MW01234152

HORTICULTURE – NEW TECHNOLOGIES AND APPLICATIONS

Current Plant Science and Biotechnology in Agriculture

VOLUME 12

Aims and Scope
The book series is intended for readers ranging from advanced students to senior research scientists and corporate directors interested in acquiring in-depth, state-of-the-art knowledge about research findings and techniques related to plant science and biotechnology. While the subject matter will relate more particularly to agricultural applications, timely topics in basic science and biotechnology will also be explored. Some volumes will report progress in rapidly advancing disciplines through proceedings of symposia and workshops while others will detail fundamental information of an enduring nature that will be referenced repeatedly.

The titles published in this series are listed at the end of this volume.

Horticulture – New Technologies and Applications

Proceedings of the International Seminar on New Frontiers in Horticulture, organized by Indo-American Hybrid Seeds, Bangalore, India, November 25–28, 1990

edited by

J. PRAKASH
Indo-American Hybrid Seeds,
P.O. Box 7099,
Bangalore 560 070,
India

and

R. L. M. PIERIK
Department of Horticulture,
Agricultural University,
P.O. Box 30,
6700 AA Wageningen,
The Netherlands

SPRINGER SCIENCE+BUSINESS, MEDIA, B.V.

Library of Congress Cataloging-in-Publication Data

International Seminar on New Frontiers in Horticulture (1990 : Bangalore, India)
Horticulture, new technologies and applications : proceedings of the International Seminar on New Frontiers in Horticulture / organized by Indo-American Hybrid Seeds, Bangalore, India, November 25-28, 1990 ; edited by J. Prakash, R.L.M. Pierik.
p. cm. -- (Current plant science and biotechnology in agriculture ; 12)
ISBN 978-94-010-5401-0 ISBN 978-94-011-3176-6 (eBook)
DOI 10.1007/978-94-011-3176-6
1. Horticulture--Technological innovations--Congresses. 2. Horticulture--Congresses. 3. Plant biotechnology--Congresses. I. Prakash, J. (Jitendra) II. Pierik, R. L. M. III. Indo-American Hybrid Seeds. IV. Title. V. Series.
SB317.53.I57 1990
635--dc20 91-17527

ISBN 978-94-010-5401-0

Printed on acid-free paper

TABLE OF CONTENTS

Preface

In November 1990 Indo-American Hybrid Seeds (IAHS), one of the largest and very innovative horticultural enterprises of its kind in India, celebrated its silver jubilee year in the town of Bangalore, India. On the occasion of this silver jubilee of IAHS an International Seminar on 'New Frontiers in Horticulture' was organized from 25-28th of November 1990 at the Ashok Radisson Hotel in Bangalore. IAHS was almost fully responsible in terms of organization and financially for this International Seminar. Assisted by an International Scientific Advisory Board, the organizing committee, all members of the company IAHS, really did a great job. I would like to thank in particular Mr. Mammohan Attavar (the company's founder) and Mr. Sri N.K. Bhat (partner of the company), respectively chairman and treasurer of the organizing committee, for their organizational and financial support in organizing this conference. Very special words of thanks go to my colleague editor, Dr. Jitendra Prakash, Secretary Organizing Committee and Director of Biotechnology - IAHS, who was really the spill in the whole organization of our very successful conference.

The International Seminar was proceeded by the inauguration of the Silver Jubilee Research & Development Block of IAHS in Bangalore by His Excellency the Governor of the State Karnataka, Shri Bhanu Pratap Singh. During the inauguration the Chief Minister of Karnataka Shri S. Bagarappa and Dr. M.S. Swaminathan addressed the audience. Renowned agricultural scientist and President of the National Academy of Science of India, Dr. M.S. Swaminathan suggested in his keynote address a five-pronged strategy for horticulture to take advantage of recent advances in plant science: the strengthening and streamlining of the Indian National Horticultural Development Board, the establishment of a national research and information centre for the horticulture trade, ecological horticulture estates, nutrition gardens and horticultural seed villages.

The International Seminar was divided into 6 large sections:

1. Advances in production technology.
2. Plant genetic manipulation.
3. Plant cell and tissue culture.
4. Greenhouse technology.
5. Plant health and crop protection.
6. Post harvest and landscape horticulture.

In each section key note lectures, lead papers and contributory papers were presented; these are included in the proceedings. Apart from these lectures, panel discussions workshops and poster sessions were included in the congress but they are not included in the proceedings of this seminar.

Finally I would like to thank all people who contributed to the success of this seminar, in particular quite a number of staff members of IAHS.

During this seminar scientists from India and abroad exchanged the ideas and views that will help in understanding complex biological problems which in turn will fulfil the aim for plant improvement and development of plant products for the good of humanity.

R.L.M. Pierik

System Analysis of Yield Trials can Raise Efficiency of Breeding for Yield

D.H. Wallace
Professor of Vegetable Crops and Plant Breeding
Department of Plant Breeding and Biometry
Cornell University, Ithaca, New York 14853-1902

Improvement of efficiency of breeding for higher yield of seed crops can be achieved, based on relationships that arise from partitioning of the available photosynthate.

Effects by partitioning of photosynthate.

If a large proportion of the photosynthate is partitioned to already existing organs of yield, these organs will grow at a rapid rate, which will cause them to develop to harvest maturity in a short time (Fig. 1). For a seed crop, yield arises from growth of the buds, pods and seeds. Any photosynthate that is partitioned to these organs of yield will be unavailable to support continued growth of vegetative organs. This competitive deprivation of photosynthate will reduce the rate of vegetative growth, which will reinforce the early maturity caused by rapid growth of the reproductive organs.

Partitioning a large proportion of the photosynthate to the existing organs of yield will give the highest yield for a short growing season, because it results in rapid growth of the reproductive organs, which is a rapid rate of accumulation of yield, and it will also result in a high harvest index. A growing season is that time across which the environment facilitates plant growth.

The proportion of the photosynthate that is partitioned to vegetative organs will support growth and development of additional vegetative organs, if its quantity is more than required for maintenance respiration of the already existing vegetative organs.

If a large proportion of the photosynthate is partitioned to the vegetative organs, this will result in rapid addition of nodes accompanied by increase in the number of leaves and leaf area (increased capacity for photosynthesis). Also, each new node will provide an additional site where additional organs of yield can arise at a later time. The continued vegetative growth and development will extend the time required to develop to harvest maturity. The already existing organs of yield will be competitively deprived of any photosynthate that goes to vegetative growth, so they will grow slowly. This will reinforce the extension of the days to maturity caused by the continued growth and development of vegetative organs. Initial but not indefinite extension of the duration of partitioning a large proportion of the photosynthate toward (continued) vegetative growth will result in the highest yields for a longer growing season, because the additional nodes will provide positions on the plant for a larger number of reproductive organs.

J. Prakash and R. L. M. Pierik (eds.), Horticulture – New Technologies and Applications, 1–11.

These organs may grow slowly but they will accumulate yield across a longer time.

The concept presented in the two previous paragraphs is bluntly repeated as follows. For each cultivar, the rate of the ongoing accumulation of yield controls the days the cultivar requires to develop to harvest maturity. This concept, which is new and also unfamiliar to most plant scientists, precedes and is therefore more fundamental than the following correct and universally accepted phenomenon. The number of days the cultivar needs (uses) to develop to harvest maturity strongly controls the yield, by determining the adaptation of the cultivar to the growing season duration. A cultivar is adapted, and its yield can be high relative to other cultivars, if the time the cultivar needs and uses to mature corresponds with the length of the growing season. Yield will be reduced if the time the cultivar requires is either longer or shorter than the growing season duration.

Three genetically-controlled components of yield

Each major physiological component of yield results from interaction and integration of many preceding biochemical and physiological processes. These gene-controlled traits are always further controlled by the environment. A major component is one that directly results in yield accumulation, rather than doing so distantly as do the biochemical and physiological processes that are integrated into the major component.

There are three genetically-controlled, major physiological components of yield. They are: 1) the accumulated biomass; 2) the harvest index (the proportion of the biomass that has been partitioned to the yield); 3) the time required to develop to harvest maturity. The biochemical and physiological processes that result in these three components integrate all the gene activities of the cultivar.

Competition for and partitioning of the photosynthate (as described above, and see Fig. 1) causes the biomass to become negatively correlated with the harvest index, and causes the harvest index to become negatively correlated with the days to maturity. Biomass and days to harvest maturity (the duration of plant growth) become positively correlated.

A 4th major component is physical rather than biological.

Maximization of yield requires that the time the cultivar uses to develop to harvest exactly match the growing season duration. The growing season duration plus all environmental factors associated with it are the 4th physiological component of yield (Fig. 1). They are physical factors, without control by a genotype. However, they modulate all the gene activities that result in the expressed level of each of the three genetically controlled components of yield.

Temperature and moisture establish the rates of plant growth and development, in addition to establishing the beginning and end of the possibility for plant growth. Thus, they establish the duration of the growing season. Simultaneously, the same temperature and moisture exert additional control over the yield by their modulation of the gene activities. Additionally, the daylength alters plant development by modulating the activity of the photoperiod genes. Responses to all environmental factors and the potential for yield depend on the genotype. Depending on which stage(s) of development the cultivar is at when the irregular adequacies to stresses by temperature and moisture occur, these two environmental factors plus the daylength, which varies systematically, alter the expressed levels of all three of the genetically-controlled major components of yield.

The focus of this paper is on effects by daylength and temperature on the partitioning of the photosynthate, on the consequent effects on the number of days the cultivar requires to develop to harvest maturity, and on the effects upon the adaptation and yield of the cultivar of the correspondence of the time to maturity with the growing season duration.

Relatively few maturity genes.

The partitioning and its effect on the days to maturity expressed by the cultivar, which is the cultivar's adaptation to the growing season, is controlled by the sensitivity of a few photoperiod gene(s), plus by a few maturity genes of other classes . The other classes control the number of nodes on the plant shoots, the earliest possible node to flower, and the rate of node development. Collectively, these traits constitute the plant habit. For some species there is also an effect on days to flowering by vernalization, which requires low temperature.

The number of maturity genes with control over partitioning and therefore over the days to flowering and maturity is far fewer than the number of genes that control the accumulated biomass. Virtually all genes affect the net biomass, including the maturity genes. Maturity genes partially control the net accumulation of biomass through the alternative pathways of partitioning. The first pathway is by direct feedback control over the magnitude of the vegetative growth and development. This control results from the competitive partitioning of the available photosynthate to continued vegetative growth and development, or alternatively, to the continued growth and development of the already initiated reproductive organs, i e. to the accumulation of yield (Fig. 1). The second pathway of feedback is indirect, through the hormonal signals that will or will not hasten senescence, respectively, as the yield (the reproductive organ) approaches or else remains short of development to harvest maturity.

Measuring the physiological components of yield

A yield system analysis (YSA) has been developed (Wallace and Masaya, 1988). Applied to each yield trial, YSA quantifies the variation in control by genotype over each of the three genetically controlled components of yield. The yield, aerial biomass, time to flowering, and the time to harvest maturity are measured, followed by calculations of rates, harvest index and the duration of seedfill. This quantifies each of the three genetically controlled major components of yield, from either two or three perspectives (Table 1).

A complete YSA measures the aerial biomass. Root biomass is not measured, because it would be time consuming and expensive beyond the potential benefit. Measurement of root biomass is not essential, because for most cultivars, the aerial biomass will be correlated with the root and total biomass. Based similarly on expense plus benefit from correlation, leaves that abscise can also be ignored. Fresh weights can suffice for plant species with fruits and other organs with high moisture content. The requirement for improving efficiency of breeding for higher yield is just identification of the genotypes with the highest and near-highest yields, biomasses, and harvest indices, plus acquisition of understanding of effects on levels of these three traits (all of which involve only measures of biomass) by the time the genotype requires to develop to harvest maturity, plus by the correspondence of this time with the growing season duration.

Each YSA is completed by calculating the correlation, across all of the genotypes tested within the yield trial, between each possible pairing among the nine outputs from the yield system (Table 2). The correlations quantify how each of the major physiological components of yield changes in relation to changes of each of the other components.

The variation in level of each trait within one yield trial, as measured by YSA, quantifies the variation caused by the tested genotypes within the specific environment (site-season) in which the yield trial was conducted.

Correlation between yield and days to maturity.

YSA of multiple yield trials at the same site has indicated that correlation between yield and days to maturity is often significant and positive for a site. Just as often, however, it is significant and negative. Consequently, across multiple seasons, the average correlation is close to zero (Table 2). This is explained as follows.

Because of variation in moisture and/or temperature regimes, almost every site has relatively short and intermediate to relatively long growing seasons. For this reason, plus attendant variations of the environmental factors associated with the growing season, breeding materials are tested across multiple seasons before being recommended as cultivars to be used by farmers. Usually, two or more cultivars will be recommended. Some will have relatively early maturity for the site, and some will have intermediate and late maturities. Early maturing cultivars will tend to give the highest yield for short seasons, and the later maturing ones will tend to do so for the longer seasons.

Correlation between time and harvest index.

The aerial biomass was measured for only 34 of the 52 yield trials. Correlation between days to maturity and harvest index was calculated for each of the 34 yield trials. It was negative for 32. The average of the 34 correlations was -0.45 (Table 2). The probability is 0.0001 that the true average correlation is zero or positive. The negative correlation agrees with the partitioning of the photosynthate between the existing organs of yield vs. continued vegetative growth being a control over the number of days to maturity. Supporting this additionally, the yield accumulated per day of seedfill and per day of plant growth were highly and positively correlated with the harvest index.

Improving efficiency of breeding for higher yield.

Efficient breeding for higher yield requires simultaneous selection for all three of the genetically-controlled physiological components of yield: the biomass, harvest index, and days to maturity. Simultaneous selection is required because of the negative correlation between harvest index and days to maturity, plus the negative correlation between harvest index and the accumulated biomass.

As practiced in the past, selection for only yield will continue to improve it primarily by raising the harvest index. Gain in harvest index has occurred and will continue because the control over the partitioning and its effect on the days to maturity is by far fewer genes (has higher heritability) than the genetic control over the accumulated biomass which is by virtually all genes. This includes the partial control through the genes that control the partitioning and maturity (Fig. 1).

Continuing to select for only yield will gradually reduce the genetic variability available for improving the harvest index. Gradually, a yet higher harvest index will become detrimental to yield, because the larger biomass partitioned to the yield cannot be supported by the vegetative structure. As this detrimental status approaches, continued selection for yield only will shift to becoming indirect selection for higher biomass, rather than the previous indirect selection for higher harvest index, because improvement of partitioning will no longer be the pathway to higher yield which has the highest heritability. Improvement will be dependent upon the lower heritability for increase of the biomass.

Due to the correlations which the partitioning causes among the major physiological components of yield, selection for only higher biomass will lead to later maturity and lower yield. Selection for only harvest index will lead to earlier maturity and lower yield.

YSA will contribute toward improved efficiency of breeding for higher yield, because it identifies for the site of each yield trial the genotypes with superiority for yield and/or with superiority for each major physiological component of yield. Therefore, YSA of successive yield trials at a site, i.e. across the environmental variability of successive seasons at the site, will facilitate recurrent selection for yield and for each of yield's genetically-controlled major physiological components. A superior expression indicates that the genotype has superior adaptation to the growing season duration and the environmental factors of the site-season. The recurrently selected superior genotypes will include the full range of genotypes having demonstrated adaptation at the site for yield and/or for each of the three major components of yield.

The recurrently selected genotypes should be recurrently intercrossed. This will merge the large range of genetic variability with adaptation to the site into a common gene pool. A common pool with much genetic variability is a requirement for maximized genetic advance. The recurrent selection and intercrossing will increase the proportion of segregants from the crosses that will have higher genetic potential for yield. The selection for use as parents of genotypes representative of the full range of physiological pathways to yield with adaptation to the site is the major contribution that YSA can make toward improved efficiency of breeding for higher yield.

Selection within segregating populations

Enhancement by YSA through effectiveness of selection among segregants from crosses will come more slowly than enhancement from the improved selection of parental germplasm. This is because, during the early segregating generations, each plant must be considered as a separate genotype, and it may continue to segregate. YSA of all or even only some individual segregants is likely to be too expensive relative to the benefits. Early generation selection will be least effective when there is large variation of the duration of the growing season and/or of the environment associated with it from one season to the next comparable season at the site. Improvement of early generation selection will arise, however, from improved insight about the ideotype that can give the highest yield. With YSA, each yield trial will improve understanding of the best ideotype(s). Each YSA will elucidate, for the site of the yield trial, both the physiological and the morphological traits that are associated with high yield.

Environmental effects on yield and its components.

The effects by the environments will be quantified, each time the yield and/or its components are compared across two or more seasons at the same site, and/or across different sites.

Genotype x daylength x temperature effects.

It is always the genotype interacting with the environment that establishes each of the major components and the resulting yield. The genotype x environment (G x E) interactions are made more complex, because the same environment also establishes the growing season duration. Lack of understanding of the the effects by the G x E interaction on the biomass, harvest index, days to maturity and yield is a major constraint to efficient breeding.

A new statistical analysis is based on a model with Additive Main effects and Multiplicative Interaction effects (Zobel et al., 1988). Analysis of variance is used to quantify the main effect by each genotype and the main effect by each environment. The ANOVA is followed with a principal component analysis of the sums of squares.of its G x E interaction. The PCA quantifies the deviation from the grand mean, that is caused by the interaction of each genotype with the set of environments, and is caused by the interaction of each environment with the set of genotypes. The information derivable from AMMI analyses is illustrated in Figs. 2 & 3).

In Fig. 2, the main effect on the days to flowering due to each of 15 environments is plotted against the G x E effects on the same days due to the same environment. Short daylength of tropical locations causes low average days to flowering. Simultaneously, interaction of the short daylength with the set of genotypes tested causes a negative effect on the deviation from the grand mean days to flowering across all the data. Long daylengths of temperate locations cause both a larger average days to flowering, and a positive effect on the deviation from the grand mean due to the interaction of the long daylength with the genotypes.

Further understanding of the G x E interaction arises by considering the effect by temperature. At short-day tropical locations, each higher temperature causes earlier flowering (Fig. 2). On the contrary, at long-day temperate locations, each higher temperature causes later flowering. This genotype x daylength x temperature interaction, in conjunction with associated studies (Wallace and Enriquez, 1980; Wallace, 1985; Muhammad, 1983; Gniffke, 1985; Yourstone,1988), reveal that a change of temperature always and simultaneously tends to change the days to flowering in opposing directions. A higher temperature tends to give earlier flowering because it causes development of the successive nodes to occur in less time. This effect toward earlier flowering is through acceleration of the rate of vegetative development by the higher temperature. The same higher temperature simultaneously delays the node to flower, i.e. enlarges any delay of flowering caused by photoperiod gene(s) functioning in a daylength that is non promotive of flowering. This is enlargement by the temperature of the effect on the rate of reproductive development through the photoperiod gene(s) and daylength .

In response to temperature. the genotype x daylength x temperature interactions result in a U-shaped curve of the days to flowering. That is, each genotype will flower earlier at some intermediate temperature than at either a lower or higher temperature. The smallest days to flowering on the curve is an optimum temperature for flowering. The optimum temperature for flowering will be lowered by a genotype that is more highly photoperiod sensitive and/or by a longer daylength. The optimum temperature for flowering will be highest if the genotype is photoperiod insensitive and/or the daylength is short.

Predictions for one genotype within one environment.

Quantification of the main and G x E effects allows prediction of the yield for each combination of one genotype within one environment. If YSA has been conducted, the biomass, harvest index, days to flowering, and days to maturity can also be statistically predicted for each genotype within each environment. For example, the bean yield for each genotype at the mean temperature of each of five elevations can be predicted from the data of an AMMI analysis (Fig. 3). The predicted yield is: 1) the average of the genotype across all of the environments, plus; 2) the average of the environment across all of the genotypes; plus 3) (the effect on the deviation from the grand mean by the genotype, multiplied by the effect on the deviation from the grand mean by the environment); minus 4) the grand mean of all the data because it was included twice (in the average of both the genotype and the environment). With YSA of a set of genotypes at multiple site-seasons, such predictions can be made for each of the two or three perspectives (Table 1) for each of the three, major, genetically controlled, physiological components of yield. From the patterns of response, predictions can also be made for sites with intermediate environments, or for cultivars with intermediate genotypes.

YSA alone provides information needed to improve the efficiency of breeding for higher yield at the site of the yield trial. Improved efficiency at many sites will raise the average yield of a region. Higher yield for many regions will raise average yield of a continent, etc. Yield advances can become even more rapid and larger if the YSA data from multiple site-seasons are subjected to AMMI analysis. The AMMI will quantify the G x E interaction effects on both yield and its major physiological components. YSA plus AMMI analysis will advance progress additionally, because the information they provide will strengthen the linkages of applied plant science with molecular biology and biotechnology.

Literature

Gniffke P.A. (1985) Studies of phenological variation in the common bean (*Phaseolus vulgaris* L.) as modulated by mean temperature and photoperiod. PhD. Thesis. Cornell University. Ithaca, New York. (Diss. Abst. 46:9 p. 2698B).

Muhammad A.F.H. (1983) The effects of temperature and daylength on days to flower and maturity in dry beans (*Phaseolus vulgaris* L.). Ph.D. Thesis. Cornell University. Ithaca, NY. (Diss. Abstr. 44:08 p2294-B

Wallace D.H. (1985) Physiological genetics of plant maturity, adaptation, and yield. Plant Breeding Reviews 3:21-167.

Wallace D.H. and Enriquez G.A. (1980) Daylength and temperature effects on days to flowering of early and late maturing beans (*Phaseolus vulgaris* L.). J. Amer. Soc. Hort. Sci. 105:583-591.

Wallace D.H. and Masaya P.N. (1988) Using yield trial data to analyze the physiological genetics of yield accumulation and the genotype x environment interaction effect on yield. Annu. Rep. Bean Improvement Coop. 31:vii-xxiv.

Yourstone K.S. (1988) Photoperiod and temperature interaction effects on time to flower and its components in bean (*Phaseolus vulgaris* L.). PhD Diss., Cornell University. Ithaca, New York. (Diss. Abst. 49:08 p. 2957-B).

Zobel R.W., Wright M.J. and Gauch H.G. (1988) Statistical Analysis of a yield trial. Agron J 80:388-393.

Table 1. NINE OUTPUTS FROM THE YIELD SYSTEM OF A CULTIVAR

Output	Interpretative perspective
Four direct measurements within each yield trial	
Yield	The economically important output
Aerial biomass	The overall photosynthetic efficiency
Days to flowering	Time used for development to flowering
Days to harvest maturity	Time used to develop to harvest maturity
Five calculations from the four direct measurements	
Days of seedfill	Time used for actual yield accumulation
Yield/day to maturity	Efficiency of yield accumulation }*The rate of*
Yield/day of seedfill	Efficiency of yield accumulation }*partitioning*
Biomass/day of plant growth	Efficiency of photosynthesis
Harvest index	Endpoint efficiency of partitioning to yield

Table 2. THE AVERAGE CORRELATION ACROSS 20 TO 51 YIELD TRIALS BETWEEN EACH PAIR AMONG THE NINE YIELD-SYSTEM OUTPUT TRAITS. The yield trials were conducted in 12 US states, 2 Canadian provinces, plus Puerto Rico, Peru, Colombia and Belgium.

	Days used for development to indicated stage or for that stage			Rate of the process		Biomass and its partitioning to vegetative vs. reproductive growth		
	Flow	Matu	SdFi	Yd/D	Sf/D	Bi/D	Biom	HaIn
YIELD	0.07	0.01	-0.01	0.93	0.84	0.72	0.71	0.54
DAYS TO FLOWER		0.66	-0.11	-0.01	0.07	0.11	0.31	-0.17
DAYS TO MATURITY			0.57	-0.27	-0.16	-0.07	0.34	-0.45
DAYS OF SEEDFILL				-0.14	-0.31	-0.19	0.04	-0.23
YIELD/DAY TO MATURITY					0.92	0.76	0.54	0.67
YIELD/DAY OF SEEDFILL						0.73	0.59	0.69
BIOMASS/DAY							0.86	0.21
TOTAL BIOMASS								-0.05
HARVEST INDEX								---

Fig. 1. **Four physiological components of yield.** Three are controlled by genetics and environment. The fourth is the duration of the growing season plus the attendant environment. All effects on yield by the duration and environment of the growing season are implemented through the genetic activities of the plant. These effects are symbolized by the grey fill of the boxes representing the three genetically controlled components.

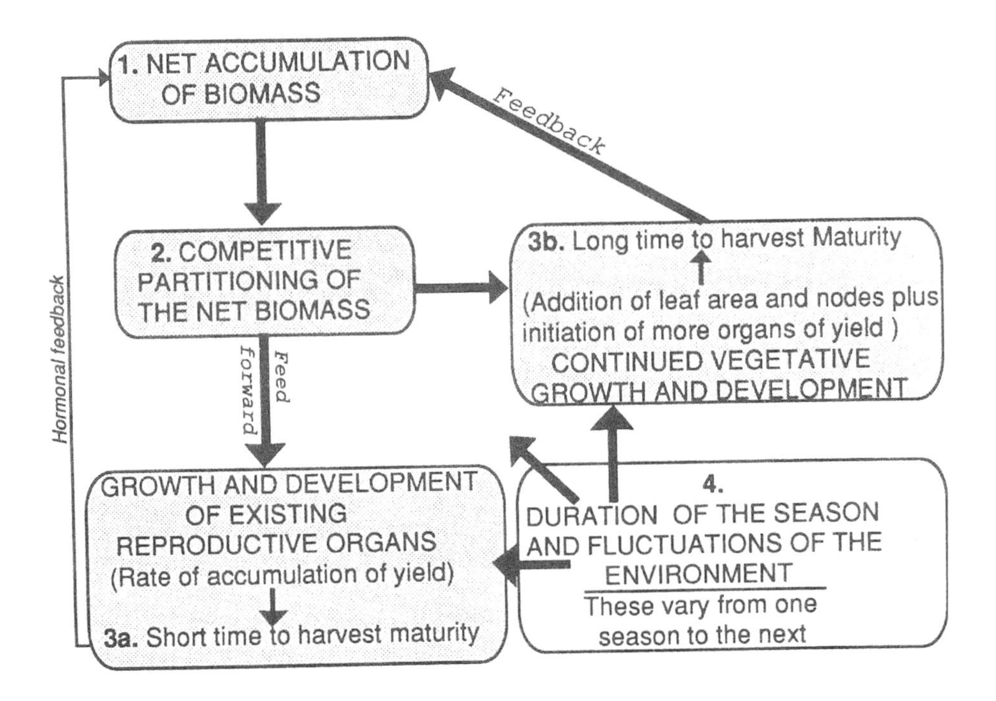

Fig. 2. The average days to flowering at each of 15 environments across a set of genotypes plotted against the effect (days to flower$^{0.5}$) on the deviation from the grand mean days caused by the same environment. (From Yourstone, 1988).

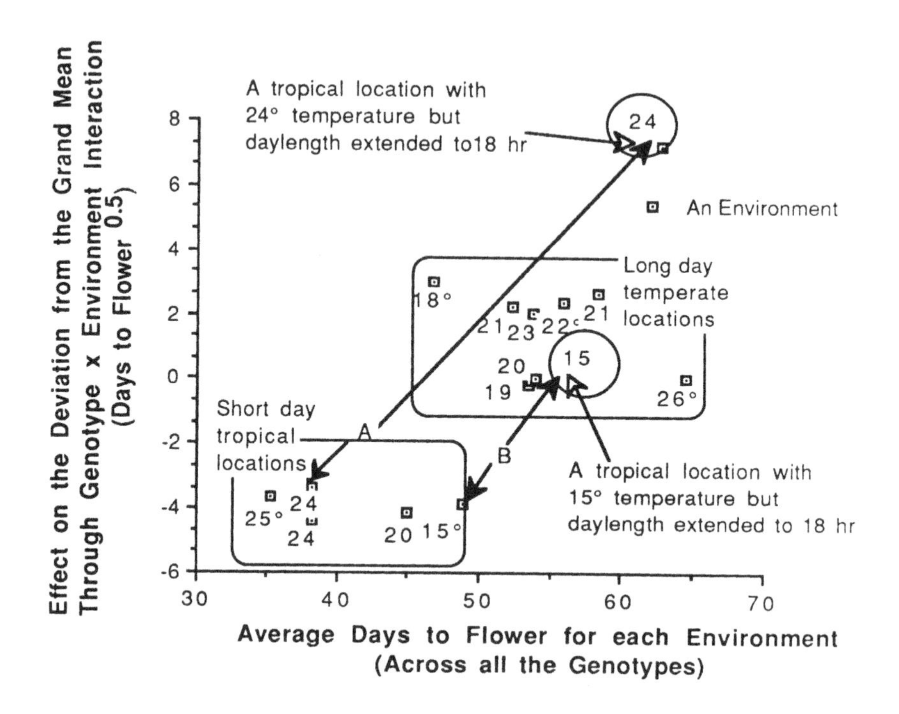

Fig. 3. The average yield at each of 5 mean temperatures and for each of 10 cultivars plotted against the effect ($yield^{0.5}$) on the deviation from the grand mean yield caused by the same environment or by the same genotype.

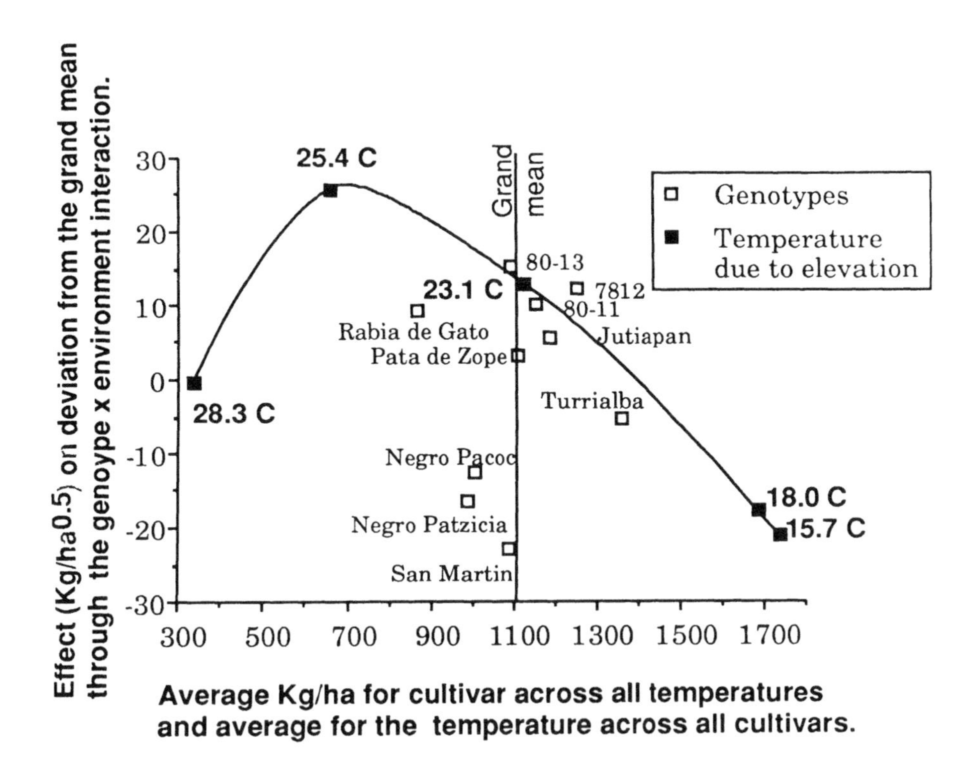

Research needs in the production of high quality seeds

FRANCIS Y. KWONG
PanAmerican Seed Company, West Chicago, IL. USA

The recent trend towards singulated seedling production in professional nurseries has created the demand for a new class of seed products. Ideally, these seeds should be of a high degree of uniformity, both physically and physiologically, so that they can be sown mechanically and germinate quickly in cultural regimes that are fully regulated. The production of these products is a new challenge to the seed industry. It requires integrated efforts in plant breeding, seed parent culture and seed conditioning.

Improving the physical uniformity of a seed lot can be done as an extension of the routine seed cleaning procedure (van der Burg and Hendriks, 1980), by utilizing proper grading equipment to obtain smaller fractions of narrow size range in a seed lot. However, this approach does not always render the desired improvement in seed performance. For example, we compared the effects of tighter size separation on two seed lots of Impatiens (Impatiens walleriana), both having initial germination of 85%. For one lot, size separation resulted in better germination in about half of the seed. The larger seeds in fractions A and B performed better than the smaller ones in fractions C and D. For the second seed lot, no significant improvement was observed (Table 1).

J. Prakash and R. L. M. Pierik (eds.), Horticulture – New Technologies and Applications, 13–20.

Table 1. Effect of size separation on germination of Impatiens seed.

A. Lot 0625, initial germination 85%

Fraction	weight(gms)	% by weight	% germination
A	216	19	92
B	429	37	87
C	383	33	83
D	122	11	74

B. Lot 1086, initial germination 85%

Fraction	weight(gms)	% by weight	% germination
A	20	9	88
B	49	21	87
C	55	24	86
D	56	24	85
E	34	15	83
F	17	7	76

These results are not surprising, considering that germination performance is dependant on the physiological state of the seed, which may or may not be correlated with its physical size. In order to achieve high and uniform germination, seed production efforts should concentrate on obtaining physiologically uniform products.

A commercial seed production process generally involves: (1) raising the parent plants, (2) pollination, by hand if it is a hybrid crop, (3) seed harvest, drying, and cleaning, and (4) seed conditioning. In principle, any manipulation on the mother plant or the developed seed can have a potential impact on the resulting seed quality. However, our knowledge of the exact relations of the various cultural measures in seed production to the quality of the product is fragmentary. We conducted experiments studying the effects of some production factors in the germination performance of flower seeds. Three examples of this work are given in this discussion.

1. Effects of the timing of pollination on seed quality

The seed is the result of a successful pollination. There is increasing evidence that seed performance is dependant not only on the environment under which the seed matures, but it may also relate to the physiological state of the gametophyte at the time of fertilization (Mulcahy and Mulcahy, 1975; Ter-Avanesian, 1978; Mulcahy et al., 1982; Winsor et al., 1987). The readiness of the flower for the fertilization process to begin, and thus the physiological age of the pollen at the time of fusion, can be influenced by the timing of pollination. We studied the effects of altering the timing of pollination on the resulting seed quality in Impatiens.

The Impatiens flower is protandrous. Subsequent to the dehiscence of the anther column, the gynoecium is exposed. During its development, the stigma tip flares open from a single point to a five-lobed, star shaped surface. This process occurs over a one week period. Distinct visual stages of stigma development, namely slightly open, 1/4 open, 1/2 open and fully open, can be identified easily. We used these visual indicators as a measure for the physiological age of the stigma. Flowers of different stigma age were pollinated with the same source of fresh pollen on the same day, the seed pods were hand harvested individually just before they opened naturally. After drying, the one hundred seed weight for each treatment was determined, and germination tests were conducted on filter papers in the laboratory, using conditions recommended by AOSA. The results are shown in Table 2.

Table 2. Effects of pollinating Impatiens of different stigma age on seed quality

variety No.	stigma appearance	100 seed wt (grams)	% germination
7649	closed	44	65
	slightly open	42	75
	1/4 open	40	75
	1/2 open	44	82
	fully open	40	56
9328	closed	52	56
	slightly open	57	75
	1/4 open	58	65
	1/2 open	63	86
	fully open	61	55

Pollination at all stigma ages was successful, and there was little difference in the resulting seed size. But the germination capacity of the seeds varied. The best timing for pollination was found to be when the stigma was half open. Pollinating the flowers earlier or later than this stage resulted in inferior quality seed.

2. Effects of developmental problems on seed quality.

Various developmental problems are observed in the course of seed maturation. Some of these problems are subtle enough that the harvested seed do not differ in size, shape, or weight from the normal ones, making mechanical separation difficult. As the proportion of such malformed seeds in a given seed lot increases, the overall quality of that lot decreases. Geranium (Pelargonium x hortorum) green seed represents one of such developmental problems. We attempted to obtain a better understanding of the nature of green seed in this crop.

Green seeds are often noticed in commercial geranium seed production. The percentage of green seeds in a given crop ranges from very few to over 10%. These seeds have green cotyledons, whereas cotyledons of normal seeds are white. Green seeds did not germinate well (Table 3). The germinating problem is unlikely to be due to a defunct embryo, since most of the excised embryos developed normally when placed on a Marashige and Skoog basal salt agar medium.

Table 3. Comparison of geranium green and normal seed.

seed color	% germination in laboratory test	% excised embryos developing normally on MS medium
Green	10	76
White	90	100

We extracted the green pigment from the seed with an acetone-ethanol mixture (80/20, v/v) and measured its absorption spectrum. It has the typical absorption characteristics of chlorophyll, with peaks around 430nm and 650nm. We then traced the presence of chlorophyll in the course of geranium seed development (Table 4). An increase in chlorophyll content was observed throughout the growth phase of the seed. After reaching full size, degradation of chlorophyll took place and the ripe seed turned white. The problem in green seeds is that this degreening process does not occur. The relationship between the degreening process and the germination capability of the seed is still unexplained.

Table 4. Geranium seed development

days after pollination	size	color
7	2 mm	white with green tip
14	3 mm	light green
21	4 mm	bright green
30	4 mm	white

3. Effect of seed drying on seed quality

Proper seed drying is of utmost importance to maintaining high seed quality (Roberts, 1981). Some practical guidelines on how seeds should be dried are available (Brandenburg et al., 1961). However, there is little information on the changes in seed properties during the drying process. This information is required if we were to fine-tune the drying conditions in order to achieve higher seed quality. We examined the drying of Impatiens seeds in some detail.

Mature Impatiens seed pods were hand-harvested just before they opened. They were brought into a laboratory with uncontrolled environmental conditions for observation. Ambient temperature ranged from 18°C at night to 26°C daytime during the observation period, and the relative humidity ranged from 52% to 100%. As soon as the seed pods opened naturally (day 0), samples were drawn on a daily basis for seed moisture content and germination testing. The results are shown in Table 5. On day 0, when released from the seed pod, the seed had a 42.5% moisture content. Within a day, it lost moisture rapidly and dried down to about 10%. It germinated at 56% at that time. Thereafter, the seed moisture content stabilized at about 13%, with a germination rate of about 70%.

Table 5. Impatiens seed drying under ambient conditions

Days from pod opening	% seed moisture content	% germination
0	42.5	-
1	9.5	56
2	13.8	56
3	12.5	68
4	13.0	72

In the same experiment, some of the seed at day two was placed in a controlled drying chamber. The drying conditions were set at 40% RH and 35°C. Samples were removed from this chamber periodically for moisture content and germination evaluations. The results of this study are shown in Table 6.

Table 6. Impatiens seed drying at 40% RH and 35°C.

Hours of drying	% seed moisture content	% germination
0	13.5	-
6	4.6	60
15	3.0	60
24	3.1	64
44	5.0	78
60	4.6	80
72	4.6	72

Under these drying conditions, seed moisture dropped from 13.5% to 4.6% within six hours, and down to 3% within a day. Subsequently, it absorbed moisture again until reaching an equilibrium at a little under 5%. Interestingly, the increase in germinability of the seed was found to coincide with the increase in seed moisture. Since the air moisture level was maintained throughout the experimental period, the shifts in the seed moisture contents are likely to be the results of changes in seed property. The exact nature of these changes, and their relation to the gain in germinability are to be studied further.

These three examples should serve to illustrate how the minute details in a seed production routine could impact on the final quality of the product. The closer to perfection in seed performance we try to achieve, the more of such small details we have to attend to throughout the production precess. Our problem is that we do not, at present, have all the knowledge that would enable us to do so efficiently. Some of the more obvious areas for further research include:

a) nutritional requirements at different stages of plant development
b) pollination ecology
c) what determined seed maturity?
d) imposition and alleviation of dormancy
e) the dynamics of seed drying and storage

The seed product in the market is the result of a long chain of cultural manipulation and seed treatments. While proper pollination and seed development pre-determine the maximum seed vigor, postharvest handling procedures are related to seed deterioration. All these steps have to be optimized before further physical grading and seed enhancement treatments can bring about the desired seed performance demanded by critical growers.

References

Brandenburg N.R., Simons J.W. and Smith L.L. (1961) Why and how seeds are dried. USDA Yearbook of Agriculture 9161:295-306.

Burg van der W.J. and Hendriks R. (1980) Cleaning flower seeds. Seed Sci Technol 8:505-522.

Mulchay D.L. and Mulchay G.B. (1975) The influence of gametophytic competition on sporophytic quality in Dianthus chinensis. Theo. Appl. Genet. 46:277-280.

Mulchay G.B., Mulchay D.L. and Pfahler P.L. (1982) The effect of delayed pollination in Petunia hybrida. Act. Bot. Neerl. 31:97-103.

Roberts E.H. (1981) Physiology of ageing and its application to drying and storage. Seed Sci. Technol. 9:359-372.

Ter-Avanesian D.V. (1978) The effect of varying the number of pollen grains used in fertilization. Theo. Appl. Genet. 52:77-79.

Winsor J.A., Davis L.E. and Stephenson A.G. (1987) The relationship between pollen load and fruit maturation and the effect of pollen load on offspring vigor in Cucurbita pepo. Amer. Naturalist 129:643-656.

Breeding for quality and resistance to fusarial wilt in gladiolus

S.S. NEGI, S.P.S. RAGHAVA, C.I. CHACKO, T.M. RAO
AND N. RAMACHANDRAN
Indian Institute of Horticultural Research,
Hessaraghatta, Bangalore-560089, India

Key words: Breeding,fusarial wilt,gladiolus,resistance,spike quality

Introduction

Gladiolus is one of the most important bulbous flower crops. It is grown in many parts of the world for cut flowers or for garden displays. The major limitation in commercial cultivation of gladiolus is a wilt disease caused by *Fusarium oxysporum* Schl. f. sp. *gladioli* Snyd. & Hans. This disease was described as early as 1909 from California, but in India, the disease was noticed only in 1969 in a severe form (Singh, 1969). Though some of the cultivars are tolerant to this disease, they do not produce good quality spikes (Jagadish Chandra *et al.*, 1985). Hence, work on breeding gladiolus was initiated at the Indian Institute of Horticultural Research, Hessaraghatta, Bangalore(India) in 1982 to develop cultivars possessing resistance to fusarial wilt disease as well as producing good quality spikes. This paper reports the results obtained from this investigation.

Methods

Five commercial cultivars, namely, 'Beauty Spot', 'Jo Wagenaar', 'Tropic Seas', 'Watermelon Pink' and ' Wild Rose',susceptible to fusarial wilt were crossed with four wilt tolerant cultivars viz., 'Australian Fair', 'Lady John', 'Mansoer' and '*Psittacinus* hybrid'. Nearly 1000 hybrid seedlings were raised. The hybrids were evaluated for resistance to fusarial wilt under field conditions for four years. Eighteen field resistant hybrids were then tested by growing them in soil sick with *Fusarium oxysporum* for two years. *Fusarium oxysporum* was isolated from infected corms of gladiolus and its pathogenecity was confirmed by artificially inoculating the gladiolus cultivar 'Friendship' which has been reported by McClellan and Pryor (1957) to be susceptible to fusarial wilt. Inoculation was done by mixing boiled sorghum grains colonised by *F.oxysporum* at the rate of 50 g per 5 kg

J. Prakash and R. L. M. Pierik (eds.), Horticulture – New Technologies and Applications, 21–25.

of potting mixture contained in each pot. Ten corms of each hybrid were tested by planting one corm in each pot and sufficient moisture was maintained in the pots. The disease incidence was recorded based on foliar yellowing 80 days after planting. The hybrids were classified as resistant (0% infection), tolerant (1-25% infection), moderately tolerant (26-50% infection), susceptible (51-75% infection) and highly susceptible (76-100% infection). Resistant/ tolerant hybrids were evaluated in the field for various vegetative and floral characteristics for two years.

Results

Of the eighteen hybrids tested after artificial inoculation with *Fusarium oxysporum*, one hybrid was found to be resistant, ten were tolerant, four were susceptible and three were highly susceptible to fusarial wilt disease. Evaluation of the resistant and tolerant hybrids in the field showed that the resistant hybrid '82-11-90' and two tolerant hybrids viz., '82-7-59' and '82-18-16' had good quantitative (Table 1) and qualitative (Table 2) characteristics. The resistant hybrid '82-11-90' (Figure 1) is from a cross of 'Beauty Spot' x 'Psittacinus hybrid'. It produced 85.2 cm long spikes each with 16 florets having reddish purple colour with yellow and purple blotch. The tolerant hybrid '82-7-59' ('Watermelon Pink' x 'Lady John') produced spikes which were 103.4 cm long having 18 bright red coloured florets per spike. Another tolerant hybrid '82-18-16'(Figure 2) developed by crossing 'Watermelon Pink' x 'Mansoer' had good quality spikes which were 98.2 cm long with 18.4 florets per spike. Florets of this hybrid were round and open-faced, pink in colour having yellow blotch. The resistant hybrid '82-11-90' and the tolerant hybrid '82-18-16' yielded two marketable spikes per corm, while the tolerant hybrid '82-7-59' yielded only one marketable spike per corm.

Table 1. Quantitative characteristics of gladiolus hybrids resistant and tolerant to fusarial wilt

Hybrid	No.of marketable spikes per corm	Spike length (cm)	No.of florets per spike	No.of florets remain open at a time	Floret size (cm)
82-11-90 (Resistant)	2.0	85.2	16.0	4.3	7.6
82-7-59 (Tolerant)	1.0	103.4	18.0	5.3	10.8
82-18-16 (Tolerant)	2.2	98.2	18.4	5.0	9.4

Table 2. Qualitative characteristics of gladiolus hybrids resistant and tolerant to fusarial wilt

Hybrid	Floret place-ment	Floret type	Floret colour*
82-11-90 (Resistant)	Good	Open-faced	Red (53.D) with Yelllow (10.C) and Red-Purple (60.B) blotch
82-7-59 (Tolerant)	Good	Open-faced	Red (45.B) with two lower tepals Red (45.A) having Yellow (12.D) lines on lower tepals
82-18-16 (Tolerant)	Good	Round and open-faced	Red (48.B) with Red (48.A) margin having Yellow (2.D) blotch

*Based on Royal Horticultural Society colour chart in association with the Flower Council of Holland

Data on corm and cormel production of the resistant and tolerant hybrids are presented in Table 3. The resistant hybrid '82-11-90' and the tolerant hybrid '82-18-16' yielded more than two daughter corms per corm, whereas the tolerant hybrid '82-7-59' produced one daughter corm per corm. Cormels produced per corm was maximum (30.5) in the tolerant hybrid '82-18-16' followed by the tolerant hybrid '82-7-59'(22.1) and the resistant hybrid '82-11-90'(18.7). Size of the daughter corms was biggest (6.8 cm) in the tolerant hybrid '82-18-16', while size of the cormels was biggest (1.6 cm) in the resistant hybrid '82-11-90'.

Discussion

In gladiolus, worst disease is caused by Fusarium oxysporum f. gladioli (Magie, 1953). Most of the commercial cultivars of gladiolus are susceptible to fusarial wilt disease (Wilfret, 1974). In order to minimise the disease incidence, the farmers have to spend a lot of money for fungicidal treatment. Hence, it is very important to develop fusarial wilt resistant/tolerant cultivars with good quality spikes. In the present study, level of resistance among the hybrids tested varied from resistant to highly susceptible. Similar results in gladiolus were reported by Jones and Jenkins (1975). Different hybrids became symptomatic at different stages

Figure 1. Spikes of resistant hybrid '82-11-90'('Beauty Spot' x '*Psittacinus* hybrid')

Figure 2. Spikes of tolerant hybrid '82-18-16'('Wild Rose' x 'Mansoer')

Table 3. Corm and cormel production of resistant and tolerant hybrids of gladiolus

Hybrid	Corms produced per corm		Cormels produced per corm	
	No.	Size (cm)	No.	Size (cm)
82-11-90 (Resistant)	2.1	5.4	18.7	1.6
82-7-59 (Tolerant)	1.2	5.6	22.1	1.2
82-18-16 (Tolerant)	2.8	6.8	30.5	1.4

of growth period. Palmer and Pryor (1958) also observed similar variation among the 160 gladiolus cultivars tested for resistance to *F. oxysporum*. Because of the heterozygous nature of gladiolus, progeny from the same cross showed different levels of resistance to fusarial wilt as well as for quality of spikes. In the present investigation, one hybrid '82-11-90' was found to be resistant and two hybrids, '82-7-59' and '82-18-16' were found to be tolerant to fusarial wilt disease. These hybrids are being tested at different agroclimatic conditions. Based on their performance results, they can be released for commercial cultivation. These hybrids and a tolerant cultivar 'Dr.Magie' (Wilfret,1986) are being used in further breeding programme to develop superior cultivars resistant to fusarial wilt disease.

Acknowledgement

The authors are thankful to the Director,Indian Institute of Horticultural Research,Bangalore,India for providing necessary facilities to carry out this investigation.

References

Jagadish Chandra K., Negi S.S., Raghava S.P.S. and Sharma T.V.R.S. (1985) Evaluation of gladiolus cultivars and hybrids for resistance to *Fusarium oxysporum* f. sp. *gladioli*. Indian J. Hort. 42(3&4) : 304-305.

Jones R.K. and Jenkins Jr. J.M. (1975) Evalluation of resistance in *Gladiolus* sp. to *Fusarium oxysporum* f. sp. *gladioli*. Phytopathology 65 : 481-484.

Magie R.O. (1953) Some fungi that attack gladiolus. In: Stefferud A. ed., Plant diseases - the year book of agriculture 1953, pp. 601-607. United States Department of Agriculture, Washington, D.C.

McClellan W.D. and Pryor R.L. (1957) Susceptibility of gladiolus varieties to *Fusarium*, *Botrytis*, and *Curvularia*. Plant Dis. Rep. 41 : 47-50.

Palmer J.G. and Pryor R.L. (1958) Evaluation of 160 varieties of gladiolus for resistance to *Fusarium* yellows. Plant Dis. Rep. 42 : 1405-1407.

Singh R.N. (1969) A vascular disease of gladiolus caused by *Fusarium oxysporum* f. *gladioli* in India. Indian Phytopath. 22 : 402-403.

Wilfret G.J. (1974) Gladiolus breeding. Plants Gard. 30(1) : 35-38.

Wilfret G.J. (1986) 'Dr. Magie' gladiolus. HortScience 21(1) : 163-164.

Reproductive Biology of Potato : Basis for 'True Seed' Production

H.Y.MOHANRAM DEPARTMENT OF BOTANY UNIVERSITY OF DELHI,DELHI-110007, INDIA
ASHA JUNEJA, DEPARTMENT OF BOTANY, DAYALBAGH EDUCATIONAL INSTITUTE, DAYALBAGH, AGRA-282005, INDIA

Key Words
Genotype TPS-3; Solanum tuberosum ssp. tuberosum;grafting; hand pollination; apical pores; longitudinal dehiscence slits; Zonocolporate; prolate; synchronous development; embryonate; non-embryonate seed; placental outgrowth; embryo-types; seed-types; extensions of epidermal cuticular thickenings.

Introduction

Potato forms a major item of the diet of a large proportion of a global population. Being highly efficient, potato is claimed to be superior to every other food crop in the amount of food produced per unit area in unit time (Weerasinghe, 1983). Its production ranks fourth among the major food crops, the other three being wheat, rice and maize in the order of importance. The ratio of protein to carbohydrate is far greater in potatoes than in many cereals and in other root and tuber crops. The quality of potato protein is excellent. Potato is best suited to cool, temperate regions or to high elevations (2000 m) in the tropics.

The area of primary domestication of potato is assumed to be in the high Andean plateau of Bolivia-Peru, around Lake Titicaca in South America, where the wild and cultivated diploids are abundant and most variable.

Potato was brought to India by Portuguese explorers in the 17th Century (1615). Presently it has become an important item of food in the country. There has been a tremendous increase in potato production in the recent times. The area under potato cultivation in India is about 7,50,000 hectares (Kaul and Sukumaran, 1984; Nayar, 1985) with an annual production of 10 million tonnes. North Indian plains account for more than 75 per cent of the area. Potato is harvested from January to March in the plains. A highly perishable comodity, potato cannot be stored without refrigeration for more than two months after harvest. About 10 per cent of the tubers are lost every month by shrinkage, rotting and sprouting.

The installed refrigerated storage capacity in India is 3.5million tonnes. This storage space, therefore, can only hold less than one third the total production. About half of this quantity is required as seed tubers.

The traditional method of vegetative propagaton of potato by seed tubers has certain advantages. It ensures maintenance of quality, easy establishment, vigorous growth

J. Prakash and R. L. M. Pierik (eds.), Horticulture – New Technologies and Applications, 27–35.

and requires no transplantation. In a tropical country like India, raising potato from seed tubers poses major problems such as: (i) high cost of tubers (nearly 50 percent of the cost of potato production is the price of seed tubers), (ii) bulkiness of tubers and need for large storage space, (iii) high cost of transpo4rtation, (iv) rapid deterioration of tubers caused by storage pests and disease or untimely sprouting when tubers are not stored in cold, (v) lack of adequate cold storage facilities, (vi) high energy inputs for cold storage, (vii) transmission of virus diseases through tubers and (viii) drain on food reserves.

Production and use of true seeds have been suggested as a means of overcoming many of these problems. True seeds are small, easy to store and convenient to handle and transport. Seeds being dry are less susceptible to pests and storage diseases. Most importantly potato seeds do not transmit virus diseases.

True potato seeds (TPS) are nearly of the same size as those of tomato. A rough calculation shows that a packet of seeds weighing 100 g should be sufficient to plant one hectare of land as against two tonnes of seed tubers. The additional advantage of employing true seeds is reduction in the initial cost of planting and cultivation.

The success of raising potatoes from true seed will depend on the technology of production of true seed in sufficient quantity at a low cost, and the ability of the plants raised from seeds to give a high yeild of tubers of acceptable quality. Plants raised from true seed should be vigorous and sturdy. Some of the true seed varieties have these attributes.

One fundamental question that is justifiably posed is whether the plants raised from true seeds will ensure purity of the genome (which is an asset of vegetative propagation) or whether the progeny would show a wide range of segregation with the manifestation of disadvantageous traits. Even in arriving at the choice of the experimental material, it was necessary to ensure that the genotype chosen was appropriate for the purpose. It is recognized (as in hybrid corn) that the population raised from true seeds is assessed for homogeneity. The phenotypes are categorized into a gradation between 1-5, on the basis of examination of foliage and the general condition of the seedlings. This method of selection is found to be satisfactory and does not show segregtation of tubers either. According to gradation, phenotypes of categories 4 and 5 are rated best.

A thorough knowledge of the reproductive biology of potato is a pre-requisite for producing consistently good seeds on a large scale. This encompasses flower initiation (induction of flowrs) in the plains; flower development and opening; origin and development of ovules; pollen production, anther dehiscence, pollen discharge, pollination mechanism and stigma receptivity, pollen tube growth; fertilization and development of endosperm and embryo; fruit formation; seed development and factors regulating seed set; seed maturity; seed viability; germinability and seed longevity; seedling vigour and performance of the adult plants.

In the mountainous regions of India flowering of potato occurs normally and fruit set is fairly high. In these areas commercially feasible seed set should be predictively high. However, in the plains of India true seed production is problematic. Potato has a large number of varieties, a few of which flower in the plains and others do not. Still others flower only under centain environmental conditons. Even when flowering occurs seed production is generally inadequate for commercial purposes.

In India human population is highly concentrated in the northern plains (especially in the Indo-Gangetic region) where potatoes can be cultivated during the winter months, but storage of tubers is needed during summers. The area under potato cultivation in the hills is limited. But as potatoes flower easily in the hills, the available land can be profitably utilized for the large-scale production of true seeds.

Although plant breeders (Pal and Pushkarnath, 1942, 1944, 1951; Pushkarnath, 1942, 1953, 1957, 1961; Simmonds, 1969; Swaminathan and Howard, 1953; Swaminathan, 1954, 1955) have used sexual reproduction as a means of genetic recombination for improvement of potato, a systematic study of the reproductive events in the potato has not been attempted. The present work is a step in this direction.

Understandably selection in potato has been for tubers and not for fruits. Most Andigena clones flower profusely but many Tuberosum cultivars produce only a few flowers, or are shy flowering (Howard, 1978; Sinha and Pushkarnath, 1964), several other cultivars do not flower normally. Therefore, a selection has to be made for flowering and fruiting among high yielding varieties.

Switching over to potato cultivation in India by true seeds requires selected genotype with high yield and good tuber quality which can flower either in the plains on their own or which can be induced to flower by easy and inexpensive methods. Most importantly flower number, fruit number and seed set should be high. As TPS-3 (Accession No.2235) produces an abundant number of flowers, it was chosen as material for the present study for the production of true seed in the plains also. Genotype TPS-3 a tetraploid Solanum tuberosum ssp tuberosum is a F_1 hybrid of Ekishiraju X Katahdin varieties. It flowers profusely in the hills of North India and bears tubers and fruits simultaneously.

Potato is sown in the plains of India as a rabi (winter) crop. In Delhi (situated 210 to 216 M above mean sea level) potato is planted towards the end of October. Development of tubers occurs between mid December-January. Tubers attain maturity and are ready for harvest between February and March. In the hills of Himachal Pradesh potato is grown as a summer crop. It is planted during mid May. Tuberization occurs from July to mid September and harvesting is done by the end of September.

Under Delhi conditions the genotype TPS-3 flowers scantily and bud abscission is quite common. Shoots of this genotype flowers profusely when grafted on to tomato stocks. Grafted plants show reduction in bud shedding.

The flowers are perfect. Stamens and pistil are well-developed. Pollen production is abundant and fertility is high. The ovary has innumerable ovules with a good potential for seed set.

Methods

Tubers were planted in the Experimental Garden of the Department of Botany, University of Delhi in beds, each measuring 1.83m x 0,914m.

When tubers of this genotype were planted in the normal way in the plains the resulting plants bore very few flowers, which were also prone to early abscission.

Two methods were adopted in the present study: (i) planting seed tubers on bricks and (ii) grafting potato scions on tomato stocks (Carson and Howard, 1944; Crafts, 1934; Dawson, 1942; Detjen, 1943; Hartmann and Kester, 1986; and Weinheimer and Woodbury, 1966).

Brick planting method developed by Thijn (1954) was used, in which the tubers are placed over a brick and planted in the soil directly. This method promotes rooting and suupresses the formation of stolons and tuberization.

Each tuber produced 1 to 6 sprouts, of which the terminal one took the lead. Tomato plants were raised for grafting potato scions. Potato shoot tips were grafted on tomato stocks by wedge grafting technique. Flowers were subjected to the following methods of pollination : (1) Hand pollination with emasculation and bagging (i) once, (ii) twice and (iii) thrice and (2) three hand pollinations without emasculation and bagging of flowers (with allowance to open pollination). The results were compared with those of controls (open pollination).

Opening and closing of flowers and their orientation on the plant were studied under both hilly and Delhi conditions. The youngest flower buds were labelled to estimate the time taken by them to reach anthesis and fruit maturity. The date and time of hand pollination were also noted. A correlated study of the gross morphology of anther, stigma and seed as seen under the light and the scanning electron microscope at different stage of flower bud development was made.

Results

The potato scion branched profusely and produced 4 to 12 haulms. In three months time the haulms attained a length of 150 to 180 cm. The grafted potato scions also flowered profusely and abscission of flower buds was minimized.

When a flower bud reaches full development, it remains upside down through the night. The petal lobes open the next morning and the anther dehisces by means of apical pores. Next the flower orients itself at right angles to the long axis. After 2 or 3 hours the flower bends back to its original position and the corolla lobes become reflexed. Longitudinal dehiscence slits begin to appear in the middle of the anther, extending both below and above and eventually joining with the apical pore. The pollen discharge begins at anthesis and the locules are empty by next four stages in the development of flower.

The study of time course development of flower showed that a stigma is positioned deep down in the middle of the anther cone. The stigma is pushed up and at anthesis stage it reaches the level of the apex of the anther cone. This coincides with stigma receptivity, formation of apical pores in the anthers and first release of mature pollen.

The genotype is both self- and cross-compatible. Flowers are self-pollinated. No external insect vectors appear to be involved in the transfer of self pollen. Gravity and wind vibration appear to aid pollen deposition on the stigma.

The pollen are prolate and are 4 (3) (6)-zonocolporate. Some pollen also have syncolporate and parasyncolporate structures. SEM pictures have shown that the exine ornamentation is verrucate and displays a mosaic with puncta at places. A unique feature of the pollen is the presence of a colpus bridge. The colpus bridge drifts apart allowing the emergence of the germ tube. The tip of the incipient germ tube comes to be covered by an operculum. The operculum is heavily granulated and persistent.

On an average 27,180 ± 3.13 pollen and 1291.29 ± 10.81 ovules are produced per flower. Thus the ovule:pollen ratio comes to 1:21 which is rather low.

The stigma has numerous papillae. The style is solid. The pollen begin to germinate at stage of a flower next to anthesis. The pollen tubes grow through intercellular spaces of the transmission tissue of the style. During basipetal development the unitegmic ovules pass through orthotropous and hemianatropous conditions. An 8-nucleate (Polygonum type) embryo sac is formed by a stage prior to anthesis stage by which time the ovules become anatropous. Thirty six hours before anthesis stage the ovules show a basipetal gradient of several stages of megagametophyte development. However, by anthesis stage nearly all the ovules come to bear a mature embryo sac showing their synchronous developent.

TPS-3 variety of potato bears 850 to 1500 ovules per ovary. Thus the potential for seed production is very high although 62 per cent of the ovules fail to set seed. Presuming that the probable limitation to seed set was in-sufficiency of pollen, it was logical to carry out 1-3 hand pollinations. Expectedly, multiovulate systems must have a prolonged stigma receptivity to ensure high reproductive capacity.

Eight days after pollination, the ovules at stylar end enlarged. The placenta grew as outgrowths forming a collar-like structure below the enlarging ovules and become cup-shaped in the next four days. After a lapse of 30 to 32 days following pollination the seeds contain massive cel-lular endosperm with an early heart-shaped embryo. The placental outgrowths extend further to overgrow the seeds. However, the abortive seeds below remain devoid of placental outgrowth. The fruit attains maturity by 40 days after pol-lination and the placenta becomes transluscent. A wide range of embryo shapes and sizes are seen in the mature seeds.

The seeds formed may be embryonate or non-embryonate. Open pollination results in 27.75 per cent seed set (only embryonate seeds). The corresponding figures for one, two and three hand pollination with emasculation and bagging are 3.30; 7.75 and 41.26 per cent. The seed:ovule ration in a control fruit is 1:3.6. Seed set is high in the stylar end of the ovary. Ovules that fail to develop into seed are seen as a pouch of shrivelled structures at the base of the placenta. On the basis of the shape and size of the embryo, seeds contained in a ripe fruit are classified into five types (A-E). The largest embryos are grouped under A and the smallest in D (in the decreasing order of size). Type E seeds have rudimentary6 embryos or no embroys and may or may not contain emdosperm. Nevertheless, the seed coat is fully differentiated in all the five types.

Discussion

The position effect of ovules in an ovary and the entry of pollen tube into the ovule bear importance in setting of seeds (Chandra and Bhatnagar, 1974). The ovary of okra was divided from top downwards into positions 1 to 15. It was observed that ovules at the apex of the ovary (positions 1, 2) received very few pollen tubes. Maximum pollen tubes were received by ovules positioned 7 to 14 on the ovary. All the ovules inside the ovary, therefore, do not have equal potential for seed set. However, the authors have not reported the condition of ovules at different locations in the ovary at anthesis.

In genotype TPS-3 of potato (present work), however, there is a basipetal sequence of ovules in the flower bud (2 days prior to anthesis). This does not persist and at anthesis all the ovules come to bear mature embryo sacs. No ovule meiotic aberrations accounting for poor seed set could be observed.

It was explained under 'Observations' that a mature fruit bears seeds which appear nearly alike externally but show embryos of different shapes and sizes when viewed under a binocular microscope with reflected light. Following the classification of (Upadhya et al., 1981) the embryo types have been categorized as types A to E and the seeds have been named accordingly.

An analysis of the results from hand pollination experiments shows a significant increase in embryos of type A and B borne by respective seeds resulting from three hand pollinations with emasculation and bagging as compared to open pollination.

SEM pictures of the spermoderm of mature seed from A-E types show the presence of stiff, horizontal, hair-like structures, each with a curved base and a pointed tip. These are not true epidermal hairs but are extensions of the cuticular thickenings of the epidermal cells.

A correlated histological preparations of developing seed have shown downwardly projecting extensions of the tangential wall of the upper epidermal cells of the seed coat. These inward projections elongate and jet out and take a curve. When the seed is soaked in water these structures release sticky mucilage and stand erect. This is not found in any other number of Solanaceae including the genus Solanum (Bhaduri, 1935; 1936; Soueges, 1907).

It is however, a unique feature of some members of family Lythraceae (Panigrahi, 1986). Seeds produced as a result of three hand pollinations (with emasculation and bagging) show 80 per cent germination. Those borne in fruits resulting from three hand pollinations (without emasculation and bagging) have a germination percentage of 56. The percentage of germination of seeds is directly related to the types of embryo contained in them. One hundred per cent germination occurs in A and B type seeds. The percentage declines to 40 in D type seeds. As A and B seed types show a higher percentage of seed germination as well as a greater vigour of the seedlings, than other seed types the production of A and B seed types is of crucial importance in commercial production of true seeds. On the basis of the present work it is concluded that three hand pollinations with emasculation and bagging yielding nearly 70 percent of seed types A and B are most suited for raising potato from true seeds.

References

Bhaduri P.N. (1935). Studies on the female gametophyte in Solancaea. J. Indian Bot. Soc. 14: 133-150.

_______________(1936). Studies on the embryogeny of the Solanaceae. I. Bot. Gaz. 98: 283-295

Carlson, E.M. and Stuart, B.C. (1936). Development of spores and gametophytes in certain New World species of Salvia. New Phytol. 35: 68-91.

Carson, G.P. and Howard, H.W. (1944). Inheritance of the 'bolter' condition in the potato. Nature, (Lond.) 154: 829.

Chandrra, S and Bhatnagar, S.P. (1974). Influence of ovule position on pollen tube entry and seed set. Naturwissenschaften 61: 688.

Dawsin, R.F. (1942). Accumulation of nicotine in reciprocal grafts of tomato and tobacco. Am. J. Bot. 29: 66-71.

Detjen, L.R. (1943). The influence of root stocks on seeds and seedling progenies of tomato grafts. Proc. Am. Soc. Hort. Sci. 43: 147-148.

Howard, H.W. (1978). The production of new varieties: 607-646. In Harris, P.M. (ed) The Potato Crop - The Scientific Basis for Improvement. Chapman & Hall, London.

Kaul, H.N. and Sukumaran, N.P. (1984). A potato store run on passive evaporation cooling: In Nayar, N.M. (ed) Technical Bull. No.11 CPRI (ICAR) Shimla Allied Publ. Pvt. Ltd., New Delhi, India.

Nayar, N.M. (1985). A perspective view of the potato variety improvement programme of India: 189-195. In Present and Future Strategies for Potato Breeding and Improvement. Rep. 26th Planning Confer. International Potato Centre (CIP), Lima, Peru.

Pal, B.P. and Pushkarnath (1942). Genetic nature of self-and cross-incompatibility in potatoes. Nature, (Lond.) 149: 247-249.

_______________(1944). Self and cross-incompatibility in some diploid species of Solanum. Curr. Sci. 13: 235-236.

_______________(1951). Potato breeding investigations in India. Emp. J. Exp. Agric. 19: 87-103.

Panigrahi, S.G. (1986). Seed morphology of Rotala L., Ammania L., Nesaea Kunth and Hionanthera Fernandes & Diniz (Lythraceae). Bot. J. Linn. Soc. 93: 389-403.

Pushkarnath, (1942). Studies on sterility in potatoes. I. The genetics of self-and cross-incompatibility. Indian J. Genet. Pl. Breed 2: 11-36.

_______________(1953). Studies on sterility in potato. IV. Genetics of incompatibility in Solanum aracc papa. Euphytica 2: 49-58.

_______________(1957). Breeding potato varieties in South Asia. Indian J. Genet. Pl. Breed. 17: 197-211.

_______________(1961). Potato breeding and genetics in In-

dia. Indian J. Genet. Pl. Breed. 21: 77-86.
Simmonds, N.W. (1969). Prospects of potato improvement. A. Rep. Scott. Pl. Br. Sta. 48: 18-38.
Sinha, S.K. and Pushkarnath (1964). The relationship of Indian potato varieties to Solanum tuberosum subspecies an[illegible] digena. Indian Potato J. 6: 24-29.
Swaminathan, M.S. (1954). Microsporogenesis in some commercial potato varieties. J. Hered. 45: 265-272.
____________(1955). Overcoming cross -incompatibility among some Mexican diploid species of Solanum. Nature, (Lond.) 176:887-888.
Swaminathan, M.S. and Howard, H.W. (1953). The cytology and genetics of the potato (Solanum tuberosum) and related species. Bibliogr. Genet. 16: 1-192.
Upaadhya, M.D; Chandra, R and Sharma, Y.K. (1981). Potato embryo characteristics and variation. True Potato Seed Letter 2: 1.
Weerasionghe, S.P.R. (1983). The potato for the hot humid tropics: 183-184. In Paroc. Internatl. Congr. Research for the Potato in the year 2000. International Potato Centre (CIP), Lima, Peru.

RESISTANCE BREEDING UNDER COORDINATED PROGRAMME

H.S. Gill and A.S. Kataria

Project Directorate on Vegetable Research,
Division of Vegetable Crops, I.A.R.I, New Delhi-110012

Breeding of vegetable varieties resistant to diseases and pests has greater relevance in Indian conditions, since most of the vegetables are consumed in fresh form which necessitates either complete check on use of insecticides and pesticides or use of chemicals known to have low residual effects. Some viruses and soil borne pathogens are difficult to check even with the use of chemical, which makes it further important to go in for resistance breeding.

At present 24 important vegetable diseases and 10 pests have been identified for resistance breeding programmes at thirty centers under the Directorate of Vegetable Research, as given in Table 1.

Table 1. Diseases and pests included in resistance breeding programme.

Crop	Disease
1. **Brinjal**	Phomopsis blight *(Phomopsis vexans)*
	Little leaf
	Bacterial wilt (*Pseudomonas solanacearum)*
2. **Cauliflower**	Black rot *(Xanthomonas campestris)*
	Sclerotinia rot *(Sclerotinia sclerotiorum)*
	Curd and inflorescence blight *(Alternaria brassicicola)*
3. **Cabbage**	Black rot *(Xanthomonas campestris)*
4. **Chillies**	Fruit rot (*Colletotrichum spp.)*
	Leaf curl virus
5. **Cowpea**	Bacterial blight *(Xanthomonas vignicola)*
6. **Muskmelon**	Powdery mildew *(Sphaerotheca fuliginea)*
	Downy mildew *(Pseudoperonospora cubensis)*
	Viruses
7. **Okra**	Yellow Vein Mosaic Virus
8. **Onion**	Purple Blotch *(Alternaria porri)*
9. **Peas**	Powdery mildew *(Erysiphe polygoni)*
	Fusarium wilt *(Fusarium oxysporum f. pisi)*

J. Prakash and R. L. M. Pierik (eds.), Horticulture – New Technologies and Applications, 37–40.

Table 1. (Contd.)

10.	**Tomato**	Late blight *(Phytopthora infestans)* Bacterial wilt *(Pseudomonas solanacearum)* Early blight *(Alternaria solani)* Spotted wilt Leaf curl
11.	**Watermelon**	Anthracnose *(Colletotrichum spp.)* Blossom end rot

Crop		Insect Pest
1.	**Bitter gourd**	Fruit fly *(Dacus cucurbitae, D. dorsalis)*
2.	**Brinjal**	Shoot and Fruit Borer *(Leucinodes orbonalis)*
3.	**Chillies**	Thrips, mites and pod borer *(Scirto thrips)*, *(Polyphagotarsonamus latus)* and *(Heliothis armigera)*
4.	**Muskmelon**	Red Pumpkin Beetle *(Aulacophora foveicollis)* Fruit Fly
5.	**Okra**	Fruit borer *(Earis fabia, E. insula and E.vettela)* Jassids *(Amrasca biguttula beguttula and Earis spp.)*
6.	**Onion**	Thrips *(Thrips flavus)*
7.	**Tomato**	Fruit Borer *(Heliothis armigera)*

Salient findings:

The list of experiments given above include both screening and testing against diseases and pests.

Screening against diseases: The experiments conducted at various centers indicated the presence of field resistance to various diseases in a number of crops, i.e. Annamalai against little leaf in brinjal, KT-9, KT-16 in cauliflower against black rot, Arka Niketan against Purple Blotch in onion and EC 118277, Sel-2, BWR-5 and LE-79 against bacterial wilt and Ottawa 31 against late blight of tomato. *L. pimpinellifolium* (LE 904), *L. hirsutum glabratum* (LE 1117) were found to have resistance to Leaf Curl Virus at Coimbatore. In okra P-7, Sel 10, Sel-4 and Parbhani Kranti showed resistance to Yellow Vein Mosaic Virus at several locations. Minimum incidence of leaf curl disease was recorded in chilli cultivars ie. CA(P)-1068, LIC-13, X-197, C-2B Mexico and Pant C-1. In tomato screening against Leaf curl virus at Hyderabad, showed that the lines C-17 d, C-11 d, LE-79 and CNo. 112 were promising. A tomato cultivar, EC. 110700 recorded high degree of resistance to Spotted wilt virus at Rahuri.

Screening against pests: Screening of numerous local and exotic vegetable cultivars against pests has so far eluded the authentic results. A number of lines ie, EMS-8 (okra) against jassids, LIC-45 against thrips and LCA -235 against mites in chillies, Plaza local, VL-1 against thrips in onion, and E-3, E-12, E-15 against leaf minor in peas exhibited low degree of incidence of pest under field conditions but these results need further confirmation. Three lines of brinjal, ie. Punjab Barsati (7-3), Pant Samrat, SM 17-4 tested at PAU Ludhiana, were found to carry high degree of resistance to Fruit and Shoot borer and jassids. Among the various breeding lines of okra, a cross PS x Sel-1 showed resistance to cotton jassids at Ludhiana. Another cross, IC 12930 x Sel. 6-2 F_3 x EMS-8 FS-7 was also found to be free from virus incidence and had low borer infestation. In bittergourd a prickly variety Phule BG-4 recorded comparative resistance to fruit fly at Rahuri. In muskmelon two crosses of EC 163888 (phut) x Haramadhu and Pusa Madhuras recorded high degree of resistance/tolerance to red pumpkin beetle and fruit flies.

Testing: All the varieties\lines bred for resistance to diseases and pests at different centers under the project are subjected to multilocation testing. More than 60 newly developed varieties of six vegetables have been tested for 3 to 4 years at different centers. Data has been recorded on percentage of disease incidence and productivity of the resistant cultivars compared with standard susceptible checks for respective diseases. Eight vegetable varieties resistant to some important diseases were identified for recommendation to farmers from the trials concluded during 1990. The recommendations are given in table 2.

Table 2. Resistant Vegetable Varieties Identified During 1990.

Crop/variety Source	Recommended for zone	Resistance to disease	Source of resistance	Av. yield and disease intensity (Three year's average)
Brinjal **BB-7** (Bhubneshwar)	II, V	Bacterial wilt	Gopa local (1)	Av Yield- 422 Q/ha Av. Disease intensity 4.8%
Brinjal **BWR-12** (Hesserghatta)	VIII	Bacterial wilt	Arka sheel (1)	Av. Yield- 425 Q/ha Av. Disease intensity 1.6%

Table 2. Contd.

Okra **P-7** (Ludhiana)	For all Zones	Yellow Vein Mosaic Virus	*A. manihot* ssp. *manihot* (2)	Av. Yield- 425 Q/ha Av. Disease intensity 0.5%.
Okra **PB-57** (Parbhani)	For all Zones	-do-	*A. manihot* ssp. *manihot* (3)	Av. yield-120 Q/ha Tall plants,Dark green leaves and fruits slightly narrow at the base.
Okra **Sel-10** (Hesserghatta)	For all Zones	-do-	*A. manihot* ssp. *tetraphyllus* (4)	Av. yield 115 Q/ha Highly resistant
Peas **PRS-4** (Kalianpur)	IV,VI VII	Powdery mildew	Rachna (1)	Av. yield 100 Q/ha Resistant to Powdery mildew.
Peas **JP-4** (Jabalpur)	IV, VIII	-do-	6588-1 (1)	Av. yield 103 Q/ha Resistant to Powdery mildew.
Tomato **BT-1** (Bhubneshwar)	For all Zones	Bacterial wilt	BWRS-1 (1)	Av. yield 489 Q/ha Av. disease intensity 1.45%.

Some of the resistant vegetable varieties have also been identified earlier besides the ones given in the table above which have already become very popular with the growers. Cauliflower Pusa Shubhra and Pusa Snowball K-1 and cabbage Pusa Mukta from IARI, were identified for resistance to black rot with high yielding ability. Watermelon, Arka Manik from IIHR for resistance to Powdery mildew, Cowpea, Pusa Komal for resistance to Bacterial blight and Pusa Purple Cluster for resistance to bacterial wilt were identified for release by the Project Directorate and some of them have been released by central variety release committee. Multilocational trials on testing of resistant varieties against phomopsis blight of brinjal and black rot of cabbage were carried out for 3 to 4 years and were concluded during 1990, after making suitable recommendations. A new trial was initiated this year on testing of early pea varieties for resistance against Fusarium wilt.

REFERENCES

1. H.S. Gill, *et al.* (1990) Annual Report (Part 1) (1988-89 and 1989-90) Project Directorate on Vegetables. pp 273-292.
2. Thakur, M.R. (1986). Veg. Sci **13**(2): 311-315.
3. Nerkar, Y.S. and Jambhale, N.D. (1985). Indian J. Genet. **45**(2): 261-70
4. Dutta, O.P. (1984). Annual Rep. IIHR, Bangalore. pp 43.

Quality improvement of Punjab grapes

S.S. CHEEMA, A.S. BINDRA AND W.S. DHILLON
Department of Horticulture
Punjab Agricultural University
Ludhiana-141 004, India

Key words - Grapes, dormax, gibberellic acid, ethephon, berry thinning, shot-berries

Introduction

Punjab is emerging as an important grape growing state with more than 2000 hectares under this fruit. Perlette, the dominant table grape cultivar of this region has the desirable characters like seedlessness, early ripening and heavy yields. It has certain drawbacks too, such as over compact clusters, uneven sized berries and high proportion of shot-berries. To avoid damage from pre-monsoon showers and to capture early market, grape growers of Punjab have a tendency to start harvesting the crop before it is fully ripe. Such low quality fruits fail to get remunerative price and causes a distaste for Punjab grapes. As the entire grape production of this area is used for table purpose only, clusters with uniformly bold sized berries having high T.S.S. are desired.

Dormancy breaking chemicals like hydrogen cynamide are known to start early bud burst and advance fruit ripening. Reduction of crop load by berry thinning leads to improvement in berry size and quality and is being practised for table grapes in horticulturally advanced countries. Studies were, therefore, undertaken to enhance fruit ripening and improvement of berry quality by using chemicals and mechanical methods.

Method

Studies were carried out on full grown, horizontal trellis trained Perlette grapevines, spaced at 3m x 3m, Dormax (50% H_2CN_2) @ 1.0, 1.5 and 2.0% was sprayed on dormant grapevines on 25 December, 1 January and 7 January, 1989-90

J. Prakash and R. L. M. Pierik (eds.), Horticulture – New Technologies and Applications, 41–44.

immediately after pruning. Some of the Dormax (2.0%) treated vines were berry thinned by removing 25, 50, 75% berries about one week after fruit-set. Similarly to improve berry quality, few Dormax (2.0%) treated vines were superimposed with 600 ppm (ai) ethephon (2-chloroethyl-phosphonic acid) at the time of colour break stage. In a separate experiment flower thinning was done by 30, 40, 50 ppm GA_3, 1000, 1500, 2000 ppm (ai) Sevin and 5, 10, 20 ppm (ai) ethephon. The experiments were laid out according to C.R.D. design. Observations on bud sprouting, flowering, fruit ripening and its quality were recorded. Standard methods were used to analyse physico-chemical fruit parameters.

Results

Dormax (2.0%) sprayed on 25 December proved to be most effective for advancing fruit ripening among all the dates and concentrations. This treatment advanced bud burst by 30 days. Flowering was earlier by 20 days and the fruit ripened eight days earlier than the untreated vines (table 1). The quality of the fruit on this date was much superior over control.

Gibberellic acid (40 ppm) gave the desirable flower thinning and the results of this chemical only are presented in table 2. This treatment resulted in reducing berry number per unit of the lateral significantly over the control. It also improved berry size, fruit quality and reduced shot-berries. Mechanical berry thinning improved berry size and quality but the results varied according to severity of berry thinning (table 3). The highest berry weight (2.56 g) and T.S.S. (19.8%) as well as least acidity (0.60%) were obtained in the 75% berry removal treatment, but this reduced the bunch weight drastically. It is interesting to note that 25% berry removal improved the bunch weight in addition to improving fruit quality as compared to unthined vines.

Ethephon (600 ppm) sprayed at colour break stage improved berry T.S.S. and reduced juice acidity considerably. However, berry as well as cluster weight were not affected (table 4).

Discussion

Dormax advanced the bud break and fruit ripening by breaking the bud dormancy. Its action is through the reduction of catalase activity (Shulman et al. (1985). The results are in line with the findings of McColl (1986) who observed advanced fruit ripening in table grapes with cynamide treatment in central Australia. Gibberellic acid is an inhibitor of pollen germination and thus affected flower thinning. Flower thinning improved berry size and quality by reducing the competition for photosynthates and making these available for the remaining berries. Similar observations were reported by Bravdo *et al.* (1985). Ethephon released the ethylene which is a known fruit ripening hormone which enhanced fruit ripening by increasing T.S.S. and reducing juice acidity. The findings corroborate the work of Weaver and Pool (1971).

References

Bravdo, B., Hepner, V., Loinger, C., Cohen, S. and Tabacman, N. (1985) Effect of crop level and crop load on growth, yield, must and wine composition and quality of 'Cabernet Sauvignon'. Amer. J. Enol. Viticult. 36 : 125-131.

McColl, C.R. (1986) Cynamide advances the maturity of table grapes in central Australia. Aust. J. Exp. Agric. 26:505-509.

Shulman, Y., Nir, G. and Lavee, S.(1985) Oxidative process in bud dormancy and the use of hydrogen cynamide in breaking dormancy. 5th International Symposium on 'Growth regulators' in fruit production in Rimini (Italy), Sept. 2-6, 1985.

Weaver R.J. and Pool R.M. (1971) Effect of (2-chloroethyl) phosphonic acid (ethephon) on maturation of *Vitis vinifera* L. J. Amer. Soc. Hort. Sci., 96 : 725-727.

Table 1. Effect of Dormax (50% H_2CN_2) on fruiting in 'Perlette' grapes

Treatments	Characters				
	Bud burst	Full bloom	Harvesting	TSS (%)	Acidity (%)
Dormax (2%)	24 Jan	16 Mar	26 May	14.4	0.70
Control	23 Feb	6 Apr	3 Jun	10.2	1.00
Difference (days)	30	20	8	-	-

Table 2. Effect of chemical treatments on flower thinning in 'Perlette' grapes

GA_3 (ppm)	Berries/cm Lateral(no.)	Berry weight (g)	Shot-berries (%)	TSS (%)	Acidity (%)
30	4.0	1.6	14.5	15.2	0.74
40	3.6	1.8	13.0	15.4	0.70
50	2.9	1.9	12.8	15.0	0.72
Control	5.5	1.3	26.4	14.2	0.74
L.S.D.(0.05)	1.1	0.3	3.1	0.82	N.S.

Table 3. Effect of berry thinning on fruit quality in 'Perlette' grapes

Berry thinning (%)	Bunch weight (g)	Berry weight (g)	TSS (%)	Acidity (%)
25	345	1.92	18.0	0.64
50	172	2.22	19.4	0.63
75	115	2.56	19.8	0.60
Control	332	1.51	15.2	0.82
L.S.D.(0.05)	56	0.21	0.7	0.07

Table 4. Effect of ethephon on fruit quality in 'Perlette' grapes

Ethephon (ppm)	Bunch weight (g)	Berry weight (g)	TSS (%)	Acidity (%)
600	348	1.56	18.4	0.63
0	340	1.48	15.2	0.82
L.S.D. (0.05)	N.S.	N.S.	1.1	0.09

N.S. = Non significant

Exploitation of alien genes for yellow vein mosaic resistance in okra

B.S.DHANKAR, B.S.SAHARAN, V.S.HOODA AND P.SINGH
Department of Vegetable Crops, Haryana Agricultural University, Hisar-125 004, India

KEY WORDS
Yellow vein mosaic, resistance, A. esculentus, A. manihot, A. tetraphyllus.

INTRODUCTION

The cultivated okra species Abelmoschus esculentus (L.) Moench is grown as a vegetable crop in the tropical and subtropical regions of Asia, Africa, America and the temperate regions of the Mediterranean. In India, the infection of yellow vein mosaic (YVM), a white fly (Bemisia tabaci Gen.) transmitted virus is the main production constraint during rainy season. The extent of loss in yield varies from 50 to 90 % depending on the crop stage when it is infected by the virus (Sastry and Singh, 1974). The resistance to YVM could not be found in the cultivated species. However, the sources of resistance have been identified in the related wild species (Sandhu et al., 1974; Arumugam et al., 1975; Thakur, 1976; Ugale et al., 1976). YVM resistance in the related wild species is controlled by two complementary dominant genes (Thakur, 1976) and a single dominant gene (Jambhale and Nerkar, 1976).

Recently, as a result of success achieved in India in transferring YVM resistance genes from the wild species into the cultivated species YVM resistant varieties such as P7 (A.esculentusxA. manihot ssp. manihot) (Thakur, 1986), Parbhani Kranti (P.Kranti) (A. esculentus X A. manihot) (Jambhale and Nerkar, 1986) and Selections 4 and 10 (A. esculentus x A. manihot ssp. tetraphyllus) (Dutta, personal communication) have been developed. Our main objective in the present investigation was fruther exploitation of these alien genes in an intervarietal hybridization programme to improve upon the existing cvs.

METHODS

At Haryana Agricultural University, Hisar a YVM resistance breeding programme was initiated during 1985 using these varieties as the sources of resistance and simple crosses were made between P 7 and P. Kranti as males and 7 susceptible commercial cultivars as females (identity withhold) for production of 14 F1 hybrids.

The F1 and subsequent generations were grown in the field under epiphytotic conditions created by growing rows of Pusa Makhmali (highly susceptible to YVM) inbetween and around the progenies. In order to advance the generations, progenies were also grown in the disease free spring-summer season. Pedigree method of selection was adopted. The number of plants and families selected in each generation was as suggested for this crop (Swarup, 1977). The selection in the segregating generations was adopted for resistance to YVM, earliness, number of fruits, smoothness and fruit colour in the rainy

J. Prakash and R. L. M. Pierik (eds.), Horticulture – New Technologies and Applications, 45–48.

season crop.

We have developed 9 resistant strains of okra, incorporating YVM resistance of P7 and P.Kranti. These strains in F 7 and F 8 generations were evaluated in a replicated yield trial for their performance against check varieties (P7, P.Kranti and Pusa Sawni) during rainy season in 1989 and 1990. The trial was conducted in randomised block design with three replications. Each plot contained 72 plants of the test strain at a spacing of 45x 25 cm. Infector rows of Pusa Makhmali were grown inbetween the genotypes and around replications to avoid plant escape from the disease. Two initial insecticidal sprays were done against leafhopper to boost up plant growth. Records on YVM occurrance (% plants infected) were made at 30, 60 and 90 days after sowing.

RESULTS

All the resistant strains except HRB 12 did not show YVM symptoms over the years (Table 1). The resistant check P7 had less number of susceptible plants than P.Kranti. The cultivar Pusa Sawni a commercially grown variety had 91 and 100 % infected plants at 90 days after sowing.

The strains were significantly early in flowering (days to first flowering) and took 40 to 41 days as compared to 46 and 48 days in the resistant checks (Table 2). The differences in the plant height between strains and the checks were not significant. On the basis of mean performance, highest branch number was observed in HRB 12-1, HRB 55, HRB 54 and HRB 9-1, however, some of these strains showed wide variation between the years. The resistant checks and HRB 12-1 showed stability for this character.

Highest mean performance for yield over checks was exhibited by HRB 55 (86.6 q/ha) and HRB 9-2 (82.4 q/ha) and these strains maintained their superiority in the individual year as well (Table 3). The strains HRB 12-1 and HRB 54 were unstable for yield in different years. Among checks, P.Kranti and Pusa Sawni also changed their positions for this character in individual year. The strains HRB 55 and HRB 9-2 produced 8 fruits per plant as compared to 6 and 5 in P.Kranti and P7, respectively (Table 3). The weight per fruit recorded was more than 8 g in HRB 10-1, HRB 10, HRB 55 and 8 g in HRB 9-2 (all statistically at par) as compared to 6.4, 6.6 and 6.8 in Pusa Sawni, P7 and P.Kranti, respectively. High yield in the strains HRB 55 and HRB 9-2 was due to high number of fruits and higher fruit weight over the checks.

TABLE 1 Performance of YVM resistant strains and checks against YVM infection

Strain/variety	percent infected plants					
	30 days		60 days		90 days	
	1989	1990	1989	1990	1989	1990
HRB 9-1	0	0	0	0	0	0
HRB 9-2	0	0	0	0	0	0
HRB 10	0	0	0	0	0	0
HRB 10-1	0	0	0	0	0	0
HRB 12	0	0	1.0	0	2.0	0
HRB 12-1	0	0	0	0	0	0
HRB 54	0	0	0	0	0	0
HRB 55	0	0	0	0	0	0
HRB 55-2	0	0	0	0	0	0
P.Kranti	1.0	0	3.0	2.0	8.0	5.9
P7	0	0	1.5	1.0	6.0	4.8
Pusa Sawni	36.0	24.0	62.0	48.0	100.0	91.0

TABLE 2 Days to first flowering, plant height(cm) and number of branches/plant of YVM resistant strains and checks

Strain/Variety	Days to first flowering			Plant height(cm)			Branches/Plant		
	1989*	1990	Mean	1989*	1990	Mean	1989*	1990	Mean
HRB 9-1	40.7	41.9	41.3	71.8	89.6	80.7	4.5	3.2	3.8
HRB 9-2	39.3	40.1	39.7	74.1	90.3	82.2	3.7	2.4	3.0
HRB 10	41.0	42.1	41.5	72.5	98.1	85.3	2.6	2.8	2.7
HRB 10-1	41.3	41.1	41.2	77.4	96.1	86.7	3.3	2.7	3.0
HRB 12	39.7	40.8	40.2	85.9	103.8	94.8	3.7	2.5	3.1
HRB 12-1	40.0	40.2	40.1	81.6	109.8	95.7	4.0	4.0	4.0
HRB 54	40.0	40.3	40.1	77.5	93.3	85.4	5.5	2.3	3.9
HRB 55	39.3	40.0	39.6	78.3	99.5	88.9	5.4	2.5	3.9
HRB 55-2	39.7	39.9	39.8	75.1	94.1	84.6	4.5	2.4	3.4
P.Kranti	47.6	48.3	47.9	90.8	104.9	97.8	3.2	3.2	3.2
P7	45.7	46.5	46.1	77.2	84.2	80.7	2.9	2.8	2.8
Pusa Sawni	44.0	45.4	44.7	81.0	88.1	84.5	3.9	3.3	3.6
L.S.D.(0.05)	1.62	2.58		NS	NS		1.77	0.85	

*Severe leafhopper attack

TABLE 3 Fruit yield and its components of YVM resistant strains and checks

Strain/Variety	Fruit yield (q/ha)			Fruits/Plant			Weight/Fruit(g)		
	1989*	1990	Mean	1989*	1990	Mean	1989*	1990	Mean
HRB 9-1	42.9	88.8	65.8	4.0	8.7	6.3	4.3	11.4	7.8
HRB 9-2	63.3	101.4	82.4	5.8	9.8	7.8	4.3	11.6	7.9
HRB 10	47.9	93.5	70.7	3.6	8.5	6.1	5.5	12.2	8.8
HRB 10-1	40.1	101.1	70.6	2.9	9.2	6.1	5.4	12.3	8.8
HRB 12	55.3	79.8	67.5	6.0	8.1	7.0	3.7	10.1	6.9
HRB 12-1	25.5	92.2	58.8	2.3	9.1	5.7	4.4	11.3	7.8
HRB 54	32.4	98.6	65.5	3.5	9.5	6.5	3.9	11.5	7.7
HRB 55	71.1	102.1	86.6	6.0	9.7	7.8	4.9	11.7	8.3
HRB 55-2	42.7	98.8	70.7	4.0	10.2	7.1	4.3	10.7	7.5
P.Kranti	38.6	64.4	51.5	3.7	7.6	5.6	4.2	9.4	6.8
P 7	36.7	50.3	43.5	3.9	5.8	4.8	3.7	9.6	6.6
Pusa Sawni	44.6	60.5	52.5	4.3	7.7	6.0	4.2	8.7	6.4
L.S.D.(0.05)	18.14	13.33		2.69	1.13		0.94	0.97	

*Severe leafhopper attack

DISCUSSION

In vegetable crops, particularly where diseases are more devastating, resistance breeding is the only means to increase their productivity. The setback in okra cultivation in the country was due to the break down of YVM resistance in the cultivar Pusa Sawni, developed about 3 decades ago (Singh et al., 1962). Recently, the successful transfer of YVM resistance genes (limited to the wild species only) from the related wild species into the agronomically desirable okra has been considered a classical work of the Indian research workers. They incorporated YVM resistance into A. esculentus from A. manihot, A. manihot ssp. manihot and A. manihot ssp. tetraphyllus (symptomless carriers) by over-coming barriers of sterility, cross incompatibility and transmission of undesirable linked genes.

Our efforts were concentrated to further exploit the alien genes responsible for YVM resistance (which have been already transferred into the cultivated species) by undertaking an intervarietal hybridizataion programme. We achieved success in inducing earliness, increased fruit number and fruit weight which led to increased yield in the YVM resistant strains. The strains HRB 55 and HRB 9-2 were found superior to other strains and check varieties because of all these desirable characters present in them. The fruit quality of these strains is better due to smooth surface, appealing green colour, low fibre content and medium seed number. These strains indicated their superiority over other genotypes in 1989, the year of severe leafhopper (Amrasca biguttula biguttula) infestation and in 1990 as well. These strains are being tested at multilocations and farmers field.

It may be concluded that YVM resistance in the related wild species of okra used as sources of resistance to develop P 7 and P.Kranti is of "symptomless carrier" type nature, therefore, it is yet to be seen that how long the resistance of HRB 55 and HRB 9-2 will last.

ACKNOWLEDGEMENTS

We are thankful to Dr. V.K.Srivastava, Professor and Head Vegetable Crops for providing facilites. Our thanks are also due to Dr. B.S.Dahiya, Professor of Seed Technology and Dr. Suringder Singh Professor of Agronomy for their help and valuable suggestions in the preparation of this manuscript.

REFERENCES

1 Arumugam R., Chelliah S. and Muthukrishnan C.R. (1975) Abelmoschus manihot a source of resistance to bhendi yellow vein mosaic. Madras Agricultural Journal 62: 310-312.

2 Jambhale N.D. and Nerkar Y.S. (1981) Inheritance of resistance to okra yellow vein mosaic disease in interspecific crosses of Abelmoschus. Theoretical and Applied Genetics 60 : 313-316.

3 Jambhale N.D. and Nerkar Y.S. (1986) 'Parbhani Kranti', a yellow vein mosaic resistant okra. HortScience 21 : 1470-1471.

4 Sandhu G.S., Sharma B.R., Singh B. and Bhalla J.S. (1974) Sources of resistance to jassids and white fly in okra germplasm. Crop Improvement 1 : 77-81.

5 Sastry K.S.M and Singh S.J. (1974) Effect of yellow vein mosaic infection on growth and yield of okra crop. Indian Phytopathology 27 : 294-297.

6 Singh H.B., Joshi B.S., Khanna P.P. and Gupta P.S. (1962) Breeding for field resistance to yellow vein mosaic virus in bhindi. Indian journal of Genetics 22 : 137-144.

7 Swarup V. (1977) Breeding Procedures for Cross Pollinated Vegetable Crops. ICAR Publication, New Delhi, India, 73 pp.

8 Thakur M.R. (1976) Inheritance of resistane to yellow vein mosaic virus in a cross of okra species, A. esculentus x A. manihot ssp. manihot, SABRAO Journal 8 (1) : 69-73.

9 Thakur M.R. (1986) Breeding for disease resistance in okra. Vegetable Science 13 (2) : 310-315.

10 Ugale S.D., Patil R.C. and Khupse S.S. (1976) Cytogenetic studies in the cross between A. esculentus and A. tetraphyllus. Journal of Maharashtra Agricultural University. 1 (2-6) : 106-110.

USE OF VESICULAR - ARBUSCULAR MYCORRHIZA (VAM) AS BIOFERTILIZER FOR HORTICULTURAL PLANTS IN DEVELOPING COUNTRIES

S.B. SULLIA, DEPARTMENT OF BOTANY, BANGALORE UNIVERSITY, BANGALORE 560 056.

1. INTRODUCTION

Mycorrhiza is the product of an association between a fungus and plant root. Vesicular-arbuscular mycorrhiza (VAM) is formed by the symbiotic association between certain phycomycetous fungi and angiosperm roots. The fungus colonizes the root cortex forming a mycelial network and characteristic vesicles (bladder-like structures) and arbuscules (branched finger-like hyphae). The mycelia are aseptate or septate ramifying intercellularly thus causing little damage to tissues. The arbuscules are the most characteristic structures, formed intracellularly and probably having an absorptive function. The vesicles are terminal swellings of hyphae formed inter and intracellularly having a storage function. There are six genera of fungi belonging to Endogonaceae which have been shown to form mycorrhizal associations: Glomus, Gigaspora, Acaulospora, Entrophospora Sclerocystis and Scutellospora. These are mainly identified by their characteristic spores and sporocarps which are formed mostly in the soil surrounding the roots and rarely inside the roots. The identification of VAM fungi directly from roots has been difficult. One of the striking features of VAM fungi is their very wide host range which includes angiosperm species belonging to almost all the families. Even the roots of some aquatic plants are colonized by VAM fungi.

2. WHAT IS SPECIAL ABOUT VAM?

Though the mycorrhizae have been known for a century now, mycorrhizal research gained impetus only recently with the demonstration that VA-mycorrhizal fungi increase the efficiency of plants in phosphate utilization (Baylis, 1967; Gray and Gerdemann, 1969; Mosse et al., 1973). Since the 1970s, agricultural scietists have been trying to use VAM fungi for better utilization of phosphorus by plants in poorly endowed soils. The VAM mycelia are capable of extending the region of absorption reaching away from the phosphorus depletion zone which forms around absorbing roots. These fungi also make available to plants insoluble phosphates such as rock phosphate.

It has been shown in recent years that the VAM fungi confer on host plants several other benefits in addition to enhancement of phosphate uptake. These benefits are biological control of root diseases (Shonbeck, 1979), nodulation and nitrogen fixation in legumes (Crush,1974; Mosse et al.,1976) hormone production (Schultz et al.,

J. Prakash and R. L. M. Pierik (eds.), Horticulture – New Technologies and Applications, 49–53.

1979; Allen *et al*., 1980), drought resistance (Powell and Bagyaraj, 1984) and increased uptake of several elements such as N, K, Zn, Mg, Ca and S (Gray and Gerdemann, 1979; Hayman, 1982). VA mycorrhizae have, therefore, found use in agriculture, horticulture afforestation and reclamation of deserts.

3. SOIL MICROORGANISM AND VAM FUNGI

Interactions between the general soil microbiota and VAM fungi have been studied extensively (reviewed: Bagyaraj, 1984). MOre recent works hae shown suppression of soil-borne plant pathogens due to prior inoculation of plants with VAM fungi especially with reference to wilt and root rot pathogens. Significant reduction of infection by *Rhizoctonia solani* due to mycorrhizal inoculations has been shown in several hosts. Seed pelleting with sporcarps of the VAM fungi gave maximum protective effect against*Rhizoctonia solani*. Control of *Aphanomyces euteiches*, the root rot pathogen of peas has been shown. However, most of the studies on biological control of root diseases have been conducted under controlled conditions in glasshouses and the results obtained may not be necessarily reproducible under field conditions.

4. INOCULUM PRODUCTION

The mass production of VAM inoculum for commercial purpose is beset with practical problems that are yet to be fully overcome. One of the major problems is that of growing VAM fungi in aexnic cultures. These fungi need host roots for growth and remain as dormant spores in the absence of the host. The axenic culture of *Glomus aggregatum* has been reported recently (Janardhanan *et al*., 1990) but the method is tedious and the growth and sporulation in the medium is poor. Initial growth of the fungus on the medium requires stimulation from the roots of axenically cultured plants of the host palmarosa (*Cymbopogon martinii* var. *motia*). With our present knowledge it may be safer to consider VAM fungi as obligate symbionts and this fact has restricted researches on the genetics and molecular biology of this group of fungi. The existing methods of raising bulk inoculum are based on (i) plants grown on solid substrates, (ii) plants grown in hydroponic cultures and (iii) root organ cultures in agar media.

4.1. *Inoculum on plants grown on solid substrates*

Solid substrates for raising plants used till recently include soil, sand, soil-sand mixture and synthetic inert materials such as vermiculite, perlite and soilrite. Vermiculite has been most useful because of its layered structure which favours sporulation of VAM fungi (Mosse, 1990). It is necessary to begin inoculum production with a single species free from contaminants. Spores obtained from soil, identified ans surface-sterilized with 2% chloramine T and 200 mg L streptomycin or 2% sodium hypochlorite have been used as initial inoculum. Sterilization of the solid substrae is done (i) by autoclaving at 15 psi for one hour, (ii) by steaming 2 to 3 times (iii) by fumigation with methyl bromide and (iv) by gamma irradiation. Nutrient media provided in the substrate should not contain large amounts of availabe phosphorus which may suppress colonization of the fungus. However, the medium should be fertile enough to allow the growth of the plant (menge, 1984). A wide variety of host plants have been used for inoculum production, e.g., citrus, sorghum, coleus, onion, pepper, corn, clover, groundnut, cotton and guinea grass. Grasses have been most useful because of the plentiful adventitious roots produced at lesser depths in the substrate.

Several new substrates have been recently introduced for VAM inoculum production and it is claimed that these are superior to vermiculite or perlite. "Calcined montmorillonite clay" marketed under the trade name "Turface" proved to be a good substrate for plant growth and abundant mycorrhizal colonization. A more recent introduction is the "Light aggregated clay particles (LECA) ", also called "Blahton". The advantage of these clay based substates is that the physical structure of the porous material and the absorptive capacity of the clay function as a slow release mechanism for nutrients particularly phosphorus (Mosse, 1990).

4.2 Nutrient Flow cultures

Circulating hydroponic cultures have been used for inoculum production eliminating the necessity of soil or other solid substrates. Inoculum in this case becomes purely root based. In nutrient flow technique (NFT) plant roots grow between plastic sheets or capillary matting. A weak nutrient solution is pumped into the inclined plane where it flows down over the roots as a film and into a container from where it is recyled. Root mats several centimetres thick build up under such conditions. By this method, root mats of beans (*Phaseolus vulgaris*) containing upto 60% mycorrhizal roots with attached floating films of mycelium, spores and sporocarps have been obtained.

4.3 Root Organ Cultures

VAM colonization of root organ cultures of clover in modified White's medium was reported. However, the method did not hold promise for large scale inoculum production as the root growth remained relatively restricted. By using roots of *Convolvulus sepium* transformed with Ri-T-DNA (root inducing plasmid genes), carried by *Agrobacterium rhizogenes*, the growt potential of roots greatly increased (Mosse, 1990)

5. CONCLUSIONS

Introduction of new substrates such as expanded clay aggregates has made the inoculum packets less bulky and lighter reducing transport costs. If root organ culture method is further improved using genetic engineering techniques, there is hope for still better inoculum package that could also be trusted for its purity. Use of VAM inoculum may minimize fertilizer inputs. though VAM incolum is claimed to be a biofertilizer cum biopesticide, its potential as a biopesticide is limited to the containment of root pthogens, and not foliar pathogens. The protective effect of VAM against virus and bacterial diseases has not been proved. Exaggerated claims on the benefits of VAM, therefore, may be misleading and may divert the attention of the users from the real scope of VAM use.

6. REFERENCES

Allen M.F., Moore T.S. and Christensen M. (1980) Phytohormone changes in Bouteloua gracilis infected by vesicular-arbuscular mycorrhizae. I. Cytokinin increases in the host. Can. J. Bot. 58:371-374 .

Bagyaraj D.J. (1984) Biological interactions with VA Mycorrhizal fungi. In: Powell C. Ll. and Bagyaraj D.J. eds., VA Mycorrhiza, pp 131-153, CRC Press, Florida.

Baylis G.T.S. (1967) Experiments on the ecological significance of phycomycetous mycorrhizas. New Phytol. 66: 231-243.

Baylis G.T.S. (1972) Fungi, phosphorus and evolution of root systems. Search 3 : 257-258.

Crush J.R. (1974) Plant growth responses to vesicular-arbuscular mycorrhiza. VII. Growth and nodulation of some herbage legumes. New Phytol. 73:743-749.

Gray L.E. and Gerdemann J.W. (1969) Uptake of phosphorus-32 by vesicular-arbuscular mycorrhizae. Plant Soil 30:415-422.

Gray L.E. and Gerdemann J W. (1973) Uptake of sulphur-35 by vesicular-arbuscular mycorrhizae. Plant soil 39: 787-689.

Hayman D.S. (1982) The physiology of vesicular-arbuscular endomycorrhizal symbiosis. Can.J. Bot. 61:944-963.

Janardhanan K.K.,Gupta M.L. and Husain A. (1990) Axenic culture of a vesicular-arbuscular mycorrhizal fungus. Curr. Sci. 59: 509-513.

Menge J.A. (1984) Inoculum production. In:Powell C.Ll. and Bagyaraj D.J. Eds., VA Mycorrhiza, pp 187-203, CRC Press, Florida.

Mosse B., Hayman D.S. and Arnold D.J. (1973) Plant growth responses to vesicular-arbuscular mycorrhiza V. Phosphate uptake by three plant species from P-deficient soils labelled with P-32. New Phytol. 76:331-342.

Powell C. Ll. and Bagyaraj D.J. (1984) VA-Mycorrhizae: Why all the interest? In: Powell C.Ll. and Bagyaraj D.J. Eds., VA Mycorrhiza, pp 1-3, CRC Press, Florida.

Schultz R.C., Kormanik P.P., Bryan W.C. and Brister G.H. (1979) Vesicular-arbuscular mycorrhizae influence growth but not mineral concentrations in seedlings of eight sweetgum families. Can. J. For. Res. 9:218-223.

Shonbeck F. (1979) Endomycorrhiza in relation to plant diseases. In:Schippers B and Gams W.Eds., Soil-borne Plant Pathogens, pp 271-280, Academic Press, London.

Foliar feeding of zinc and iron in peach

G.P.S. GREWAL, A.S. DHATT AND P.P.S. MINHAS
Department of Horticulture
Punjab Agricultural University
Ludhiana-141 004, India

Key words - Peach, leaf Zn and Fe

Introduction

The low-chill peach cv. Flordasun is extensively grown in the Punjab (India) for fresh fruit consumption. It is budded on the seedlings of Sharbati/wild peach. The trees are vigorous, the new shoots grow more than a metre and both the vegetative and fruit growth and maturation occurs in the summer (April/May). The trees exhibit visual symptoms of zinc and iron deficiency during this period which if not corrected, affect the fruit quality as well as the next year cropping adversely. An attempt was made to increase the leaf contents of zinc and iron by foliar sprays of their salts and chelates and to correct the leaf chlorosis.

Methods

The full grown trees of peach cv. Flordasun were sprayed with aqueous solution of $ZnSO_4$ and $FeSO_4$ (0.4% or 0.8% + 0.4% lime) alone or in combination; Chelfer (0.4% Aries Agro-vet India), Tracel-2 1.6% (Rallies India); Bhavtone (1.6% Ram Kishan Middha), or 0.2% Na Fe EDTA and tubewell water. The spray was given to a slight run-off in end of April when the chlorosis had just started to appear. The mid-shoot levels from current season's growth were collected during mid May and analysed for total Zn and Fe contents (ppm dry wt. basis) following standarised leaf washing, sample preparation and analysis procedures for micro-nutrients.

Results

A perusal of the data in table 1 reveals that all the spray materials increased the leaf Zn and/or Fe content

J. Prakash and R. L. M. Pierik (eds.), Horticulture – New Technologies and Applications, 55–57.

over the untreated trees. The maximum leaf Zn (38 ppm) was recorded with sprays of 0.8% $ZnSO_4$ + 0.4% lime over the control (12 ppm Zn). It was better than all the other treatments including Tracel-2 and Bhavtone 0.2%.

Na Fe EDTA spray was more effective for increasing leaf Fe contents as compared to Ferrous Sulphate (0.4% and/or 0.8%), Chelfer 0.4%, Tracel-2 and Bhavtone. However, a combined spray of $ZnSO_4$ 0.8% + $FeSO_4$ 0.8% neutralized with 0.4% quick lime most effective for increasing the leaf contents of both the elements in Flordasun peach, than single application of $ZnSO_4$ and $FeSO_4$ over the control. All the treatments were partially effective in alleviating the chlorotic condition of leaves.

Discussion

The increase in total leaf Zn and Fe was recorded following their spray applications showing thereby, that peach leaves are able to absorb zinc and iron from inorganic salt mixtures as well as chelated form. The leaf chlorosis was also corrected since the spray solutions were freshly prepared before application and no typical antagonism between absorption of zinc and iron is visible from the results. The beneficial effects of Zn and Fe sprays on peach have already been reported from the Punjab (Singh *et al*., 1986).

References

Singh, K., S.S. Grewal and A.S. Dhatt. (1986) Effect of foliar sprays of iron and zinc on leaf chlorosis and fruit quality in peach (*Prunus persica* Batsch.). Haryana J. Hort. Sci. 15 : 4-8.

Table 1. Effect of foliar sprays of Zn and Fe salts/ chelates on leaf Zn and Fe content in peach

Treatments	Leaf Zn (ppm)	% increase over control	Leaf Fe (ppm)	% increase over control
0.4% $ZnSO_4$	31.0	158.33	-	-
0.8% $ZnSO_4$ + 0.4% Lime	38.0	216.66	-	-
0.4% $FeSO_4$	-	-	170.0	70.0
0.8% $FeSO_4$ + 0.4% Lime	-	-	190.0	90.0
0.4% $ZnSO_4$ + 0.4% $FeSO_4$	33.0	175.0	170.0	70.0
0.8% $ZnSO_4$ + 0.8% $FeSO_4$ + 0.4% Lime	39.0	225.0	204.0	104.0
0.2% NaFe EDTA	-	-	200.0	100.0
Chelfer 0.4%	-	-	180.0	80.0
Tracel-2 1.6%	22.0	83.3	165.0	60.0
Bhavtone 1.6%	26.0	116.6	145.0	45.0
Untreated Control	12.0	-	100.0	-

Conservation of wild plants of Horticultural importance from Tumkur District, Karnataka

C.G.KUSHALAPPA AND V.BHASKAR
Department of Farm Forestry, University of Agricultural Sciences, GKVK, Bangalore-560 065

Key words: Conservation, Dryland Horticulture, Germplasm, Tumkur

Plants are part of our human heritage. This biological treasure is under constant threat of extinction due to many reasons. Never before the rate of extinction of ancestral plants has been this high. To meet increasing demands of our population there is a conflict between agricultural modernization to optimise production and the preservation of indigenous agriculture along with the genetic diversity found in these areas associated with agricultural origins and development.

Germplasm is the source of genetic potential of living organisms. The potential represented in a gene pool is the foundation for our biological renewable crops in agriculture, horticulture and forestry. So it is essential that wild germplasm should be preserved for the survival of mankind.

In the recent days there has been growing interest among horticulturists to grow new and native species as many of our plants have become common. South Indian Flora is ancient and has large number of endemic plants. There are very few attempts made in India to introduce wild plants and most of these efforts are with respect to orchids from N.E. and Himalayan forests. For plant introduction to dryland horticulture one needs to introduce plants from areas which have similar climatic and edaphic conditions so that plants adapt themselves well. Very little is known on the potential of wild plant genetic resources in dry plains of tropics. Karnataka State possesses nearly 1,39,360 sq.km. of semiarid area. Tumkur, a typically dry district receiving less than 750 mm annual rainfall, was selected for the present study to explore the plant weath and to identify the species that could be introduced for dryland horticulture and forestry. The results of the botanical exploration carried out between 1985-88 revealed that even though the district falls under dry plains it is quite rich in botanical wealth. 2090 collections were made and 950 species have been identified which is significant for a

J. Prakash and R. L. M. Pierik (eds.), Horticulture – New Technologies and Applications, 59–62.

dry district.

From the collection made the wild germplasms which could be introduced to dryland horticulture are listed below:

Wild fruit yielding plants

The following fruit yielding trees occur in wild and their germplasms could be exploited for introduction and further improvement.

Anona squamosa (Custard apple): The rocky hills of the district have many varieties which could be used in improvement of this minor fruit.
Buchanania lanzan (Maradi): The seeds contain a white endosperm that is used as a condiment.
Diospyros melanoxylon (Tendu or Tupre): Edible fruit yielding tree, better known for its use in beedi making.
Feronia elephantum (Wood apple): Yields fruits which are used in juice making and the juice is medicinal. A new rare variety with large oblong fruits has been identified and efforts are made for conserving and propagating this tree.
Grewia tiliifolia (Thadaslu): Tree yields dark red attractive sweet fruits used in juice making.
Emblica officinalis (Goose berry): A large germplasm is available which could be used for introduction and improvement.
Xeromphis spinosa ('Negare'): A throny shurb with attractive orange coloured fruits which are not eaten by birds. The outer peel is bitter and inner pulp is acidic and seed has white endosperm.
Spondias anacardium ('Amate'): Its unripe fruits are used in pickling.
Semicarpus anacardium (Indian marking nut tree or 'Karigeru'): Locally, the fleshy part of the false fruit is dried in sun on rocks and the dried fruits are made into beads and sold in market. The fruit is a nut and it is used for making ink used in marking cloths by laundrymen. The nut has been reported to have a chemical used in cancer curing.
Syzygium jambolana (Jamoon): A very attractive sweet fruit and medicinal also. The forests in the district have wide variety of these trees.
Tamarindus indica (Tamarind): A important tree for dryland, horticulture. The villages in the district have many varieties which could be identified and used in improvement.

Medicinal plants

This district has one of the richest storehouses of wild medicinal plants. Sidderabetta, Devarayanadurga, Madhugiri, Thimmalapura are some of the well known forest areas, where medicinal plants are concentrated. In a earlier study conducted by Yoganarashiman et al. (1982) nearly 143 plants of medicinal importance were listed from the district. The present study reports further additions to the plants of medicinal value. They are: Abutilon indicum, Abrus precatorius, Aloe vera, Asparagus racemosus, Boswellia serrata, Careya arborea, Chloroxylon swietenia, Coleus spicatus,

Commiphora mukul, Dioscorea oppositifolia, D.pentaphylla, Gloriosa superba, Gymnema sylvestre, Helicteres isora, Holarrhena antidysenterica, Jatropha curcas, Litsea deccanensis, Martynia annua, Memecylon umbellatum, Nicandra physalodes, Phyllanthus embilica, Sarcostemma acidum, Sida cordifolia, Strychnos nux-vomica, Terminalia chebula, T.bellirica, Tylophora indica, Vitex negund, Withania somnifera, Wrightia tinctoria, Zingiber zerbmbet.

Plants for Floriculture and Landscaping

The district has many wild plants which could be introduced to dryland, horticulture and landscaping. The authors have reported 7 species of orchids of which 3 are epiphytic and 4 land orchids (Kushalappa and Bhaskar, 1989).

Epipsytic orchids: Luisia zeylanica, Vanda tessellata, Vanda testacea.
Land orchids: Habenaria longicorniculata, Habenaria marginata, Habenaria rariflora, Aphyllorchis montana.
Other flowering and foliage plants: Capparis grandiflora, Caralluma umbellatum, Ceropegia spiralis, Clematis gouriana, Crinum asiaticum, Didymocarpus tomentosa, Hibiscus abelmoschus, Ipomoea bona-nox, Ipomoea hispida, Leonotis nepetifolia, Martynia annua, Zephyranthus sp.
Trees for landscaping: Butea monosperma, Cochlospermum religiosum, Firmiana colorata, Gardenia gummifera, Gardenia latifolia, Hymenodictyon excelsum, Lagastrocmia parviflora, Ochna gambllei, Shorea talura, Sterculia villosa, Sterculia urens, Terminalia paniculata, Wrightia tuinctoria.

From the field studies undertaken in the district it is clear that the district is rich in wild plants which have a very great potential for introduction to dryland horticulture. There are sites in the district which have very rich botanical assemblage. These microsites have conditions that are favourable to support rich ground flora with many rare and diverse plant species. From the observations made, it is also evident that the plant resources in the district have developed, evolved and localized under the prevailing microhabitat environments.

But as in any other forest ecosystem the wild plant genetic resources are endangered due to many factors and the forests are being degraded and the weeds like Lantana and Dichrostachys are replacing the native ground flora.

In a effort to conserve the plants of horticultural importance the following measures have been suggested.

1. In-situ conservation: In the first phase, the forests in Sidderabetta and Laxmankere, the two microsites with very rich assemblage of medicinal plants are being protected with the help of the local Forest Department staff.

2. Ex-Situ conservation: The endangered plants of horticultural importance from Tumkur District have been identified and these plants are raised in botanical gardens and propagating these plants in the University Forest Nursery. The use of micropropagation technique for their multiplication is also planned.

3. Re-creation of certain microsites and extending natural habitats through protection and enrichment planting is recommended.

Thus, the forests in Tumkur district is a source of many wild plants which find a place in dry land horticulture. But as these valuable resources from wild are endangered an effort is made to conserve and propagate the wild horticulturally important plants for future. If these potential wild plants are not saved from destruction it is going to affect our efforts to increase production and crop improvement. This would make us depend on exotics for our future horticultural needs.

References

Kushalappa,C.G. and Bhaskar,V. (1989) - Orchid flora of Tumkur district, Karnataka State. Myforest, **25**(1):61-66.

Yoganarasimhan,S.N., Togunashi,V.S., Keshavamurthy,K.R. and Govindaiah (1982) - Medico-botany of Tumkur district. J.Econ.Tax Bot., **4**(1):15.

Potential of seed storage at ambient temperature using organic liquids to avoid refrigeration

D.K. PANDEY
Seed Storage Laboratory, Division of Plant Genetic Resources, Indian Institute of Horticultural Research, Hessaraghatta Lake Post, Bangalore-560089, India

Key words : ageing, dehydrogenase activity, field emergence, French bean, germination, longevity, membrane integrity, propanediol.

Introduction

Water, temperature and oxygen are factors that most influence seed ageing. Ageing involves changes in macromolecules - nucleic acids, proteins and lipids (Priestley, 1986), the extent of which may determine the vigour and viability of a seed lot. The presently recommended and practiced germplasm conservation of desiccation tolerant (orthodox) seeds for long term requires drying of seeds to low moisture (ca < 5%) and storage at low temperature (<-18 °C) which make the method very expensive (International Board for Plant Genetic Resources, IBPGR, 1985). At any given temperature the lower the seed moisture the greater is the longevity down to species specific low moisture content limit to logarithmic relationship between seed moisture content and longevity (Ellis et al., 1989).

Presently emphasis is being placed on exploring use of ultra dry conditions as a cost effective method for storage of seeds at ambient temperature avoiding refrigeration (IBPGR, 1985). However, this attractive suggestion suffers from lack of suitable techniques for ultra drying the seeds and subsequent handling and storage of very dry seeds (Ellis et al., 1986). The present investigation reports possibility of using organic liquid medium for overcoming the limitations inherent with the use of ultra dry storage for long term conservation of seeds at ambient temperature avoiding refrigeration.

J. Prakash and R. L. M. Pierik (eds.), Horticulture – New Technologies and Applications, 63–68.

Materials and Methods

French bean (*Phaseolus vulgaris* L.) cv. Selection-9 seeds 20 in each of those with moisture content 10% (fresh weight) and those dried to 5% moisture content were hermetically sealed in lamenated aluminium foil packets in sufficient replications and incubated at 58 °C for accelerating ageing. Batches of 20 seeds with 9% and 10% moisture content each were immersed in 20 ml propanediol hermetically sealed and incubated at 58 °C and at ambient temperature 28 ± °4 C. The seeds from all treatments were sampled at intervals. Moisture contents in the seeds immersed in propanediol at 58 °C and at the ambient temperature were determined by the method described by the International Seed Testing Association (ISTA, 1985). The seeds which imbibed the propanediol were counted on each sampling. The seeds immersed in propanediol and those with 5% moisture content were equilibrated to 10% moisture before assessment of the applied treatments.Viability was tested as emergence upto 16 days (ISTA, 1985) under field conditions by planting seeds 3 cm deep in earthenware pots and vigour was determined as time taken for 50% emergence (Pandey, 1988). Membrane integrity deterioration as manifested in increase in electrolyte efflux into imbibition medium (Parrish and Leopold, 1978) was determined by the method described earlier (Pandey, 1989) except that the steeping period was 18 hours. Method for dehydrogenase activity determination in embryonic axes has been described earlier (Pandey, 1989) and was used with slight modification by incubating 20 axes in 4 ml of 1% 2, 3, 5 - triphenyl tetrazolium chloride in dark for 6 hours. All determinations were repeated at least thrice and the data were statistically analysed.

Results

The data (Figure 1) show that ageing of seeds as revealed by loss of vigour, viability, membrane integrity and dehydrogenase activity in embryonic axes with period of incubation was consistent in both the seeds with 10% moisture content hermetically sealed and immersed in propanediol.

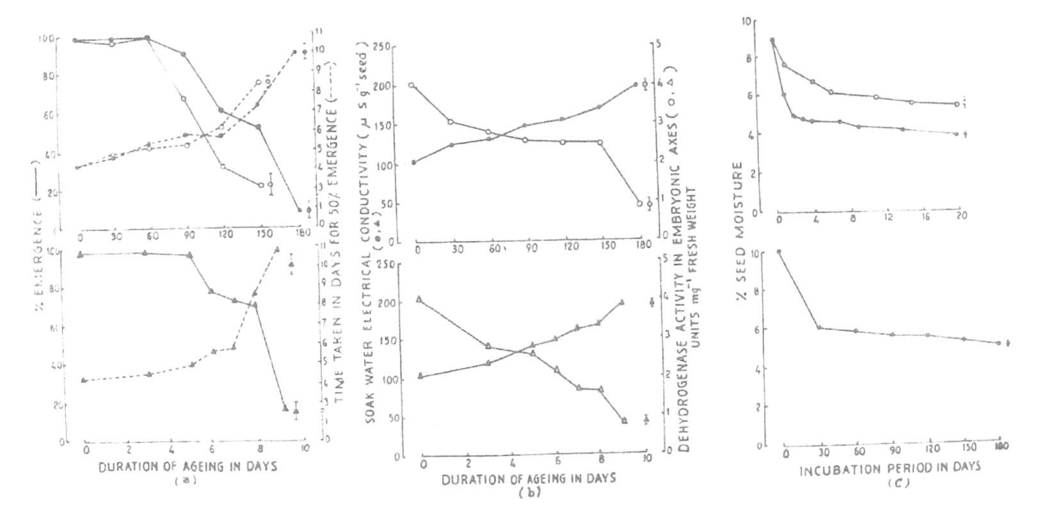

Figure 1. Effect of immersing seeds in propanediol (propdiol) and hermetic storage (HS) on ageing at 58 °C and on moisture content (MC) at 58 °C and 28 °C -(a) seeds with 10% MC in propdiol • or 10% MCHS ▲ and with 5% MCHS o; (b) seeds with 10% MCHS Δ, ▲ or in propdiol o, •; and (c) MC % in propdiol at 58 C • and at 28 ± 4 °C o.

However, immersion in propanediol enhanced seed longevity by about 18 fold. Vigour and viability loss in the seeds with 5% moisture content was comparable to the seeds immersed in propanediol. Propanediol quenched seed moisture rapidly at 58 C and slowly at the ambient temperature. The lower the initial seed moisture, the lower was the equilibrium moisture content of seeds in propanediol.

Discussion

Ageing of seeds with 10% moisture was rapid at 58 °C. Dynamic quenching of seed moisture and probably by eliminating oxygen from seed environment by propanediol probably retarded the rate of ageing by reducing deterioration of macromolecules as manifested in far less rapid loss of vigour, viability, membrane integrity and dehydrogenase activity in embryonic axes. Propanediol enhanced longevity by about 18 fold. The efficacy of immersing seeds in propanediol in retarding the rate of ageing was comparable to the extent that was achieved by reducing initial seed moisture to about 5%.

Exposing seeds to 60 °C enhances ageing (Roberts and Abdalla, 1968; Roberts, 1972). Investigations have shown no effect of temperature between 30-60 °C and below this range on the relation between chromosomal aberrations and seed viability (Rao et al., 1987). Ageing of French bean seeds, 10% moisture content, at 58 °C resembles with natural ageing at ambient temperature 28 ± 4 °C in its manifestations such as membrane integrity disruption, dehydrogenase activity decline in embryonic axes and loss of vigour and viability (Pandey, personal communication). Longevity of the French bean cv. Selection-9 seeds with 10% moisture content at ambient temperature 28 ± 4 °C was about 3 years (Pandey, 1989), against 10 days at 58 °C. Thus, 10 days accelerated ageing at 58 °C corresponds to 3 years natural ageing. Therefore 180 days accelerated ageing in propanediol will probably correspond to about 54 years provided natural ageing takes course similar to accelerated ageing at 58 °C or vice versa, an increase in longevity by about 18 fold. This figure appears to be very high unless experimentally proved at ambient temperature. However, such a potential of seed storage at ambient temperature using suitable organic liquids to avoid refrigeration exists since i. like propanediol, organic liquids may quench seed moisture at ambient temperature, ii. by using liquid medium air, thus oxygen is eliminated from seed environment, iii. it appears possible to design an embedding medium which would not be imbibed by seeds, iv. additives (Benson, 1990) can be supplied in the liquid medium for further retarding the rate of ageing.

Interestingly, immersing seeds in a derivative of polyhydric alcohol (Himalayan Colours, New Delhi, India) at 58 °C extended longevity of chilli, pea, maize, okra, gram, wheat, onion, and lentil etc. all crops studied to an extent of 120 to 310 days depending on the species was comparable to longevity of the seeds with moisture content in ultra dry range - 2-3% - against merely 10-15 days in the seeds with about 10% moisture content. None of the crop seeds studied imbibed the chemical and immersion in it did not have any adverse effect on the seeds. These findings, further strengthen that immersing desiccation tolerant (orthodox) seeds in a suitable organic liquid medium which would ultra dry the seeds (preferably on single placing), eliminate oxygen from the seed environment and possibly should have adjuvents to desirably release free radical scavengers, antioxidant and sequestrants, etc. into seeds appears to be a cost effective promising possibility for long term conservation of seeds at ambient temperature avoiding refrigeration.

Acknowledgements

The investigation is contribution No. 113/90 of the Institute. I thank the Director for providing facilities and Mr V.R. Srinivasan for helping in preparing the paper on 'Word Perfect' software.

References

Anonymous (1985) Cost effective long-term seed stores. IBPGR, Rome, 38 pp.

Anonymous (1985) International rules for seed testing - International Seed Testing Association. Seed Sci. Technol. 13: 356-513.

Benson E.E. (1990) Free radical damage in stored plant germplasm. IBPGR, Rome, 128 pp.

Ellis R.H., Hong T.D. and Roberts E.H. (1986) Logarithmic relationship between moisture content and longevity in sesame seeds. Ann. Bot. 57: 499-503.

Ellis R.H., Hong T.D. and Roberts E.H. (1989). A comparison of low-moisture content limit to the logarithmic relation between seed moisture and longevity in twelve species. Ann. Bot. 63: 601-611.

Pandey D.K. (1988) Electrolyte efflux into hot water as a test for predicting the germination and emergence of seeds. J. Hort. Sci. 63: 601-604.

Pandey D.K. (1989) Ageing of French bean seeds at ambient temperature in relation to vigour and viability. Seed Sci. Technol. 17: 41-47.

Parrish D.J. and Leopold A.C.(1978) On the mechanism of ageing in soybean seeds. Pl. Physiol. 61: 365-368.

Priestley D.A. (1986) Seed ageing - implications for seed storage and persistance in the soil. Cornell University Press, Ithaca and London, 304 pp.

Rao N.K., Roberts E.H. and Ellis R.H. (1987) Loss of viability in lettuce seeds and the accumulation of chromosome damage under different storage conditions. Ann. Bot. 60: 85-96.

Roberts E.H. (1972) In Viability of seeds. Chapman and Hall Ltd., London, 14-58 pp.

Roberts E.H. and Abdalla F.H. (1968) The influence of temperature, moisture and oxygen on period of seed viability in barley, broad bean and peas. Ann. Bot. 32: 97-117.

Chemical Weeding as a labour saving device in management of fruit nurseries

PRABHA CHALLA
Division of Physiology and Biochemistry
IIHR, Hessaraghatta, Bangalore - 560 080,
INDIA

Key words : fruit nurseries, herbicides, parameters, active ingredient (a.i)

Introduction:

Weed control in nursery management is very important to get healthy planting material. Skilled labourers are required as there is always a chance of damaging the roots and sometimes a possibility of pulling out the plant itself. Conventional method of hand weeding being labour intensive, chemical weeding by using herbicides come in handy in such situations. A survey of weed flora in nurseries maintained at Hessaraghatta and around Bangalore showed that *cynodon dactylon, cyperus rotundus, Euphorbia hirta, Parthenium hysterophorus and Mimosa pudica* are the most common weeds.

Materials and Methods:

Trials were conducted at the IIHR experimental station, Hessaraghatta with herbicides to control weeds in fruit nurseries.

The basal media consisted of soil, farmyard manure and sand mixed in equal proportions. This medium was either filled in pots or polythene bags or raised beds. After watering the medium, the stones, cuttings or seedlings were planted. Various pre-emergent herbicides like diuron, atrazine, fluchloralin, butachlor, alachlor, oxadiazon and benthiocarb were sprayed at 2.0 and 3.0 kg a.i/ha, and oxyfluorfen at 0.5 and 1.0 kg a.i/ha along with a hand weeded control. The treatments were replicated ten times. The nurseries were maintained for different periods of time depending on the crop used. The number of hand weedings also varied with the crop.

The parameters such as, seed germination, seedling mortality, stem diameter, root growth, leaf number and their weight were recorded to find out the herbicide tolerance in fruit crops.

J. Prakash and R. L. M. Pierik (eds.), Horticulture – New Technologies and Applications, 69–72.

Results:

The parameters recorded showed that papaya seedlings were highly succeptible to diuron, atrazine, and oxyfluorfen at both the concentrations used. Herbicides like fluchloralin, butachlor and alachlor were found to be effective for both weed control and seedling growth when they were incorporated in the nursery medium. It was observed that two hand weedings were necessary before transplantations to orchards.

Table-1 Economics of chemical weeding using pre-emergent herbicides in Fruit nurseries

Sl No	Crop Plant	Herbicides recommended	Concentration kg a.i/ha (No.of sprays)	Effective duration DAS*	Number of hand weedings **	Net profit over hand weeding Rs. /ha
1	Papaya	Fluchloralin	2.0 (one)	60	2	424.00
2	Mango	Diuron	2.0 (two)	180 - 190	12	1360.00
3	Banana	Diuron	2.0 (two)	180 - 190	12	1360.00
4	Pomogrante	Diuron	2.0 (two)	180 - 190	12	1360.00
5	Custard apple	Diuron	2.0 (two)	180 - 190	12	1360.00
6	Guajava	Diuron	2.0 (two)	180 - 190	12	1360.00
7	Phalsa	Diuron	2.0 (two)	180 - 190	12	1360.00
8	Citrus	Diuron	2.0 (two)	180 - 190	12	1360.00
9	Grapes	Oxyfluorfen	1.0 (one)	140 - 150	3	304.00

* DAS - Days After Spray
** 8 labourers are required for manual weeding of 1 hectare

In mango, custard apple, pomegranate, guajava, phalsa, ber and lime, diuron at 2.0 kg a.i/ha was found to be suitable as it not only controlled the dicot and monocot weeds but also improved the seedling growth. 24 months are required to transplant them to permanent orchards as they have to be grafted. Nearly 12 hand weedings are necessary for the whole period. In banana diuron was applied soon after the planting of suckers and no phytotoxic effects was observed. Diuron used twice at 2.0 kg a.i/ha proved to be highly economical. In grapes, the cuttings are sown in the nursery and it takes 5-6 months for planting the rooted cuttings in the field under bower system. The cultivars Anab-e-Shahi, Thompson seedless, Arka kanchan and Arka shyam proved to be highly resistant to diuron, atrazine and oxyfluorfen. Luxuriant root growth was seen in oxyfluorfen treatment. Atleast two to three hand weedings are necessary to maintain the grape in healthy and weed free condition.

Discussion:

Chemical weed control in fruit nurseries has not been worked out well and hence scanty information is available on the effect of herbicides on growth of young fruit seedlings. In general, effective concentrations of recommended herbicides for the mature orchards when used in nurseries, cause severe damage to young seedlings as has been demonstrated in papaya using diuron.

Using herbicides in nurseries has come to be accepted as a regular cultural practice in the fruit industry of Hawaii and other countries (Nishimoto and Yee, 1980). It is important to work out the correct dose of specific herbicides for fruit crop Nurseries under the Indian conditions. Of the several herbicides tested in fruit crop nurseries over a decade of research at the IIHR diuron, oxyfluorfen and fluchloralin are found to be the most effective herbicides. Besides controlling weeds, these chemicals improved the quality of seedlings as evidenced by various growth parameters thereby, proving their safety.

The results obtained indicated that herbicides were beneficial to fruit nurseries maintained either in glass house, green house or under tree shade in open areas. The economics calculated indicate that Rs. 300 to 1360 can be saved over the conventional hand weeding method depending on the fruit crop. Apart from saving money, one can also get healthy seedlings which prove beneficial in the long run.

Acknowledgements:

The author is thankful to the Director, I.I.H.R, Bangalore for providing necessary facilities. Help rendered by Sri SN Challa during the preparation of the paper is duly acknowledged.

Reference:

Nishimoto R.K. and Yee W.Y., 1980. A guide to chemical control in tropical and sub-tropical fruit and nut crops in Hawaii. Circular Cooperative Extension Service, University of Hawaii, No.423 (revised) 11-pp.

Role of germplasm in Citrus rootstock improvement

M.B.N.V.PRASAD, P.K.AGARWAL, S.D.SAWANT AND A. REKHA
Indian Institute of Horticultural Research,
Bangalore - 560 089, India

Key words : Citrus, rootstocks, Rangpur lime, Rough lemon, Trifoliate orange, Phytophthora nicotianae, vigor, disease resistance

Introduction

The need for evolving citrus rootstocks which are highly compatible with different scions, high yielding with good quality fruits and resistant to diseases, particularly *Phytophthora* root rot, has been widely stressed (Koltz and De Wolfe, 1958; Cameron and Frost, 1968; Prasad and Rao, 1983 and Rao and Prasad, 1983). Rangpur lime (*Citrus limonia* Osbeck.) and Rough lemon (*Citrus jambhiri* Lush.) possess the most desirable traits for a rootstock and are widely used (Cohen, 1970; Passos and Sabrinho, 1981; Agarwal, 1982 and Holtzhausen et al., 1988), but are highly susceptible to *Phytophthora* root rot (Carpenter and Furr, 1982; Prasad and Rao, 1983). Trifoliate orange (*Poncirus trifoliata* (L.) Raf.,) though resistant to *Phytophthora* (Rao and Prasad, 1983 and Spiegel-Roy, 1988), is incompatible with many citrus cultivars (Ashkenazi, 1988). Trifoliate orange has a generous donor of valuable genes, particularly the resistance to *Phytophthora* (Cameron and Frost, 1968; Prasad and Rao, 1983 and Spiegel-Roy, 1988), and combines well with Rangpur lime and Rough lemon. Hence, it is imperative to impart *Phytophthora* resistance of Trifoliate orange to Rangpur lime and Rough lemon.

A large amount of variation exists within the different citrus species. A number of strains/cultivars have been reported within Rangpur lime, Rough lemon and Trifoliate orange (Hodgson et al., 1963 and Bitters et al., 1977). Identification of superior strains for hybridization is therefore a pre-requisite. In the present studies the available strains of Rangpur lime and Rough lemon were evaluated for vigor parameters and Trifoliate orange strains were evaluated for both vigor and resistance to *Phytophthora* root rot.

Methods

The studies on vigor parameters were conducted on seven year old seedling plants in a randomized block design. Observations on plant height, canopy height, diameter and volume

J. Prakash and R. L. M. Pierik (eds.), Horticulture – New Technologies and Applications, 73–77.

and trunk diameter were recorded on nine Rangpur lime, five Rough lemon and nine Trifoliate orange strains(Table-1). One year old seedlings of the nine strains of Trifoliate orange were also screened for resistance to *Phytophthora* (Table-2) as per the procedure of Grimm and Hutchison(1973).

Results

The different strains of Rangpur lime, Rough lemon and Trifoliate orange varied significantly within each group for the vigour parameters(Table-1). Rangpur lime knors recorded highest canopy volume(15.34m^3) and trunk diameter (8.50 cm), followed by Rangpur lime-Brazil(14.75 m^3 and 8.40 cm, respectively), among the different strains of Rangpur lime. Jambhiri Kodur recorded highest canopy volume(21.86m^3) and trunk diameter(9.00 cm), followed by Julandhar Khatti(18.02 m^3 and 8.00 cm, respectively), among the Rough lemon strains. English Large Trifoliate orange recorded the highest canopy volume(3.68m^3) and trunk diameter(6.20 cm) followed by Gotha Road Trifoliate orange (3.10m^3 and 6.10 cm, respectively, while Flying Dragon Trifoliate orange recorded the lowest values for all the vigour parameters, among the different strains of Trifoliate orange.

The different strains of Trifoliate orange showed varying degrees of disease ratings(Table-2). The total disease rating on feeder and tap root ranged from 3.00 in Flying Dragon and Argentine Trifoliates to 9.50 in Large flowered Trifoliate, while there was no leaf and stem infection on any of the strains. Two strains, i.e.,Flying Dragon and Argentine Trifoliates were resistant while four were susceptible.

Discussion

The results revealed the variation among the different strains of Rangpur lime, Rough lemon and Trifoliate orange for various vigour parameters. The variation among the Trifoliate orange strains for resistance to *Phytophthora* was also clearly established.

Rangpur lime-knors and Rangpur lime-Brazil were found to be the most vigorous among the Rangpur lime strains studied. Jambhiri Kodur and Julandhar Khatti were found to be the most vigorous among the Rough lemon strains. Rootstock vigour has been reported to be a major factor in citrus fruit yields(Agarwal,1982). Hence, due importance should be given to vigour parameters while selecting parental lines particularly, seed parents, for rootstock breeding programmes. Rao and Prasad(1983) also emphasized the need for considering plant vigour parameters in citrus rootstocks in relation to *Phytophthora* root rot. Since Rangpur lime (Agarwal,1982 and Spiegel-Roy,1988) and Rough lemon(Agarwal, 1982 and Holtzhausen,1988) are the most promising citrus rootstocks, they should be used as seed parents in the hybridization so as to retain their desirable traits in the hybrids. However, in view of the strainal variation, it is imperative to use only the most vigorous strains as seed parents. The present studies identified Rangpur lime knors Rangpur lime-Brazil among Jambhiri Kodur and Julandhar Khatti as the most vigorous strains and would be ideal for use as seed parents.

The variation among the different strains of Trifoliate orange also revealed the importance of selecting pollen parents. Though English Large and Gotha Road were the most vigorous strains among the Trifoliate oranges, Flying Dragon and Argentine Trifoliate orange were the most resistant against *Phytophthora*. Flying Dragon is a very slow growing dwarf strain while Argentine Trifoliate is a medium vigour plant. The studies thus, indicated the suitability of Argentine Trifoliate orange as the pollen parent in the hybridization programme.

The role of germplasm in citrus rootstock improvement is thus highlighted in the present studies. Within the available strains, the utilization of Rangpur lime knors, Rangpur lime-Brazil, Rough lemon Kodur and Julandhar Khatti as seed parents and Trifoliate orange Argentina, as pollen parent is emphasized.

Acknowledgements:
The authors are grateful to the Director, Indian Institute of Horticultural Research, Bangalore, India, for providing the facilities.

References

Agarwal, P.K.(1982). Performance of different citrus rootstock in India - A review. *Agric. Rev.*, 3:17-34.

Ashkenazi, S.(1988). Incompatibility of some stock-scion citrus combinations in Israel. *Proc. Sixth Intl. Citrus Congr.*, 1:57--60.

Bitters, W.P., Mc Carty, C.D., and Cole, D.A.(1977). Evaluation of trifoliate orange selections as rootstocks for Washington navel and Valencia orange. *Proc. Intl. Citrus Congr.*, 1973, 2:127--131.

Cameron, J.W. and Frost, H.B.(1968). Genetics, breeding and nucellar embryony. In: Reuther, W. ed., Citrus Industry. Vol.2, pp.325--370. Univ. Calif.

Carpenter, J.B. and Furr, J.R.(1962). Evaluation of tolerance to root rot caused by *Phytophthora parasitica* in seedlings of citrus and related genera. *Phytopathology*. 52:1277--1285.

Cohen, M.(1970). Rangpur lime as a citrus rootstock in Florida. *Proc. Fla. State Hort. Soc.*, 83:78--84.

Fraser, L.(1949). A gummosis disease of citrus in relation to its environment. *Proc. Linn. Soc.*, N.S.W., 74:5--18.

Grimm, G.R. and Hutchison, D.J.(1973). A procedure for evaluating resistance of citrus seedlings to *Phytophthora parasitica*. *Plant Dis. Repr.*, 57:669--672.

Hodgson, R.W., Singh, R. and Singh, D.(1963). Some little-known Indian citrus species. Calif. Citrog., 48:211--294.
Holtzhausen, L.C., Grundlingh, J.A., Niven, P.N.F. and Maritz, J.(1988). Nine rootstocks evaluated for four navel cultivars in Eastern Cape. Proc. Sixth Intl. Citrus Congr., 1:33--45.
Koltz, L.J. and De Wolfe, T.A.(1958). Possible solution for a basic disease problem. Calif. Citrog., 43:80--85.
Passos, O.S. and Cunha Sabrinho, A.P.(1981). Citrus rootstock in Brazil. Proc. Intl. Soc. Citriculture, 1:102--105.
Prasad, M.B.N.V. and Raghavendra Rao, N.N.(1983). Reaction of some citrus rootstock hybrids for tolerance to Phytophthora root rot. Indian Phytopath., 36:726--728.
Raghavendra Rao, N.N. and Prasad, M.B.N.V.(1983). Evaluation of strains of Poncirus trifoliata and trifoliate orange hybrids for resistance to Phytophthora root rot. Scientia Hort., 20:85--90.
Spiegel-Roy, P.(1988). Citrus Breeding - Past, Present and Future. Proc. Sixth Intl. Citrus Congr., 1:9--17.

Table 1. Plant vigour parameters in different strains.

Genotype	Plant ht.(m)	Can. ht.(m)	Can. dia.(m)	Can. Vol.(m^3)	Trunk dia.(cm)
Rangpur limes:					
R. lime-knors	2.80	2.30	3.65	15.34	8.50
R. lime-Brazil	3.00	2.30	3.50	14.75	8.40
Florida Rangpur-8747	2.90	2.10	3.65	14.65	7.60
Limocryo Brazil	2.90	2.30	3.45	14.33	8.00
R. lime-8784	2.90	2.50	3.15	12.99	7.00
Brazilian R.lime	3.80	1.80	3.55	11.88	5.60
Sawaranta	2.90	2.10	3.25	11.61	7.70
R.lime-Australia	2.80	1.70	3.55	11.22	6.80
R.lime Kirumakki	2.40	1.40	3.45	8.72	7.00
C.D.at 5% level	0.30	0.36	NS	3.01	0.68
Rough lemons:					
Jambhiri Kodur	4.00	3.15	3.65	21.86	9.00
Julandhar Khatti	3.80	2.73	3.55	18.02	8.00
Pooney Jambhiri	2.90	2.67	3.17	12.01	7.10
Kata Jamir	3.23	2.67	2.72	9.08	9.10
Sohmyndong	3.03	1.95	2.90	8.92	8.60
C.D. at 5% level	0.38	0.39	0.68	5.54	0.71
Trifoliate oranges:					
English Large T.O.	2.97	2.30	1.72	3.68	6.20
Gotha Road T.O.	2.78	2.17	1.64	3.10	6.10
Rubidoux T.O.	2.75	2.18	1.43	2.38	5.70
T.Orange-Japan	2.52	1.97	1.33	2.16	5.50
Argentine T.O.	2.43	1.80	1.21	1.50	5.00
Srirampur T.O.	2.30	1.67	1.19	1.38	5.40
Christiansen T.O.	2.22	1.75	1.16	1.35	3.70
Large flowered T.O.	2.37	1.67	1.14	1.17	5.00
Flying Dragon T.O.	1.97	1.38	0.84	0.51	3.60
C.D. at 5% level	0.45	0.53	NS	1.76	1.45

Table 2. Reaction of trifoliate orange strains to screening against Phytophthora.

Genotype	Disease ratings			Rank*
	Feeder root	Tap root	Total	
English Large T.O.	5.00	4.50	9.50	S
Gotha Road T.O.	3.60	3.20	6.80	MS
Rubidoux T.O.	2.00	2.00	4.00	MR
T.orange-Japan	5.00	4.50	9.50	S
Argentine T.O.	2.00	1.00	3.00	R
Srirampur T.O.	2.00	2.00	4.00	MR
Christiansen T.O.	5.00	4.50	9.50	S
Large flowered T.O.	5.00	4.50	9.50	S
Flying dragon T.O.	2.00	1.00	3.00	R

Rank* - R = Resistant; S = Susceptible;
MR = Moderately Resistant; MS = Moderately Susceptible.

Comparative efficacy of single and split applied nitrogen on yield and quality of 'Perlette' grapes

W.S. DHILLON, A.S. BINDRA AND S.S. CHEEMA
Department of Horticulture
Punjab Agricultural University
Ludhiana-141 004, India

Key words : Nitrogen, split, Perlette, grapes

Introduction

The common fertilizer programme for a full-grown grapevine in Punjab (India) includes nitrogen (N) @ 500 g. The N fertilizer is applied in two equal parts, one each at the time of pruning (January) and after fruit set (April). However, the grapevines, especially on sandy soils, oftenly exhibit N deficiency symptoms during the summer months. The foliar sprays of 1% urea have been found useful for controlling temporary N-deficiency (Brar et al., 1988). Generally, the efficiency of quick-release fertilizers is increased with split application. The present studies were, therefore, carried-out to determine the response of 'Perlette' cv. of grapes to three doses and three split applications of N.

Methods

The experiment was conducted on 6-year full-grown, bower trained 'Perlette' grapevines planted 3m x 3m apart. The vineyard soil was loamy sand with dark-brown colour. Initially the soil had organic matter (0.48%), N (156 kg), P (26.8 kg), K (163 kg) per hectare and pH (8.5).

Nitrogen @ 500 g (recommended), 400 g and 300 g per vine in single (after pruning), two (after pruning + bud sprouting) and three (after pruning + bud sprouting + full-bloom) doses were applied. Besides these, two sprays of 2% urea were applied, one each at full-bloom and ten days after. A basal dose of phosphorus (720 g), potassium (720 g) and FYM (80 kg) were also given. The experiment was laid-out according to the randomized block design with three replications using three vines as experimental unit.

J. Prakash and R. L. M. Pierik (eds.), Horticulture – New Technologies and Applications, 79–82.

Standard methods (A.O.A.C., 1980) were used to record shoot growth, yield and various physico-chemical parameters of the fruit. Nitrogen content of the vines was studied from petiole samples collected at full-bloom, five-days each after 1st and 2nd urea sprays and at the time of fruit harvesting by Micro Kjeldahl's method (A.O.A.C., 1980).

Results

All the N treatments produced significantly higher yield over the control (Table 1). Highest fruit yield of 27.2 kg was recorded under 400 g N III treatment which was slightly more than 500 g N III application. The yield (25.0 kg) obtained by the treatment of 400 g N II was almost at par with 400 g N III. The lowest yield of 16.2 kg per vine was recorded under the control. Bunch weight was also highest under these treatments in which the yield was higher. Maximum shoot growth (176 cm) upto fruit harvest was, however, recorded under the treatment of 500 g N III, closely followed by 500 g N II and 400 g N III. These treatments produced significantly more shoot growth than the vines (128 cm), where no N was applied (Table 2). Differences were not significant within the various N treatments.

Nitrogen treatments resulted in significantly bigger berries as compared to unfertilized vines. The highest berry weight of 1.72 g (Table 1) was recorded under 400 g N III, followed by 500 g N III and 400 g N II treatments. The treatment of 300 g N III also produced berry weight upto 1.59 g. The lowest berry weight of 1.46 g was recorded under the control treatment. No significant differences in fruit quality (TSS and acidity) were noted in any of the treatments. Maximum TSS (15.2%), however, was recorded under 300 g N I. Lowest acidity of 0.72% was recorded under 300 g N I & II. A similar trend in case of TSS/acid ratio was observed under the same treatments (Table 1).

The results pertaining to petiole-N presented in Table 2 revealed that petiole-N was variable under different treatments at different stages. There was little variation in N content at full-bloom stage under different N treatments, however, these had significantly higher level over control. There was significant increase in the level of N after five-days of 1st and 2nd urea sprays. Maximum petiole-N content (2.41%) was recorded under 500 g N III after five-days of 1st spray. There was further considerable increase in the level of N after five-days of 2nd urea spray. Minimum of 1.58% and 1.57% N was recorded after five-days of 1st and 2nd spray, respectively in case of unfertilized vines. The level of N content was the lowest at the time of fruit harvesting under all the N treatments but still were significantly higher than the control vines. The lowest level of N (1.07%) was recorded, where no soil and foliar-urea application was given. The petiole-N content at the time of harvesting was the highest in 400 g N III.treatment.

Discussion

Increase in yield (Table 1) with split applied N in three doses and subsequent sprays of urea might be due to more photosynthetic activity resulted in manufacturing of greater quantity of food material (Pecznik and Meric, 1962). Also the split application of N would have enhanced the efficiency of fertilization. However, less yield due to production of slightly lighter bunches under 500 g N III as compared to 400 g N III might be correlated to the more vegetative growth which utilized the photosynthates in shoot extension. Improvement in berry weight was due to readily available-N to the vines which led to better photosynthetic activity and consequently better development of berries (Bindra and Bajwa, 1978). Considerable increase in petiole-N (Table 2) after 1st and 2nd urea spray indicate that N was quickly absorbed by the leaves. The fall in N-content at the time of harvesting was due to maximum utilization of N during vegetative and fruit development phases. Silva and Fontana (1966) studied urea absorption with radioactive-N and reported that N retained by the leaves was inversely and lograthmatically related to the N-application. These findings corroborate with the results of the present study.

References

A.O.A.C. (1980) Official methods of analysis. 12th ed., Washington, D.C., U.S.A.

Bindra, A.S. and Bajwa, G.S. (1978) Effect of nutrient spray on coulure in 'Thompson Seedless' grapes. Sci. and Cul. 44(12) : 549-50.

Brar, S.S., Bindra, A.S., Cheema, S.S. and Singh, S. (1988) Effect of urea spray on yield and quality of grapes (*Vitis vinifera* L.) cv. Perlette. Comp.Physiol.Ecol. 13(2) : 69-72.

Pecznik, J. and Meric, I.G. (1962) Results on large scale experiments on the foliar nutrition of vines. Agartud.Egypt. Mezoyzd. Kar. Kozlen. pp:319-22 (c.f. Hort.Abstr. 35: 660).

Silva, S. and Fontana, P. (1966). Foliar absorption of nitrogen by vines. Ann. Fac. Agri. S. Gurgi, Milan Co. 6 : 49-62.

Table 1. Comparative efficacy of single and split applied nitrogen on yield and quality of 'Perlette' grapes

Treatments		Shoot growth (cm)	Yield/ vine (kg)	Bunch weight (g)	Berry weight (g)	TSS (%)	Juice acidity (%)	TSS/ acid ratio
500 g N	I	157	20.2	355	1.60	14.4	0.75	19.2
	II	164	23.2	369	1.63	14.4	0.74	19.4
	III	176[h]	26.8	408	1.68	14.0[l]	0.78[h]	17.8[l]
400 g N	I	146	20.9	353	1.58	14.5	0.74	19.5
	II	157	25.0	367	1.64	14.4	0.74	19.4
	III	162	27.2[h]	412[h]	1.72[h]	14.7	0.76	19.3
300 g N	I	142	18.6	301	1.52	15.2[h]	0.72[l]	21.1[h]
	II	148	20.4	326	1.57	14.6	0.74	19.7
	III	156	24.3	364	1.59	14.8	0.72[l]	20.8
Control		128[l]	16.2[l]	225[l]	1.46[l]	14.5	0.74	19.5
L.S.D. (0.05)		28	2.3	23	0.14	NS	NS	NS

h = highest, l = lowest, NS = Non significant

Table 2. Effect of soil and foliar applied urea on petiole-N content of 'Perlette' grapes

Treatments		Petiole-N (%) content at			
		Full-bloom stage	Five-days after urea spray 1st	Five-days after urea spray 2nd	Fruit harvesting
500 g N	I	1.73	2.26	2.45	1.43
	II	1.77	2.33	2.53	1.47
	III	1.78	2.41	2.58	1.50
400 g N	I	1.71	2.23	2.44	1.40
	II	1.75	2.31	2.57	1.49
	III	1.76	2.40	2.60	1.51
300 g N	I	1.67	2.17	2.37	1.37
	II	1.71	2.25	2.46	1.40
	III	1.73	2.31	2.52	1.46
Control		1.47	1.58	1.57	1.07
L.S.D. (0.05)		0.19	0.13	0.07	0.25

Induced mutations recovered in M2 and subsequent generations in three varieties of okra(**Abelmoschus esculentus** (L.)Moench)

L. JANARDHAN RAO
Agricultural Research Institute, A.P.Agricultural University, Rajendranagar, Hyderabad - 500 030, India.

Keywords: **dES:diEthyl sulphate, EMS:Ethyl Methane sulphonate HZ: Hydrazine.**

INTRODUCTION

Induced mutational studies in Okra (**Abelmoschus esculentus** (L.) Moench), resulted in the recovery of mutants which were early, late, high yielding (Nandpuri et al.,1970), varying in fruit length (Yashvir, 1977), yellow vein mosaic resistant with more number of fruits and high yield (Thandapani et al., 1978),plants with changed height,fertility of seeds(jehangir and Chandra Sekhar, 1978) true breeding chlorina mutant (Jambhale and Nerkar, 1979), Palamatisect mutant (Jambhale and Nerkar, 1980).

The present paper deals with description of several marphologically distinct mutants isolated from M2 and M3 generations in three varieties of Okra. Some of these mutants have been directly used for general cultivation as they possessed distinct advantage over varieties grown, while others may find use in further utilization in breeding programmes.

METHODS

Pure seeds of three varieties of Okra viz., Pusa sawani, Co.1 and Shankerpally (Local) were treated with Gamma rays,with dosage of (20 Kr to 70 Kr with 10 Kr intervals (6 doses),dEs 0.25% to 0.125% with 0.025% intervals (5doses) EMS with 0.1% to 0.5% with interval of 0.1% (5 doses) and HZ 0.01% to 0.05% with 0.01% intervals (5 doses).

The plants in M2 generation were thoroughly screened for Identifying mutations effecting all the parts of the plant which were confirmed in M3 generation. All deviations from the normal plants which bred true are described in this paper.

RESULT AND DISCUSSION

A number of macro mutants (Viable mutations) with various morphological and physiological changes have been observed in both M2 and M3 generations. The seeds from these altered forms have been collected separately and were grown in succeeding generations, to confirm the true genetic nature of the altered forms. Most of these mutants bred true in the follow-

J. Prakash and R. L. M. Pierik (eds.), Horticulture – New Technologies and Applications, 83–86.

ing generations. A brief description of each of these mutants observed in M2 or M3 generations of the three varieties treated with both physical and chemical mutagents at various doses is furnished below.

Disease resistant mutants

Mutants resistant to yellow vein mosaic virus were isolated in M2 generation. These seedlings appeared to possess resistance factor under field conditions. This observation was based on the healthy appearance of these seelings in midst of other lines which were heavily infected with yellow vein mosaic virus. However no tests were made using artificial inoculation. Virus resistant lines isolated in M2 generation were mostly dwarf although tall resistant lines were also identified.

Jassid resistant mutants

In M2 generation certain plants showed jassid resistance. These jassid resistant lines were dwarf. Screening of the jassid resistant line was based on field infestation. These mutants showed lower levels of insect population compared to normal ones which had higher infestation.

Dwarf and very early mutants

Certain lines in M2 generation were very dwarf (25 to 45cm height). The flowering started at 4th node in these mutants and took 36 days to 42 days. As a consequence, mature fruits could be obtained on plants which were only 30 cm height in comparison with control 90 cm height. Fewer fruits were borne on these mutants.

Cup shaped leaf mutants

One progeny in M2 generation of Co.1 was found to be late and had cup shaped leathery leaves. The height in this mutant was comparable to control. The flowering started at 45 days after sowing. The fruits in these mutants were small with pointed tip.

Tall mutant with long stout fruits and bold seeds

This mutant had leathery, olive green leaves and long fruits which were very stout having 7 to 8 ridges with bold seeds. The fruits were heavier compared to control.

Mutants with very long fruits

In M2 generation certain plants produced extra long fruits ranging from 22 cm to 30 cm at maturity in Pusa Sawani. The number of fruits per plant was higher than the control.

One of these mutants viz., PS-10 has recorded higher yield compared to control and several other mutants of same class. This mutant has out yielded the others and is in the process of being released for general cultivation.

In the local variety, a few mutants had longer fruits than control but shorter than the ones recorded in Pusa Sawani and Co.1.

Mutants with short fruit and blunt tip "JANARDHANA"

This mutant had short fruits identified in M2 generation and bred true in M3 and subsequent generations. The plants were of medium height (60cm) and fruits had a length of about 6.0 cm to 7.0cm.The length of internodes was lesser than control and the number of internodes remained the same as that of control. The mutant exhibited high level of field resistance to yellow vein mosaic virus disease during rainy and summer seasons. This mutant was isolated in Pusa Sawani treated with dES. The fruits of this mutant remained tender for longer period,was early bearing and had more number of fruits per plant compared to control, as there were 2-3 branches from the basal portion of the main stem.

This mutantt was tested extensively and was released as "JANARDHANA". Presently this variety is being grown for canning. The advantage of this variety is that the fruits can be directly canned, without cutting as the size and shape are more desirable for canning. The mutant is also being exported as fresh vegetable as it remains tender for longer period. Also the size and shape of the fruits are greatly desired in certain markets of the world.

Mutant with small fruits and bristles

This mutant had very peculiar fruits which were about 5 cm in length with bristles and appeared like a brush. These mutants were devoid of yellow vein mosaic disease symptoms probably due to resistance. On maturity the fruits split, along the walls of locules probably facilities dispersal of seeds as in wild state.

This particular mutant gives a clue to establish the phylogenetic aspects of **Abelmoschus esculentus**. Probably the wild proginator of Okra had fruits of this nature which due to domestication gradually changed its appearance to the present day Okra.

few fertile mutants with long fruits,short fruits and fruits with bristles were observed. These mutants bred true and had field resistance to jassid infestation.

During the past two decades, a number of crop varieties in rice, sorghum, groundnut, castor and even in Okra were released for general cultivation which were evolved through mutagenesis.Many other mutants were found to be useful in further breeding programmes and have been successfully utilized for incorporation of mutant character (Varghese and Swaminathan, 1966 and Swaminathan and Siddique, 1968). Thandapani et al (1978) were successful in identifying a superior mutant which was later released as MDU.1 Okra as improved variety.

The macro-mutations identified in Okra in the present study, like the ones effecting duration and resistance to jassids and yellow vein mosaic virus disease could be successfully utilized in breeding programme where the objectives warrant such mutants. Mutants with long slender fruits can be tried in yield trials in most of the Okra growing areas as the fruits shape and size are attractive and may command good consumer preference. Also, the number of fruits borne per plant were higher than control and had higher yield potential than their

respective controls.
The mutant with small fruits (6 cm to 7 cm) with blunt tip has been tested in large areas for yield potential and their suitability for canning purpose in different containers.
This mutant is presently being exported to middle east countries as fresh vegetable besides canned ones. The mutants with stout fruits with more than five ridges and mutants which appear like brush and split locucidally need thorough study along with species related to cultivated Okra to reconstruct phylogenetic history of Okra and identify the species which could have been its proginator among the many wild species presently being collected from Western Ghats (Personal communication with **Dr. Muthu Krishnan, 1983**). Similar work has been undertaken by Swaminathan and Siddique, (1968) in rice.

Acknowledgement
The Author is thankful to **Dr. S.N. Rao**, the then Director of Research,A.P.Agricultural University for the facilities provided at Fruit Research Station, Sangareddy, A.P. India, for undertaking the present study. The Author also wishes to thank **Dr. C.A. Jagadish**, Professor, Department of Plant Breeding and Genetics, for critically going through the original manuscript.

References

Jambhale, N.D.,Nerkar, Y.S.,(1979) Inheritance of Chlorina an induced chlorophyll mutant in Okra (Abelmoschus esculentus (L.) Moench). J. Maharashtra Agric. Univ. 4(3) : 316.
Jambhale, N.D. Nerkar, Y.S. (1980) Inheritance of 'Palmatisect' an induced leaf mutation in Okra. Indian. J. Genet. 40:600-601.
Jehangir, K.S., and Chandrashekar, P.,(1978) Comparative mutagenic effects of gamma rays and Diethyl sulphate in Bhendi (Abelmoschus esculentus (L.) Moench) Madras agric. J. 65 (4) : 211-217.
Nandupuri, K.S., Sandhu, K.S., and Randhawa, K.S.,(1970) Effect of irradiation on variability in Okra (Abelmoschus esculentus (L.) Moench). J.Res.VII (2): 183-188.
Prasad, M.V.R.,(1976) Induced mutation in green gram. Indian J. Genet. 40 (1) : 172-175.
Swaminathan, MS., and Siddique, E.A.,(1968) Mutational analysis of racial differentiation in Oryza sativa L. Mutation Res. 6 : 478-481.
Thandapani, V., Soundara Pandian, G.,Jehangir, K.S.,Marappan. P.V.,Chandrasekharan, P.,and Venkatraman, N.(1978) MDU-1,A new high yielding Bhendi variety. Madras Agric.J.6:603-605.
Varghese, G.,and Swaminatnan, M.S.,(1966) Changes in protein quantity and quality associated with a mutation for amber grain colour in wheat. Curr. Sci. 35 (18): 469-470.
Yashvir, (1977) Mutagenic effects of individual and combined treatments of gamma rays and EMS in Okra (Abelmoschus esculentus (L.) Moench). J. Cytol. Genet. 9 & 10 : 93-97.

Role of viruses in plant biotechnology

M.J. Foxe,
National Agricultural and Veterinary Biotechnology Centre,
University College, Belfield, Dublin 4, Ireland

Introduction

The development of biotechnology in agriculture is seen to be a major initiative which will enhance the growth of crops and will also provide significant commercial development for companies supplying new products derived from Biotechnology. One particular area in which biotechnology is and will continue to make a significant impact, is in diagnosis and control of plant diseases. Currently plant disease caused by viruses accounts for in excess of $2 billion losses per year in the U.S. Since viruses are not amenable to direct chemical control it is important that other methods can be developed which will result in significant reduction in the impact of viruses on crop productivity. New biotechnological approaches are addressing this issue of virus diseases and are providing new and novel methods for their control through (1) the development and utilisation of rapid diagnostic tests based on immunology and on molecular probes and (2) the development of novel resistance mechanisms in plants utilising properties of the viral genome.

At a technological level significant progress has been made in developing a range of diagnostic tests for a large number of the commercially significant viral diseases and in addition methods for the incorporation of novel resistance to viruses, insects and herbicides is now at the field testing stage. The commercial success of agricultural biotechnology will depend to a significant extent upon the market uptake of both these tests and more importantly of these new plant types with these novel resistances.

In the area of virus resistance the strategies being pursued are two-fold, namely to incorporate the coat protein gene for economically important viruses into the crop plants and secondly to use satellite RNA, again incorporated into crop plants to reduce or control virus development. The commercial exploitation of these technologies is less clear as it is still not certain whether there is any significant benefit associated with the incorporation of these novel genes into crop plants when compared to existing breeding programmes where genes resistant to viruses already exist.

J. Prakash and R. L. M. Pierik (eds.), Horticulture – New Technologies and Applications, 87–94.

The approach of using coat protein mediated resistance is based on the long recognised phenomenon of cross protection in which infection of plants with a mild strain of a virus protects against infection by more virulent strains. In general cross protected plants exhibit a delay in symptom development, reduction in lesion number and reduction in accumulation of the challenge virus. Additionally the extent of protection is dependent upon how closely related the challenge strain is to the inducing strain and also on the concentration of the challenge inoculum.

There are a number of potential disadvantages associated with the use of cross protection, e.g. the mild strain used to protect some plants may result in some loss of yield or may be more virulent and spread to different hosts, or could mutate to a more severe form. The protected plants although resistant to strains may be more susceptible to synergistic effects when infected with a second non-related virus. The mechanism for the action of cross protection has not been satisfactorily explained although a number of theories have been put forward.

These theories include: (1) the capsid protein produced by the first virus infection encapsidates the RNA of the superinfecting virus strain and thus preventing its replication; (2) the superinfecting virus is prevented from infecting through a perturbation of the cell membrane by capsid or other viral protein thereby blocking virus attachment or virus encoating; (3) viral nucleic acid produced by the primary infection yields to RNA of the superinfecting strain thereby preventing its replication.

Because of the potential for use of the cross protection phenomenon, considerable scientific investigation has been undertaken in order to use genetic engineering to develop novel forms of resistance to viruses. Both genetic engineering and plant transformation techniques have been used to incorporate specific viral genes in the plants. Successful applications include the use of TMV coat protein gene (Bevan *et al.*, 1985; Powell Avel *et al.*, 1986), the coat protein genes from alfalfa mosaic virus (AMV) (Tumer *et al.*, 1987; Loesch-Fries *et al.*, 1987; van Dun *et al.*, 1987), tobacco rattle virus (TRV) (Van Dun *et al.*, 1988,a), cucumber mosaic virus (CMV) (Cuozzo *et al.*, 1988), potato virus X (PVX) (Hemenway *et al.*, 1988), potato virus Y (PVY) (Lawson *et al.*, 1990) and tobacco streak virus (TSV) (van Dun *et al.*, 1988,b). In addition plants have been engineered to express viral satellite RNAs from CMV (Harrison *et al.*, 1987) and tobacco ringspot virus (TobRV) (Gerlach *et al.*, 1987).

Coat Protein Mediated Virus Resistance

This approach has been successfully demonstrated for a range of different viruses, TMV, AlMV, TSV, CMV, PVX and PVY. This form of resistance has shown the properties of cross protection in that in each case expression of the coat protein gene from the virus protected plants against infection by the homologous virus, and disease symptoms and virus accumulation were reduced or eliminated in these engineered plants. In addition other properties of classical cross protection were evident including being less effective at high levels of inoculation and largely overcome by inoculation of viral RNA. The first and most notable success was obained with TMV where the PMV coat protein gene was stably expressed in transformed plants (Bevan *et al.*, 1985, Powell Abel *et al.*, 1986). The levels of expression of coat protein gene were approximately 0.02-0.1% of total leaf protein. Resistance in transformed plants was expressed as an absence or delay in symptom development coupled with reduced virus accumulation. The absence or delay in symptom development was dependent on inoculum concentration. This form of resistance was also expressed in tomato plants where coat protein mediated resistance was expressed against TMV and the closely related tomato mosaic virus (ToMV) (Nelson *et al.*, 1988). The transgenic tomato plants were protected against strains of TMV and ToMV with symptom development and virus accumulation reduced by comparison with control plants. Many plants escaped infection for up to 60 days after inoculation. Field tests of these genetically engineered tomato plants showed stable expression of resistance through four generations and yield characteristics similar to non treated tomato plants.

The initial success with TMV has been repeated for a number of other viruses including AlMV, PVX and PVY. AlMV is a bacilliform-shaped virus with a tripartite genome in which RNA 1, 2 and 3 and either the subgenomic RNA 4 or corresponding coat protein are required for infection. Expression of AlMV coat protein gene in tobacco has been reported by a number of workers (Loesch-Fries *et al.*, 1987; Tumer *et al.*, 1987; van Dun *et al.*, 1987). Protection against AlMV infection was observed in transgenic plants expressing the coat protein gene. Protection was effective against related strains of AlMV but not against other viruses. Plants expressing resistance had few or no symptoms on inoculated and upper leaves. In addition plants that expressed high levels of coat protein had fewer symptoms than plants expressing low levels.

Among the most important developments has been the recent demonstration of resistance to two important potato viruses, PVX and PVY. PVY is one of the most damaging potato viruses causing significant reductions in yield. In combination with PVX, PVY results in a synergistic interaction which is even more damaging than either virus alone. PVX is a flexuous rod-shaped virus containing a single genomic RNA of approximately 6,000 nucleotides in length. The coat protein gene is translated from a subgenomic RNA, is not encapsidated and is not required for infection. Hemenway *et al.* (1988), expressed the PVX coat protein gene in transgenic tobacco and obtained resistance to PVX infection. Resistance was expressed as reduced lesion numbers and delay or absence of systemic symptoms. The levels of expression of coat protein gene in transformed tobacco plants was approximately 0.02-0.1% of total leaf protein. In addition potatoes expressing the PVX coat protein gene were evaluated in the field for agronomic traits and it was demonstrated that the properties of the potato cultivars were not impaired or changed following transformation with the PVX coat protein gene (Hockema *et al.*, 1989). Similar results were obtained in studies with PVY.

PVY is a flexuous rod-shaped virus consisting of a single genomic RNA approximately 10,000 nucleotides. Subgenomic RNAs are not found in infected tissues but a polyprotein precursor is processed to yield individual proteins. Coat protein is generated by cleavage from the polyprotein by an encoded protease. Coat protein mediated resistance to PVY has been demonstrated in transformed potato (Lawson *et al.*, 1990). Transgenic potato plants cv. Russet Burbank, expressing the coat protein genes of both PVX and PVY were developed, and resistance to simultaneous infection by both viruses by mechanical inoculation was demonstrated. Resistance to PVY was observed in plants expressing the PVY coat protein gene although the level of expression of the gene was much lower than that observed for the PVX coat protein gene. The transgenic plants did not show any symptoms and did not contain any detectable virus.

When transgenic potato plants expressing both PVX and PVY coat protein genes were inoculated simultaneously with PVX and PVY the number of plants that became infected with PVX and PVY increased over single inoculation. These lines were not infected when only PVX was used as inoculum. In one case, one transgenic lines was resistant to infection by PVX and/or PVY regarless of whether either virus was inoculated alone or in combination. The lower levels of PVY coat protein expression in the transgenic potato plants could be responsible for the lower level of resistance to PVY when PVX and PVY are simultaneously inoculated. Several explanations were offered for these lower levels of expression. These included that in a mixed infection by PVX and PVY, if PVY coat protein is not expressed at sufficient levels at the sites where viral uncoating and replication take place, it might not be effective in preventing viral infection. Or that an interaction between the two coat protein genes would reduce the effective concentration of PVY coat protein making plants more susceptible to infection by PVY. Unlike coat protein mediated resistance to TMV, AlMV, PVX and CMV in which higher levels of coat protein expression were associated with higher levels of resistance to the virus, no such correlation exists in the case of PVY. For example, the most highly resistant potato line to PVY have the lowest PVY coat protein levels (Kaniewski *et al.*, 1990). Clearly further characterisation of coat protein expression is required to improve our understanding of the factors involved in resistance to viruses such as PVY.

In addition, resistance to transmission of PVY by aphids was also achieved. Field evaluation of these transgenic plants was carried out and it was demonstrated firstly that yield of virus inoculated transgenic plants was unaffected while yield of control plants was decreased, and that secondly uninoculated transgenic plants had yields as high as the control plants. These studies were carried out with the commercial cultivar Russet Burbank. The results are important as they confirm that resistance to PVX and PVY is effective in the field and can prevent yield losses due to dual infection by these viruses.

The salient features of coat protein mediated resistance are either the absence or reduction in number of lesions on inoculated leaves, a delay in or absence of systemic disease symptoms, and a reduction in virus levels in the leaves after virus inoculation. The reduction in lesion number appears to be associated with the coat protein gene producing levels of coat protein which interfere with or prevent the uncoating of the challenge virus. This is supported by the finding that in the case of TMV and AlMV coat protein mediated virus resistance is overcome by inoculation with viral RNA (Nelson *et al.*, 1987; Loesch-Fries *et al.*, 1987; van Dun *et al.*, 1987). By contrast resistance to PVX was not overcome when viral RNA was used implying that a somewhat different mechanism may be involved in this particular case (Hemenway *et al.*, 1988).

The delay or absence of systemic movement may result from reduced cell-to-cell spread of virus, decreased long distance spread of the virus or failure to initiate infection in systemic leaves. The precise roll of the coat protein mechanism in developing these forms of resistance remains to be elucidated.

Genetic engineering of virus resistance using satellite RNA

Satellite RNAs have been found associated with a number of plant viruses and have been effective in protecting plants against the effects of the virus. This phenomenon has been used to genetically engineer resistance to two viruses, cucumber mosaic virus (CMV) and tobacco ringspot Virus (TobRV), with plants expressing virus satellite RNA. In both cases the transgenic plants had reduced virus symptoms particularly systemic symptoms which was associated with reduced replication of the viral RNA (Harrison *et al.*, 1987; Gerlach *et al.*, 1987). In addition, in the case of TobRV, reduction in symptom expression in transgenic plants was also observed when antisense constructs of the satellite RNA were used. These results suggest a method of protecting plants against the effects of viruses containing satellite RNAs. However, caution must be exercised as for example, satellites that are benign in one host may be virulent in others or may mutate to more virulent forms. Nevertheless some of these difficulties may be eliminated by developing transgenic plants in which biologically active satellite RNA is maintained in the plant by transcription from the nuclear genome.

Prospects for commercialisation

Currently there are a number of key issues which will affect the commercialisation of genetically engineered plants. The major issues are regulatory approval, proprietary protection and public perception. The field testing of genetically engineered plants represents an important step in the commercialisation of plant biotechnology. The technology is developing very rapidly and issues such as regulatory costs and registration time are of increasing concern to companies developing genetically engineered crops. Regulatory approval for release of genetically modified plants varies enormously between countries. In addition the complexity of the regulatory process is such that it adds considerably to the cost of development of the plant. Regulatory standards must be developed which meet the general public's needs for a safe and reasonably priced food supply and recognise the inherent low risk of gene transfer technology and the benefits to growers, processors and consumers.

Patent protection for genetically engineered plants must be provided in order to protect the large scale investment required to produce these genetically engineered plants. Without this form of protection of intellectual property, it is very difficult to encourage continued investment in this kind of development. Currently the patenting laws governing plants differ considerably between the U.S. and Europe. These differences must be abolished. The third and possibly the most critical area for commercialisation, is that of the public perception of biotechnology. It is vital that there is public acceptance of the benefits of this new technology. Currently the major issues are those relating to agricultural surpluses, distrust of new technologies and the risks associated with release of genetically engineered plants. These issues must be adressed and the benefits to the farmer and the consumer stressed. If this is not achieved then the commercialisation of plant biotechnology becomes much more problematic.

Conclusions

The use of coat protein mediated resistance or viral satellite RNAs to develop novel forms of viral resistance have been demonstrated for a number of economically important viruses. In both cases resistance has been manifested as reduction in symptoms and delay in systemic spread. This has been associated with concomitant reduction in viral RNA replication.

The potential for exploitation of these technologies remains unclear although field testing of coat protein mediated resistance to PVX and PVY in potato has clearly demonstrated the effectiveness of this approach in reducing incidence of virus infection, while maintaining the other properties of the commercial cultivars.

A further benefit is that the level of coat protein in genetically engineered plants (0.01-0.5% of total protein) is below the levels found in plants infected with endemic. Additional potential benefits inclue reduced insecticide applications and improvement in food quality.

These same benefits must be demonstrated in crops such as vegetables, ornamentals, corn, wheat, rice and soybean if the technology is to become commercially attractive. Furthermore, genes for virus resistance must be inserted into impotant tropical crops such as cassava, sorghum, millet and cotton to enhance yield and reduce input costs. The technology must be integrated into the major crop objectives associated with local conditions and must complement and accelerate existing programmes and must be acceptable by governments and growers.

References

Bevan M.W., Mason S.E. and Goelet P. (1985) Expression of tobacco mosaic virus coat protein by a cauliflower mosaic virus promoter in plants transformed by Agrobacterium. EMBO J. 4: 1921-1926.

Cuozzo M, O'Connell K.M., Kaniewski W., Fang R.X., Chua N.H. and Tumer N.E. (1988) Viral protection in transgenic tobacco plants expressing the cucumber mosaic virus coat protein or its antisense RNA. Bio/Technology 6: 549-557.

Gerlach W.L., Llewellyn D. and Haseloff J. (1987) Construction of a plant disease resistance gene from the satellite RNA of tobacco ringspot virus. Nature (London) 328, 802-805.

Harrison B.D., Mayo M.A. and Baulcombe D.C. (1987) Virus resistance in transgenic plants that express cucumber mosaic virus satellite RNA. Nature (London) 328, 799-802.

Hemenway C, Fang R.X., Kaniewski W.K., Chua N.H. and Tumer N.E. (1988) Analysis of the mechanisms of protection in transgenic plants expressing the potato virus X coat protein or its antisense RNA. EMBO J. 7: 1273-1280.

Hoekema A., Huisman M.J., Molendijk L, van den Elzen P.J.M. and Cormelissen B.J.C. (1989) The genetic engineering of two commercial potato cultivars for resistance to potato virus X. Bio/Technology 7, 273-278.

Lawson C., Kaniewski W., Haley L, Rozman R., Newell C., Sanders P. and Tumer N.E. (1990) Engineering resistance to multiple virus infection in a commercial potato cultivar: Resistance to potato virus X and potato virus Y in transgenic Russet Burbank potato. Bio/Technology 8, 127-134.

Loesch-Fries L.S., Merlo D., Zinnen T., Burhop L., Hill K., Krahn K., Jarvis N., Nelson S., and Halk E. (1987). Expression of alfalfa mosaic virus RNA4 in transgenic plants confers virus resistance. EMBO J. 6, 1845-1851.

Nelson R.S., McCormick S.M., Delannay X., Dube P., Layton J., Anders J., Kaniewska M., Proksch R.K., Horsch R.B., Rogers S.G., Fraley R.T. and Beachy R.N. (1988) Virus tolerance, plant growth and field performance of transgenic tomato plants expressing coat protein from tobacco mosaic virus. Bio/Technology 6, 403-409.

Powell-Abel P., Nelson R.S. De, B., Hoffmann N., Rogers S.G., Fraley R.T. and Beachy R.N. 1986. Delay of disease development in transgenic plants that express the tobacco mosaic virus coat protein gene. Science 232, 738-743.

Tumer N.E., O'Connell K.M., Nelson R.N., Sanders P.R., Beachy R.N., Fraley R.T. and Shah D.M. (1987). Expression of alfalfa mosaic virus coat protein gene confers cross protection in transgenic tobacco and tomato plants. EMBO J. 6, 1181-1188.

Van Dun C.M.P. and Bol J.F. 1988. Transgenic tobacco plants accumulating tobacco rattle virus coat protein resist infection with tobacco rattle virus and pea early browning virus. Virology 167, 649-652.

Van Dun C.M.P., Overduin B., van Vloten-Doting L. and Bol J.F. (1988). Transgenic tobacco expressing tobacco streak virus or mutated alfalfa mosaic virus coat protein does not cross protect against alfalfa mosaic virus infection. Virology 164, 383-389.

Microbial Populations, Host-Specific Virulence & Plant Resistance Genes

D.W. GABRIEL, R. DE FEYTER, M. T. KINGSLEY, S. SWARUP and V. WANEY

Plant Pathology Dept., University of Florida, Gainesville, Florida, 32605, USA.

(Present address of R.D.F.: CSIRO-Division of Plant Industry, Canberra, ACT, Australia)

Key Words: Host range, virulence, specificity, avirulence, resistance, gene-for-gene.

Introduction

Plant pathogens are not found randomly distributed on all plants, but are restricted in host range to one or more host species. A phytopathogenic microbial species may have a wide or narrow host range, and some with a wide host range can attack a large number of plant species in many different plant families. However, individual strains of wide host range species may have a very limited host range, often limited to a single host species. Host-species specificity is considered to be relatively stable, and therefore useful in sub-species classifications. The subspecies ranks recognized on the basis of host-species (range) specificity are "forma specialis" for fungi and "pathovar", "pathotype" or "biovar" for bacteria. A plant species is included in the host range of a pathogen if even one variety of the species is found to be susceptible to that pathogen. If some plant varieties of a species are susceptible and some resistant to a given pathogen, the forma specialis or pathovar may be further subdivided into pathogenic "types", or more usually, "races". Race specificity is controlled by avirulence (*avr*) genes, and is a relatively unstable characteristic [Gabriel & Rolfe, 1990].

The genus *Xanthomonas* consists of plant pathogens, and some of them cause economically serious disease. The genus as a whole has a wide host range, and its members attack a large number of plant species in many different plant families. However, strains isolated from one host species are usually limited in pathogenicity to just that host species or perhaps genus or family. Our work using restriction fragment length polymorphism (RFLP) analyses of many different *Xanthomonas* strains from different hosts has revealed a clonal population structure among strains with a narrow host range, and a high degree of diversity among strains with a wide host range [Gabriel et al, 1988]. Furthermore, there is a strong correlation between clonal groups and host range at the plant species level, but not at the plant cultivar level. Single genes with high selective value on a given host may cause a given strain to be clonally amplified in direct proportion to the distribution and quantity of the host. For crop plants which are clonally propagated and widespread in distribution (cereal grains,

J. Prakash and R. L. M. Pierik (eds.), Horticulture – New Technologies and Applications, 95–105.

citrus, beans, etc.), this may result in proportionally widespread distribution of the pathogen. Since recombination rates are quite low in bacterial populations generally, other loci in the strain will remain fixed by linkage disequilibrium, and the clonality is revealed by RFLP analyses or other techniques. We hypothesize [Gabriel, 1989] that there are genes which confer parasitic fitness for specific hosts (host-specific virulence, *hsv* genes) which parallel the *Rhizobium* host-specific nodulation (*hsn)* genes [Djordjevic et al, 1987], and that these genes are responsible for the observed clonality of *Xanthomonas*.

Results and Discussion

Host-specific virulence (hsv) genes and common virulence (vir or hrp) genes.

Distinctions between host species specifying genes and common parasitism genes have long been be made. In *Rhizobium*, the ndvA & B genes, which are conserved at the family level, and the common nodulation genes (nodDABC), which are conserved at the genus level [Kondorosi et al, 1986] are essential for basic parasitism [Gabriel, 1989]. In contrast, hsn genes specify host range and are not functionally conserved at the genus level [Kondorosi et al, 1983]. At least four such genes (nodFEGH) have been identified by mutational analyses in R. meliloti [Djordjevic et al, 1987]. A region which appears to encode hsn function has been identified in B. japonicum [Nieuwkoop et al, 1987]. Mutation of the nodFE genes in either species results in a loss of host range [Djordjevic et al, 1985]. A few additional unique host-specific nodulation genes with additive effects are present in R. leguminosarum bv. viceae and in R. leguminosarum bv. trifolii [Djordjevic et al, 1987], which may account for the unique host range specificities of the two biovars.

Two important pathogenic genera of the family Pseudomonadaceae also demonstrate the existence of *hsv* genes. Such genes were cloned from a wide host-range strain of Pseudomonas solanacearum that extended pathogenicity when cloned into narrow host range P. solanacearum strains [Ma et al, 1988]. Similarly, Tn5 mutations of Pseudomonas syringae pv. tabaci were recovered which had lost pathogenicity to tobacco, but were still fully pathogenic to bean [Salch & Shaw, 1988]. More importantly, a 7.2 kb fragment associated with pathogenicity to tobacco was found conserved only in pv. tabaci and pv. angulata (pathogens of tobacco), but not in other pathovars tested. The conservation of a gene sequence among strains specific for a given host species is a useful criterion for distinguishing between host species-specific genes and virulence genes common to the species. Similarly, chemically induced, Tn*5* and spontaneous mutants of *X. translucens* were obtained that were either completely nonpathogenic or had altered host species specificity [Mellano & Cooksey, 1988]. Those which had altered host specificity were pathogenic on a more limited number of host species than the wild type.

Work in our lab has confirmed and extended the work [Mellano & Cooksey, 1988] on *X. translucens*. We used strain Xt-216, which causes leaf streak on barley, wheat oats, rye and triticale. Three thousand Tn*5*-GUS insertional derivatives of strain Xt216 were screened for loss of virulence on the five hosts. Prototrophic insertional derivatives affected in virulence on one of the hosts but not affected on the rest (eg.,

Hsv$^-$)were identified at frequencies of 0.07% (Barley$^-$), 0.04% (Wheat$^-$), 0.07% (Oats$^-$), 0.10% (Rye$^-$), and 0.07% (Triticale$^-$). All possible mono-specificities were recovered, as well as the completely Vir$^-$ class, as indicated in the table:

	WHEAT	BARLEY	RYE	OATS	TRITICALE
VIR$^-$	-	-	-	-	-
Hsv(w)$^-$	-	+	+	+	+
Hsv(b)$^-$	+	-	+	+	+
Hsv(r)$^-$	+	+	-	+	+
Hsv(o)$^-$	+	+	+	-	+
Hsv(t)$^-$	+	+	+	+	-
W.T.	+	+	+	+	+

All of the above Hsv$^-$ mutants gave a hypersensitive response (HR) on cotton, a heterologous host. We found no evidence for cultivar specificity of these mutations; that is, all Hsv$^-$ mutations were defective on all cultivars tested of the indicated host. Prototrophic mutants affected in virulence on all homologous hosts and failing to induce an HR on tobacco or cotton (eg., affected in hypersensitive response and pathogenicity, Hrp$^-$ [Lindgren et al, 1986]), were obtained at a frequency of 0.42%. Cosmid clones complementing two of the Hrp$^-$ mutations, and complementing two Hsv$^-$ mutations (Hsv^{w-} and Hsv^{b-}) were obtained and analyzed. One clone complementing an Hrp$^-$ mutation hybridized to the cloned *hrp* locus (pVir2) from *P. solanacearum* [Boucher et al, 1987], and we concluded that this clone represents a functionally equivalent *X. translucens* locus, homologous with *hrp* loci found in *P. solanacearum*. These *hrp* loci may be conserved at the family level within Pseudomonadaceae. The other three complementing clones examined do not cross-hybridize to pVir2 or to each other, and we have no evidence for linkage of the other loci. These observations suggest that host range in *X. translucens* may be conditioned by only a few genes, superimposed on common parasitism genes.

Significant progress has been made recently in understanding the selective value of specific virulence genes in pathogenic microbial populations and their role in conditioning disease [Selander, 1985]. Because recombination rates are quite low in bacterial populations generally, the number of different combinations of vir and *hsv* genes in the population may be limited. If normal ecological constraints on recombination are reduced by increased opportunity for horizontal gene transfer [Trevors et al, 1986; Schofield et al, 1987; Stotzky & Babich, 1986], the potential for pandemic or even epidemic scale outbreaks of new disease forms may be increased. In agricultural situations, increased opportunities for horizontal gene transfer may occur through the selection of self-mobilizing, broad host-range plasmids that carry resistance to antimicrobial agents. For example, grapefruit growers in Florida apply eighteen pounds of metallic copper per acre per year as a fungicide treatment [Knapp et al, 1988], causing documented selection of strains with self-mobilizing plasmids with copper resistance genes [Stall, 1986]. Epidemics of new disease forms may arise by recombination of *hsv* genes within a background of common genes. If the number of hsv genes required to alter host range is low, then the chances of a new disease form arising by horizontal transfer of one or a few genes may be unexpectedly high.

In 1984, a new epiphytotic disease form appeared on citrus in Florida and well over $25 million have been spent on failed eradication efforts. The causal organism, X. campestris pv. citrumelo, is not highly clonal, and has a wide host range, including Phaseolus vulgaris (bean) and Citrus paradisi (grapefruit). X. citri is clonal and attacks only citrus. X. phaseoli is clonal and attacks only bean [Gabriel et al, 1988; Gabriel et al, 1989]. How did the *X.c.* pv. *citrumelo* strains arise? Why were they never seen before?

We screened ca. 2000 X. campestris pv. citrumelo strain 3048 colonies on bean and citrus for spontaneous mutations to avirulence on one or both hosts. We were surprised by the high frequency of spontaneous Hsv^- mutants on citrus only, which arose at a frequency of 0.1%, while no prototrophic Hsv^- mutants of bean only were recovered. We concentrated efforts on the analysis of two of the Hsv^{cit-} mutants. Complementing clones were recovered from a wild type library, and a 3kb SstI fragment identified that complemented both mutations and evidently carries at least one *hsv* gene. As with the *X. translucens hsv* gene clones, no hybridization was detected with the *hrp* gene cluster on pVir2. We conclude that hsv genes are distinct from common parasitism (vir or hrp) genes, and that as with *X. translucens*, the host range of *X. campestris* pv. *citrumelo* is determined by at least one *hsv* gene, superimposed on basic virulence functions. The host range on citrus (and the resulting epidemic), could have arisen by horizontal transfer of the SstI fragment carrying the *hsv* gene(s). Since the region appears to be conserved among a large number of pathovars, it is also possible that the strain arose because of a newly developed and highly susceptible citrus cultivar that was beginning to be widely grown as a rootstock in Florida citrus nurseries [Gabriel *et al*, 1988].

An avirulence (avr) / virulence (vir) gene family.

Recently, we cloned a pathogenicity gene from *X. citri* by disease phenotype enhancement [Swarup et al, 1991]. A gene library of a severe disease strain (*X. citri*, causal agent of citrus canker) was mobilized into a mild disease strain (*X.c.* pv. *citrumelo*, causal agent of citrus bacterial spot), and the resulting transconjugants were screened on citrus for the appearance of the severe disease (citrus canker) phenotype. A 16 kb DNA fragment (pSS10.35) from *X. citri* was found to contain gene(s) essential for the elicitation of the raised lesion phenotype of Asiatic citrus canker. Strains from *X.c.* pathovars *alfalfae*, *citrumelo* and *cyamopsidis*, weakly compatible (opportunistic) on grapefruit (cv. Duncan) leaves and containing the clone pSS10.35 induced hypertrophy of the host cells leading to the formation of raised, canker-like lesions. Interestingly, strains of *X. malvacearum* and *X. phaseoli* harboring pSS10.35 displayed significant avirulence activity and induced hypersensitivity when inoculated on their homologous host plants. Mutagenesis of pSS10.35 with Tn5GUS showed that a 3.1 kb region was essential both for the increased pathogenicity on citrus and the avirulence activity on bean and cotton. Partial DNA sequence and restriction enzyme analyses revealed that the 3.1 kb region is highly similar to twelve *avr* genes from *X.*

malvacearum (refer below) and one from *X.c.* pv. *vesicatoria* (*avr*Bs3). Marker exchange mutagenesis of this gene results in a null phenotype on citrus and greatly reduced growth *in planta*.

Progress on *avr* gene analyses from *X. malvacearum* (Xm) was greatly delayed due to the difficulty of finding an adequate cloning vector to facilitate the work. We therefore constructed a new series of small vectors based on pSa that mobilized into Xm at frequencies of 1×10^{-6} and are maintained stably (>96% retention over 36 generations without selection pressure) [DeFeyter et al, 1990]. In the process of subcloning fragments from previously reported *avr* gene clones from Xm [Gabriel et al, 1986], we discovered that no avirulence activity was detected for any of the cosmids originally reported to carry *avr*B5. These and other considerations prompted us to remake the Xm strain H library in the newly created cosmid vector, pUFR034 [DeFeyter et al, 1990], and re-screen for avirulence. The results were that twelve new Xm *avr* genes were cloned and tested on plants, with reactions as follows:

	Acala										
Pathogen	B4	b6	b7	BIn	B1	B2	B5-22	b6,Bn 20-3	B5-82	BIn3	Ac44
N/avrB4	-	+	+	+	±	±	+	+	-	-	+
N/avrb6	+	-	+	++	±	±	±	-	+	-	+
N/avrb7	+	+	-	++	+	±	+	+	±	-	+
N/avrIn	+	+	+	-	+	+	+	+	-	-	+
N/avrB101	+	+	+	+	+	±	+	+	-	-	+
N/avrB102	+	+	+	+	±	±	+	+	-	-	+
N/C-1,C-4B	+	+	+	+	+	+	+	+	-	-	+
N/C-2	+	+	+	+	+	+	+	+	+	-	+
N/C-3	+	+	+	+	+	+	+	+	-	+	+
N/C-4	+	+	+	++	+	+	+	+	+	+	++
N/C-5,(C-3)	+	+	+	+	±	±	±	-	-	-	+
N/C-6	+	+	+	+	+	+	+	+	+	+	+
H	-	-	-	-	-	-	-	-	-	-	+
N	+	+	+	+	+	+	+	+	+	+	+

(In the above table, the first six entries represent strain N transconjugants carrying well-characterized avr genes that are plasmid-encoded. The next six entries represent strain N transconjugants carrying avr-homologous fragments presumably derived from the strain H chromosome. + = compatible; - = incompatible, HR response; ± = intermediate incompatible, HR response; ++ = more than usual water-soaking, compatible response.)

Strain Xm H appears to carry a single cryptic plasmid of 90 kb in size. This plasmid was characterized with 10 different restriction enzymes, and six of the *avr* genes found were localized on the plasmid by subcloning and Tn5-GUS saturation mutagenesis [DeFeyter & Gabriel, 1991]. All six plasmid-borne genes are larger than 3.2kb in size, all are similarly organized and all are homologous to *avr*Bs3 [Bonas et al, 1989] of *X.c.* pv. *vesicatoria.* Greater than 60% of *avr*B4 has been sequenced, and no portion of the sequence read so far is less than 90% homologous with *avr*Bs3. From 5-13 copies of the gene family were found in all Xm strains virulent on cotton; from 1-6 copies were found in many other *Xanthomonas* species and pathovars. At least six other *avr*-homologous DNA fragments were isolated from Xm strain H by colony hybridizations. Since no other cryptic plasmids have been observed with strain H, these other genes are presumably from the Xm H chromosome. Four of these fragments, which have not been well-characterized, reveal an avirulence phenotype, although all results reported above with the chromosomal fragments should be regarded as tentative, with only one replication. Two of the plasmid *avr* genes, avrb6 and avrb7, appear to enhance symptom development on compatible hosts. One of the chromosomal "avr" genes similarly appeared to enhanced symptom expression, but no avirulence phenotype was observed in Xm strain N. The other avr genes do not enhance symptom expression. When avrb6 was destroyed by marker-exchange with Tn*5*-GUS, a reduction in pathogenic symptoms was observed in Xm strain H (see table below). No such reduction was observed with three other marker-exchanged genes, including avrb7.

Marker exchanges:

Pathogen	B4	b6	b7	BIn	Ac44	
H::Tn*5*	-	±	-	-	±	(avrb6$^-$)
H::Tn*5*	-	-	+	-	+	(avrb7$^-$)
H::Tn*5*	-	-	-	+	+	(avrBIn$^-$)
H::Tn*5*	-	-	-	-	+	(avrB102$^-$)
H	-	-	-	-	+	(avrB4, avrb6, avrb7, avrBIn)

Not one of the *avr* genes, including those which are well-characterized, appear to operate in a strictly gene-for-gene manner. Each of the *avr* genes recognizes more than one R gene locus. Although genetic homology tests have not been made between each line, each of the R genes can be distinguished from the other by testing with combinations of the well-characterized plasmid avr genes. Genetic homology tests were run when there was reason to believe that certain lines contained more than one R gene, and different R genes were segregated by backcrossing to Ac44. Five of the *avr* genes each elicited a strong incompatible hypersensitive response when overexpressed in Xm N and inoculated onto cotton lines with the *B*1, *B*2, *B*5 or *B*n resistance genes.

Several of the R genes, including B_{1}, B_{2}, $B_{5\text{-}22}$, $B_{5\text{-}82}$, and B_{In3}, appeared to recognize more than one avr locus. The B5-82 and BIn-3 cotton cultivars, each shown to contain a single resistance locus, reacted strongly against at least seven of the *avr* genes. Although it is possible and perhaps likely that some of these R genes are really compound loci, it is clear from some crosses that the compound loci are different, and segregate independently in genetic crosses. The evidence presented here supports the idea that specificity is not gene-for-gene, or even gene-for-genes, but rather genes-for-genes.

The significance of genes-for-genes (plural) specificity is clear. Plant breeders have long searched for strong, "horizontal" resistance genes, and have usually discovered that "strong" resistance is vertical resistance and that "horizontal" resistance is polygenic. Nevertheless, cotton breeders have been successful in developing "immune" cotton lines that have been effective against all races of *X. malvacearum* found in North America for years [Brinkerhoff et al, 1984]. The molecular basis for this breeding success is now becoming clear: genes-for-genes recognition would make it statistically difficult (although not necessarily impossible) for the pathogen to overcome such resistance by mutations at multiple *avr* gene loci.

The six *avr* genes differ primarily in the multiplicity of a 102 bp repeated unit within each gene (refer figure, below). Random Bal I deletions of one or more repeat units from *avr*B101 generated new avirulence specificities, and also appeared to generate the increased virulence phenotype in some cases. Mutational analyses have shown that the virulence phenotype can be generated without avirulence, and that different specificities can be generated from avrB101. Specificity of the *avr* genes appears to depend in part on the number of repeat units in each gene. Conservation of this class of *avr* gene in *Xanthomonas* implies a highly conserved phenotypic function, but their "recognition" of resistance genes appears gratuitous, and not gene-for-gene specific.

It has long been recognized that so-called "defeated" R genes have residual effects. The term "defeated" is used to designate those resistance genes which have been used effectively in the field, but which are now less useful because of the prevalence of races now virulent on the R genes. There are a number of possible explanations which have been offered to explain the phenomenon [Gabriel & Rolfe, 1990]. Among the possibilities are: 1) that R genes have a general as well as specific effect; 2) that R genes have a residual effect against the original avr gene, but that is weaker than the original effect; and 3) that R genes have secondarily recognized avr gene specificity. A general effect of R genes has never been documented, and there is some evidence against this interpretation [Martin & Ellingboe, 1976]. We and others have seen no evidence for the possibility that there is a residual effect of an R locus against spontaneous mutants selected against specific R genes. The evidence presented above strongly supports the idea that at least some R genes have secondarily recognized, or gene-for-genes, specificity.

Three of the avr genes from Xm and the one from *X. citri* evidently function positively for virulence in their respective strains on homologous hosts. This virulence function is pleiotropic, and not cultivar specific. Interestingly, four different Xm avr genes (including those without evident virulence function on cotton) partially complemented the *X. citri* (hsvA::Tn5-GUS) marker-exchanged mutant Both deletion derivatives of avr101 and mutations of avrb6 have shown that the pleiotropic phenotypes of avirulence and virulence are separable functions of the same gene. The fact that this gene family is so highly conserved in *Xanthomonas* illustrates its selective value. Perhaps it should not be surprising that there would be R genes which recognize essential pathogenic functions, such as virulence. There would be a selective advantage

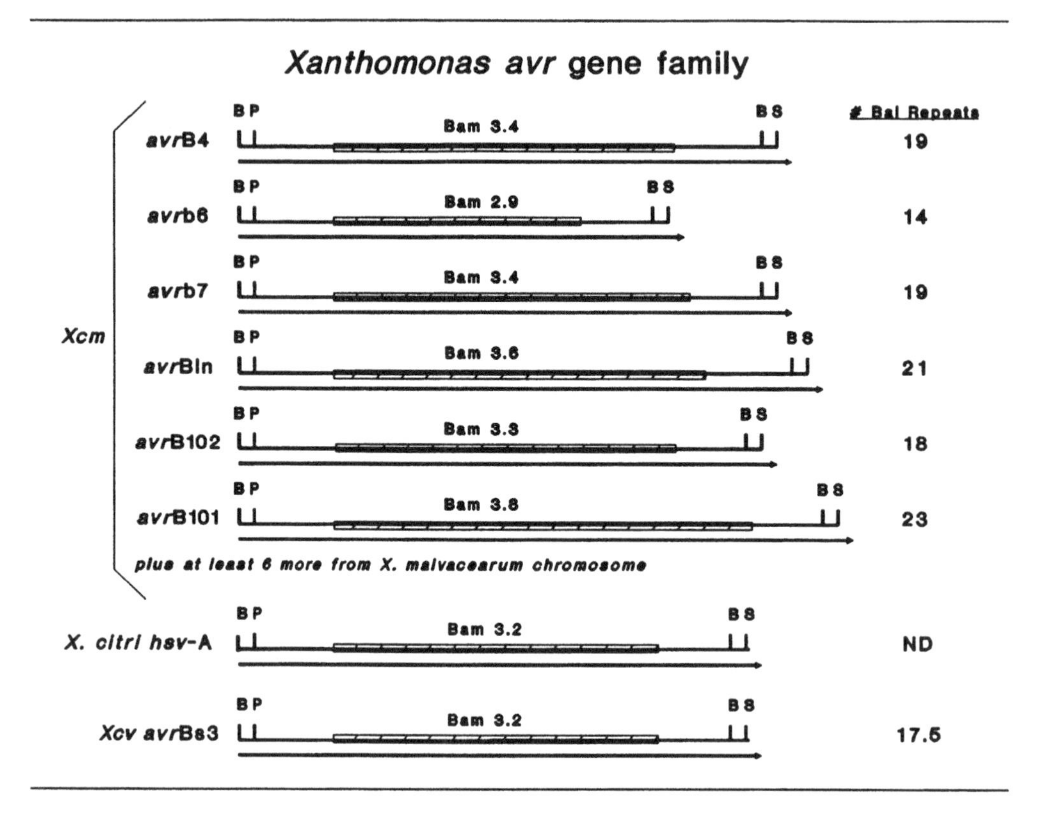

in such recognition. What is perhaps remarkable is the pathogens adaptation to this recognition. If several of the *avr/vir* genes in the family have equivalent virulence function, only one of them is essential to satisfy the selection pressure for virulence. By generating an array of genes in a family, the virulence function may be satisfied by perhaps any one of several members, leaving any "recognized" avirulence function free to mutate. The evidence presented here supports the idea that avirulence *per se* is a gratuitous function for members of this gene family.

The identification of resistance genes that react with multiple *avr* genes, and particularly with multiple *avr* genes found in other strains of the genus (such as *X. citri*), make such resistance genes obvious targets for cloning and transgenic plant transfers (say, from cotton to citrus). The demonstration that an *avr* gene from *X. citri*, when placed in *X. malvacearum*, reacts with a resistance gene from cotton provides a rational justification for the cloning of the cotton resistance gene for purposes of transfer to citrus. Such cotton *R* genes, provided they are expressed properly in citrus, could provide the first manipulable genetic control for citrus canker disease. Applications of such broad-spectrum resistance genes in other plants against other *Xanthomonas* diseases seems likely.

References

Bonas U., Stall R.E. and Staskawicz B. (1989) Genetic and structural characterization of the avirulence gene, avrBs3 from Xanthomonas campestris pv. vesicatoria. Mol. Gen. Genet. 218: 127-136.

Boucher C.A., Van Gijsegem F., Barberis P.A., Arlat M. and Zischek C. (1987) Pseudomonas solanacearum genes controlling both pathogenicity on tomato and hypersensitivity on tobacco are clustered. J. Bacteriol. 169: 5626-5632.

Brinkerhoff L.A., Verhalen L.M., Johnson W.M., Essenberg M. and Richardson P.E. (1984) Development of immunity to bacterial blight of cotton and its implications for other diseases. Plant Disease 68: 168-173.

DeFeyter R. and Gabriel D.W. (1991) At least six avirulence genes are clustered on a 90-kilobase plasmid in Xanthomonas campestris pv. malvacearum. Molec. Plant-Microbe Interact. (submitted).

DeFeyter R., Kado C.I. and Gabriel D.W. (1990) Small, stable shuttle vectors for use in Xanthomonas. Gene 88: 65-72.

Djordjevic M.A., Schofield P.R. and Rolfe B.G. (1985) Tn5 mutagenesis of Rhizobium trifolii host-specific nodulation genes result in mutants with altered host-range ability. Mol. Gen. Genet. 200: 463-471.

Djordjevic M.A., Gabriel D.W. and Rolfe B.G. (1987) Rhizobium the refined parasite of legumes. Ann. Rev. Phytopath. 25: 145-168.

Gabriel D.W. (1989) The genetics of plant pathogen population structure and host-parasite specificity. In: Kosuge T. and Nester E.W. ed., Plant-Microbe Interactions: Molecular and Genetic Perspectives, Vol 3.,pp. 343-379. Macmillan Publishing Co., New York.

Gabriel D.W. and Rolfe B.G. (1990) Working models of specific recognition in plant-microbe interactions. Annu. Rev. Phytopathol. 28: 365-391.

Gabriel D.W., Burges A. and Lazo G.R. (1986) Gene-for-gene recognition of five cloned avirulence genes from Xanthomonas campestris pv. malvacearum by specific resistance genes in cotton. Proc. Natl. Acad. Sci. USA. 83: 6415-6419.

Gabriel D.W., Hunter J., Kingsley M., Miller J. and Lazo G. (1988) Clonal population structure of Xanthomonas campestris and genetic diversity among citrus canker strains. Mol. Plant-Microbe Interact. 1: 59-65.

Gabriel D.W., Kingsley M.T., Hunter J.E. and Gottwald T.R. (1989) Reinstatement of Xanthomonas citri (ex Hasse) and X. phaseoli (ex Smith) and reclassification of all X. campestris pv. citri strains. Int. J. Syst. Bacteriol. 39: 14-22.

Knapp J.L., Tucker D.P.H., Noling J.W. and Vandiver V.V. (1988) Florida citrus spray guide. Florida Agricultural Extension Service Circular 393N : 14-22.

Kondorosi A., Banfalvi Z., Broughton W.J., Kondorosi E., Pankhurst C.E. and Randhawa G.S. (1983) Analysis of symbiotic nitrogen fixation genes of Rhizobium meliloti. In: Puhler A. ed., Molecular genetics of the bacteria-plant interaction.,pp. 55-63. Springer-Verlag, New York.

Kondorosi A., Horvath B., Rostas K., Gottfert M., Putnoky P., Rodriguez-Quinones F., Banfalvi Z. and Kondorosi E. (1986) Common and host-specific nodulation genes of Rhizobium meliloti. Third International Symposium on the molecular genetics of plant-microbe interactions, July 27-31, page 88. McGill University, Montreal.

Lindgren P.B., Peet R.C. and Panopoulos N.J. (1986) Gene cluster of Pseudomonas syringae pv. "phaseolicola" controls pathogenicity of bean plants and hypersensitivity on nonhost plants. J. Bacteriol. 168: 512-522.

Martin T.J. and Ellingboe A.H. (1976) Differences between compatible parasite/host genotypes involving the Pm4 locus of wheat and the corresponding genes in Erysiphe graminis f. sp. tritici. Phytopathology 66: 1435-1438.

Ma Q.S., Chang M.F., Tang J.L., Feng J.X., Fan M.J., Han B. and Liu T. (1988) Identification of DNA sequences involved in host specificity in the pathogenesis of Pseudomonas solanacearum strain T2005. Molec. Plant-Microbe Interact. 1: 169-174.

Mellano V.J. and Cooksey D.A. (1988) Development of host range mutants of Xanthomonas campestris pv. translucens. Appl. Environ. Microbiol. 54: 884-889.

Nieuwkoop A.J., Banfalvi Z., Deshmane N., Gerhold D., Schell M.G., Sirotkin K.M. and Stacey G. (1987) A locus encoding host range is linked to the common nodulation genes of Bradyrhizobium japonicum. J. Bacteriol. 169: 2631-2638.

Salch Y.P. and Shaw P.D. (1988) Isolation and characterization of pathogenicity genes of Pseudomonas syringae pv. tabaci. J. Bacteriol. 170: 2584-2591.

Schofield P.R., Gibson A.H., Dudman W.F. and Watson J.M. (1987) Evidence for genetic exchange and recombination of Rhizobium symbiotic plasmids in a soil population. Appl. Environ. Microbiol. 53: 2942-7.

Selander R.K. (1985) Protein polymorphism and the genetic structure of natural populations of bacteria. In: Ohta T. and Aoki K. ed., Population genetics and molecular evolution,pp. 85-106. Japan Scientific Societies Press, Tokyo.

Stall R.E. (1986) Plasmid-specified host specificity in Xanthomonas campestris pv.vesicatoria. Proceedings VI International Conference on Plant Pathogenic Bacteria, June 2-7, 1985,pp. 1042-1050. Martinus Nijhoff, Dordrecht.

Stotzky G. and Babich H. (1986) Survival of, and genetic transfer by, genetically engineered bacteria in natural environments. Adv. Appl. Microbiol. 31: 93-138.

Swarup S., DeFeyter R., Brlansky R.H. and Gabriel D.W. (1991) A host-specific virulence locus from Xanthomonas citri enables strains from several pathovars of X. campestris to induce canker-like lesions on citrus. Phytopathology (submitted).

Trevors J.T., Barkay T. and Bourquin A.W. (1986) Gene transfer among bacteria in soil and aquatic environments: a review. Can. J. Microbiol. 33: 191-8.

Genetically Engineered Resistance in Plants Against Viral Infection

D.D. SHUKLA, K.H. GOUGH, XIAO XIAOWEN, M.J. FRENKEL and C.W. WARD
CSIRO, Division of Biomolecular Engineering, 343 Royal Parade, Parkville, Victoria, 3052, Australia

Introduction

Plant viruses cause serious economic losses in crop plants throughout the world. Crop losses due to plant viruses have been estimated at $2 billion annually in the United States alone including $50 million worth of tomato, $95 million worth of wheat and $30 million worth of potato (Ralf, 1988). In general plant viruses have been difficult to control. Traditional approaches include insecticidal sprays to kill vectors or cultural practices. Examples of the latter are late or early planting of crops to avoid seasonal migration of vectors, crop rotation or cross-protection. Breeding resistant cultivars would be the favoured option since conventional plant breeding methods have played an important role in crop improvement for many decades. Genes for a variety of useful traits including resistance to diseases have been successfully transferred from noncultivated plant species and genera to cultivated crops using this approach (Goodman et al., 1987). However, classical plant breeding methods have several limitations. For instance, they are laborious and time-consuming; tightly-linked genes with undesirable traits may not be separated even after several back crossings; the desired resistance is not always available or may be available only in plants that are genetically incompatible; or the resistance may be multigenic and difficult to transfer (Goodman et al., 1987). The advent of recombinant DNA techniques and the dramatic progress made in the development of gene transfer and plant regeneration technologies have now made it possible to transfer desired genes not only between plant families but also from bacteria and other organisms into plants. These findings have provided an unparalleled opportunity to improve crop plants since they have overcome most of the drawbacks associated with the use of conventional breeding methods (Grumet, 1990).

The genes for resistance against diseases can come from either plants or pathogens, but it is much easier to identify and isolate such genes from pathogens because of the small genome size compared to that of plant species. Furthermore, plant genomes contain large stretches of repetitive DNA that makes mapping of their resistance genes difficult (Grumet, 1990).

Compared to fungal and bacterial plant pathogens, plant viruses have much simpler genome structures and the genomes of many plant viruses have now been sequenced. Plant virus genomes are multigenic with each gene coding for proteins which play several specific roles in the life cycle of viruses (Goldbach and Wellink, 1988; Shukla et al., 1991). Using molecular genetic approaches, it is now possible to interfere with the structure and inturn the function of these genes and block virus replication at critical points in the virus life cycle. Additionally, virus genomes contain several potentially useful genes which can be used to confer resistance to

J. Prakash and R. L. M. Pierik (eds.), Horticulture – New Technologies and Applications, 107–113.

plants against viruses (Beachy et al., 1990). A number of approaches to increase resistance in plants against viral infections using different viral genes have been reported in the past five years (Grumet, 1990; Beachy et al., 1990). This paper will describe these new developments.

Coat Protein-Mediated Protection

Beachy and co-workers (Powell-Able et al., 1986) were the first to show that transformed tobacco plants expressing the coat protein gene of tobacco mosaic virus were highly resistant when challenged with the virus. This approach has now been extended to several plant virus groups including alfalfa mosaic virus (Loesch-Fries et al., 1987; Tumer et al., 1987; Van Dun et al., 1988b), potexvirus (Hemenway et al., 1988; Hoekema et al. 1989; Lawson et al., 1990), cucumovirus (Cuozzo et al., 1988), tobravirus (Van Dun and Bol, 1988), ilarvirus (Van Dun et al., 1988a), potyvirus (Stark and Beachy, 1989; Lawson et al., 1990), carlavirus (Mackenzie and Tremaine, 1990) and luteovirus (Kauchuk et al., 1990) groups. These virus groups differ from each other in several properties including particle morphology, genome organization and mode of transmission, suggesting that the coat protein-mediated protection may be applicable to most plant viruses. The presence of coat protein in transgenic plants does not appear to affect their performance in glasshouse or field in terms of vigour or yield (Nelson et al., 1988). Initially it appeared that coat protein-mediated protection mimiced the classical cross-protection where only closely related strains of the one virus are protected (Sherwood and Fulton, 1982). However, recent work of Beachy and co-workers has shown that coat protein-mediated protection can confer broad-spectrum resistance against heterologous plant viruses (Stark and Beachy, 1989; Neijidat and Beachy, 1990). These workers showed that transgenic tobacco plants expressing the coat protein of soybean mosaic potyvirus, a non-pathogen in tobacco, display a significant level of protection when challenged with potato virus Y and tobacco etch potyviruses (Stark and Beachy, 1989). A similar high level of protection in transgenic tobacco plants was also obtained with tobamoviruses if the sequence identity of the coat protein of the challenging viruses and the protecting coat protein was 60% or higher (Neijidat and Beachy, 1990). Thus, it appears that genetically engineered coat protein-mediated protection does not mimic the mechanism/s involved in classical cross-protection, where only related strains are cross-protected.

Sattelite RNA-Mediated Protection

Sattelite RNAs are small molecules (300 to 400 nucleotides long) that are not part of the viral genome but are incapsidated by coat protein of viruses and are dependent on products of the viral genome for their replication. In this respect, sattelite RNAs are considered viral parasites. The presence of sattelite RNAs with viruses has been found to reduce the severity of symptoms in virus-infected plants. This property makes sattelite RNAs potential candidates for virus control (Ralf, 1988). In fact, in the Peoples Republic of China, sattelite-containing strains of cucumber mosaic virus have been sprayed onto field crops to protect them against the severe effects of sattelite-free strains of the virus (Robinson, 1988). Recently when tobacco plants were transformed with DNA copies of sattelite RNA, together with suitable sequences to promote transcription into RNA, challenge infection with a sattelite-free strain of cucumber mosaic virus led to rapid amplification of the sattelite RNA, together with amelioration of symptoms and a reduction in virus content compared to untransformed plants (Harrison et al., 1987). Another successful example is the application of the sattelite RNA protective mechanism against infection of tobacco ringspot virus (Gerlach et al., 1987). By incorporating the DNA sequence of tobacco ringspot virus sattelite RNA into the tobacco plant genome, a high degree of protection from infection by the pathogenic virus was recorded in transformed plants. However, this approach is not applicable to all plant viruses because only a few viruses have sattelites that can be used as protective agents. Moreover, the effect of sattelite RNA's is not always predictable since they can mutate and make the symptom worse (Grumet, 1990).

Viral Genome-Mediated Protection

The complete genome of a mild strain of tobacco mosaic virus has been used to transform tobacco plants to confer resistance against the virus (Yamaya et al., 1988). The resulting transgenic plants were either symptomless or displayed only mild symptoms. When these plants were challenge inoculated with a sever strain of the virus they did not develop sever symptoms. However, this approach has potential limitations, for example, the mild strain can still reduce the yield and quality of crop and it can mutate to a sever strain. Small fragments of plant virus genome have also been used to confer resistance to plant viruses. When cDNA fragment corresponding to the 3' -end of turnip yellow mosaic virus was incorporated into plant genomic DNA, replication of the virus was greatly inhibited, whereas other RNA fragments that did not contain the 3' terminal region had no effect on replication (Morch et al., 1987).

Antisense RNA-Mediated Protection

Antisense RNA is a natural mechanism for gene regulation in bacteria and has been extended to animal and plant kingdoms (Ecker and Daves, 1986; Bryant, 1988). Antisense RNA is complimentary to messenger RNA and is presumed to exert its effect at least partly by forming a stable base-paired structure with the mRNA and thus blocking its translation (Bryant, 1988). The antisense RNA approach has recently been applied to plant viruses but with only limited success (Rezian et al., 1988; Cuozzo et al., 1988; Hemenway et al., 1988, Powell et al., 1988a). In direct comparisons, the coat protein-mediated approach gave much higher level of protection at high inoculum levels than did antisense RNA which was effective only at very low inoculum levels (Cuozzo et al., 1988; Hemenway et al., 1988).

Ribozyme-mediated Protection

The term ribozyme refers to the ability of a type of RNA molecule to cleave target RNA molecules. The first example of the mechanism of ribozyme action was the self cleavage of RNA from sattelite tobacco ringspot virus and avocado sunblotch viroid (Buzayan et al., 1986; Hutchins et al., 1986). It was observed that oligomers of both these molecules, when stored, frozen and then thawed and refrozen, broke down to form RNA monomers. The cleavage was not random but occurred at specific positions (after GUC triplets) to form monomers. Around each cleavage site a conserved nucleotide sequence, which folds to give a hammerhead-like structure, was found (Symons, 1989). In the viroid and the sattelite RNA, the cleavage occurred only intramolecularly. However, it has now been demonstrated that ribozymes can be designed which will cleave RNA intermolecularly (Uhlenbeck, 1987; Haseloff and Gerlach, 1988). This approach has the potential to inactivate any unwanted gene including those from viruses infecting plants. The use of this approach has been demonstrated in vitro and in vivo in mammalian (Cameron and Jennings, 1989) and plant viral system (R.N. Nelson, personal communication). Although the approach is very effective in vitro, its effect in vivo has been disappointing so far.

Nonstructured Viral Gene-Mediated Protection

As mentioned previously viral genomes are multigenic with each gene product playing specific roles in the life cycle of viruses. Apart from the coat protein gene it is possible that resistance against viruses can be conferred by using the nonstructural genes of viruses. Recently the gene for a 54 kDa product of tobacco mosaic virus, a putative component of the replicase complex, has been expressed in tobacco plants (Golemboski et al., 1990). The transformed plants accumulated a 54 kDa gene sequence-specific RNA transcript of the expected size, but no protein product was detected. Such plants were found to be completely resistant when challenged with either virion RNA or virus at high inoculum level (Golemboski et al., 1990). Although the level of protection obtained using this system is much higher than the other approaches reported so far, it suffers from the drawback that only closely related strains can be protected using this approach.

Antibody-Mediated Protection

Recently Hiatt et al (1989) have described the successful expression of functional mouse monoclonal antibodies in tobacco plants. Complementary DNAs derived from a hybridoma messenger RNA were used to transform tobacco plants. Plants expressing single light or heavy immunoglobulin chains were crossed to yield progeny in which both chains were expressed simultaneously. The antibodies derived from the transformed plants were similar in their reactivity to the hybridoma-derived antibody.

It is well known that antibodies can neutralize the infectivity of plant and animal viruses (Van Regenmortal, 1982). Thus, it is possible that the expression of mRNA from virus neutralizing monoclonal antibodies in transgenic plants might be able to prevent infection from plant viruses. The antibody-mediated approach may have distinct advantages over the other methods of genetically engineered resistance. Broad spectrum antibodies have been produced to several plant virus groups including potyviruses (Shukla and Ward, 1989), potexviruses (Koenig and Lesemann, 1978), luteoviruses (Gerlach et al., 1990), and it is known that a number of viruses from the same plant virus group can infect a particular crop, for instance potyviruses (Shukla and Ward, 1989). This implies that in such situations a single antibody gene set would be required to render a plant species resistant to infection by several viruses. In addition, the same single antibody gene set could be used to transform other plant species to confer resistance to infection by several members of the same virus group.

Concluding Remarks

In spite of the dramatic developments in plant molecular biology and plant transformation technologies in the past decade, no report on the successful cloning of virus resistance genes or any other disease resistance genes from plants is presently available due to the difficulties associated with the mapping and tagging of such genes. However, these same developments have enabled molecular biologists, in the past five years, to clone and use several genes from plant viruses to confer resistance against viral infection. This has been possible because of the simple genome structures of viruses. Some of these approaches, viz. coat protein-mediated, sattelite RNA-mediated and nonstructural gene-mediated resistance have already been shown to render high levels of protection. Further refinements in the application of other methods, such as antisense RNA and ribozymes, could increase the effectiveness of these techniques from the present low level to high level of protection. No report is presently available on the application of antibody-mediated protection to control plant viruses but several laboratories including our own are currently engaged in this area. The next five years should reveal the use of many more novel genetic engineering approaches for the control of plant viruses based on the sequences of viral and other genes.

Acknowledgements

D.D. Shukla is grateful to the Indo-American Hybrid Seeds, Bangalore, India for funding his visit to the "International Seminar on New Frontiers in Horticulture". The authors thank Dr R. Nelson of the Samauel Noble Foundation, Oklahoma, U.S.A. for information on the use of ribozyme approach.

References

Beachy R., Loesch-Fries S. and Tumer N.E. (1990) Coat protein mediated resistance against virus infection. Ann. Rev. Phytopathol. 28: 451-474.

Bryant J.A. (1989) Antisense RNA makes good sense. Tibtech. 7: 20-21.

Buzayan J.M., Gerlach W.L. and Bruening G. (1986) Non-enzymatic cleavage and ligation of RNAs complimentary to a plant virus sattelite RNA. Nature (London) 323: 349-353.

Cameron F.H. and Jennings P.A. (1989) Specific gene suppression by engineered ribozymes in monkey cells. Proc. Natl. Acad. Sci. USA 86: 9139-9143.

Cuozzo M., O'Connell K.M., Kaniewski W., Fang R.X., Chua N.H. and Tumer N.E. (1988) Viral protection in transgenic plants expressing the cucumber mosaic virus coat protein or its antisense RNA. Bio/Technology 6: 549-557.

Ecker J.R. and Davis R.W. (1986) Inhibition of gene expression in plant cells by expression of antisense RNA. Proc. Natl. Acad. Sci. USA 83: 5372-5376.

Gerlach W.L., Llewellyn D. and Haseloff J. (1987) Construction of a plant disease resistance gene from the sattelite RNA of tobacco ringspot virus. Nature (London) 328: 802-805.

Gerlach, W.L., Martin R.R., Keese P.K., Young M.J. and Waterhouse P.M. (1990). Evolution and molecular biology of luteoviruses. Ann. Rev. Phytopathol. 28: 341-363

Goldbach R. and Wellink J. (1988) Evolution of plus-stranded RNA viruses. Intervirology 29: 260-267.

Golemboski D.B., Lomonossoff G.P. and Zaitlin M. (1990) Plants transformed with a tobacco mosaic virus nonstructural gene sequence are resistant to the virus. Proc. Natl. Acad. Sci. USA 87: 6311-6315.

Goodman R.M., Hauptli H., Crossway A. and Knauf V.C. (1987) Gene transfer in crop improvement. Science 236: 48-54.

Grumet R. (1990) Genetically engineered plant virus resistance. Hort. Sci. 25: 508-513.

Haseloff J. and Gerlach W.L. (1988) Simple RNA enzymes with new and highly specific endoribonuclease activities. Nature (London) 334 : 585-591.

Harrison B.D., Mayo M.A. and Baulcombe D.C. (1987). Virus resistance in transgenic plants that express cucumber mosaic virus sattelite RNA. Nature (London) 328: 799-802.

Hemenway C.,Fang R.X., Kaniewski W.K., Chua N.H. and Tumer N.E. (1988) Analysis of the mechanism of protection in transgenic plants expressing the potato virus X coat protein or its antisense RNA. EMBO J.7: 1273-1280.

Hiatt A., Cafferkey R. and Bowdish K. (1989) Production of antibodies in transgenic plants. Nature (London) 342: 76-78.

Hoekema A., Huisman M.J., Molendijk L., Van Der Elzen P.J.M. and Conelisen B.J.C. (1989) The genetic engineering of two commercial potato cultivars for resistance to potato virus X. Bio/Technology 7: 273-278.

Hutchins C.J., Rathjen P.D., Forster A.C. and Symons, R.H. (1986) Self cleavage of plus and minus RNA transcripts of avocado sunblotch viroid. Nucleic Acids Res. 14: 3627-3640.

Kawchuk L.M., Martin R.R. and McPherson J.C. (1990) Resistance in transgenic plants expressing the potato leafroll luteovirus coat protein gene. Mol. Plant-Microbe Interact. 3 (in press).

Koenig R. and Lesemann D.E. (1978) Potexvirus group. CMI/AAB Descript. Plant Viruses, no. 200.

Lawson C., Kaniewski W., Haley L., Rozman R., Newall C., Sanders P. and Tumer N.E. (1990) Engineering resistance to mixed virus infection in a commercial potato cultivar: Resistance to potato virus X and potato virus Y in transgenic Russet Burbank. Bio/Technology 8: 127-134.

Loesch-Fries L.S., Merlo D., Zinnen T., Burhop L., Hill K., Krahn K., Jarris N., Nelson S. and Halk E. (1987). Expression of alfalfa mosaic virus RNA 4 in transgenic plants confers virus resistance. EMBO J. 6: 1845-1851.

MacKenzie D.J. and Tremaine J.H. (1990) Transgenic Nicotiana debneyii expressing viral oat protein are resistant to potato virus S infection. J. Gen. Virol. 71: 2167-2170.

Morch M.D., Joshi R.L., Deniel T.M. and Haenni A.L. (1987) A new sense approach to block viral RNA replication in vitro. Nucleic Acids Res. 15: 4123-4130.

Neijidat A. and Beachy R.N. (1990) Transgenic tobacco plants expressing a coat protein gene of tobacco mosaic virus are resistant to some other tobamoviruses. Mol. Plant-Microbe Interact. 3: 247-251.

Nelson, R.S., McCormic S.M., Delaney X., Dube P., Layton J., Anderson E.J., Kaniewska M., Proksch R.K., Horsch R.B., Rogers S.G., Fraley R.T. and Beachy R.N. (1988) Virus tolerance, plant growth, and field performance of transgenic tomato plants expressing coat protein from tobacco mosaic virus. Bio/Technology 6: 403-409.

Powell-Abel P., Nelson R.S., De B., Hoffman N., Rogers S.G., Fraley R.T. and Beachy R.N. (1986) Delay of disease development in transgenic plants that express the tobacco mosaic virus coat protein gene. Science 232: 738-743.

Powell P.A., Stark D.M., Sanders P.R. and Beachy R.N. (1989) Protection against tobacco mosaic virus in transgenic plants that express tobacco mosaic virus antisense RNA. Proc. Natl. Acad. Sci. USA 86: 6449-6952.

Ralph W. (1988) New approaches to controlling plant viruses. Rural Res. 138: 4-8.

Rezaian M.A., Skene K.G.M. and Ellis J.E. (1988) anti-sense RNAs of cucumber mosaic virus in transgenic plants assessed for control of the virus. Plant Mol. Biol. 11: 463-471.

Robinson D.J. (1988) Plant virus diseases and the application of biotechnology. Impact Sci. Soc. 38: 193-201.

Sherwood J.L. and Fulton R.W. (1982) The specific involvement of coat protein in tobacco mosaic virus cross protection. Virology 119: 150-158.

Shukla D.D., Frenkel M.J. and Ward C.W. (1991). Structure and function of the potyvirus genome with special reference to the coat protein coding region. Can. J. Plant Pathol. (in press).

Shukla D.D. and Ward C.W. (1989) Structure of potyvirus coat proteins and its application in the taxonomy of the potyvirus group. Adv. Virus Res. 36: 273-314.

Stark D.M. and Beachy R.N. (1989) Protection against potyvirus infection in transgenic plants : evidence for broad spectrum resistance. Bio/Technology 7: 1257-1262.

Symons R.H. (1989) Self cleavage of RNA in the replication of small pathogens of plants and animals. TIBS 14: 445-450.

Tumer N.E., O'Connell K.M., Nelson R.S., Sanders P.R., Beachy R.N., Fraley R.T. and Stark D.M. (1987) Expression of alfalfa mosaic virus coat protein gene confers cross-protection in transgenic tobacco and tomato plants. EMBO J. 6: 1181-1188.

Uhlenbeck O.C. (1987) A small catalytic oligonucleotide. Nature (London) 328: 596-600.

Van Dun C.M.P. and Bol J.F. (1988) transgenic tobacco plants accumulating tobacco rattle virus coat protein resist infection with tobacco rattle virus and pea early browning virus. Virology 167: 649-652.

Van Dun C.M.P., Overduin B., Van Vloten-Doting L. and Bol J.F. (1988a) Transgenic tobacco expressing tobacco streak virus or mutated alfalfa mosaic virus coat protein does not cross protect against alfalfa mosaic virus infection. Virology 164: 383-389.

Van Dun C.M.P., Van Vloten-Doting L. and Bol J.F. (1988b) Expression of alfalfa mosaic virus cDNA 1 and 2 in transgenic tobacco plants. Virology 163: 572-578.

Van Regenmortal M.H.V. (1982) Serology and immunochemistry of plant viruses. Academic Press, New York, USA.

Yamaya J., Yoshioka M., Meshi T., Okada Y. and Ohno T. (1988) Cross protection in transgenic tobacco plants expressing a mild strain of tobacco mosaic virus. Mol. Gen. Genet. 315: 173-175.

Genetic engineering and transformation of monocots for crop improvement

ARUN P. ARYAN AND THOMAS W. OKITA,
Institute of Biological Chemistry, Washington State University,
Pullman, WA 99164-6340, USA

Keywords: Genetic engineering, monocot transformation, crop improvement

Introduction

The most significant development in the advancement of agricultural technology is the stable introduction and expression of foreign genes into plants. In the past, the introduction of novel genes into crop plants was limited by the sexual compatibility of the donor and host species. With the tools of recombinant DNA technology and plant tissue culture, genes governing desirable traits can be isolated from any organism, manipulated as required and incorporated into economically important crop plants. Thus genetic engineering technology, in concert with modern methods of plant breeding, will have a significant impact on crop improvement and productivity. In this article we will briefly describe some of the strategies involved in genetic manipulation and transformation of monocot crop plants and our current efforts to engineer rice and wheat.

I. Strategies in improvement of monocot cereals

Although our primary research projects are focussed on the cereals, rice and wheat, and not on horticultural plants, the basic approaches and steps involved in the improvement of these crops are essentially the same. They consist of the following:

1. *Identification and isolation of agronomically important gene(s).*
2. *Structural and functional analysis of these genes.*
3. *Possible manipulation of the promoter and/or coding region.*
4. *Introduction of the characterized/modified native or foreign gene(s) into plants by transformation.*

In the past decade, a number of genes governing economically important traits have been identified which are potential targets of manipulation for crop improvement. Such genes include ones that confer resistance to insects, diseases, herbicides and environmental stress in crop plants (Gasser and Fraley, 1989). In our laboratory the major emphasis centers on the genes encoding the rice and wheat seed storage proteins. The storage proteins are the predominant proteins in seed tissue and hence dictate the overall level and quality of the seed proteins. In general these proteins are deficient in several amino acids essential for human growth and development (Larkins, 1981). As these cereals are used mainly as food, our long term goals are to understand the structure-function relationships of the transcriptional regulatory elements and coding sequences of the storage protein

J. Prakash and R. L. M. Pierik (eds.), Horticulture – New Technologies and Applications, 115–121.

genes. With this knowledge, rational attempts can then be made to manipulate the promoter region of these genes for enhanced protein productivity and to mutagenize the protein sequence for a better balance of amino acids.

Once the gene of interest has been isolated and its structure determined the next important step is to identify the various cis-acting regulatory sequences of the promoter. A common approach for promoter analysis is to construct a series of deletions, from both the 5' and 3' ends, which are then fused upstream to a reporter cartridge such as chloramphenicol acetyl transferase (CAT), neomycin phosphotransferase II (npt II), or beta-glucuronidase (GUS) genes which encode easily assayable enzyme activities. These deletion constructs, in appropriate vectors, are then introduced into plants or protoplasts for their stable and transient expression, respectively. Alternatively, the use of the biolistics gun affords a unique opportunity to study the transient expression of any gene in the desired intact tissue. The regulatory sequence motifs that control transcription can be identified from the expression levels of the reporter gene driven by the series of deleted promoter fragments in various plant tissues/protoplasts.

Genes for gliadin, a prolamine class which constitutes up to 50% of total endosperm protein, have been isolated from wheat seeds and their detailed structure analyzed (Reeves and Okita, 1987). Similarly, gene copies representative of the three classes of the rice glutelin multigene family have been isolated and characterized (Okita *et al.*, 1989 a & b). To analyze the regulatory sequences of the wheat gliadin gene, deletion constructs of the promoter with a CAT reporter gene were introduced into plant protoplasts. Since the gliadin promoter was very weakly expressed by itself in wheat, rice or tobacco protoplasts, a hybrid promoter technique using the *nos* promoter elements was utilized (Aryan *et al.*, 1991). Using this approach two regions of the gliadin promoter were identified which modulated the levels of transcription. The first region, -218 bp to -142 bp from the translational start contains a C and A rich motif known as the CACA box. The deletion of this region of the gliadin promoter significantly lowered the expression of CAT activity. Similarly a second region, from -592 bp to -448 bp, was found to modulate promoter function (Aryan *et al.*, 1991).

The presence of cis-acting regulatory elements directly implies the involvement of trans-acting factor(s) which interact with these sequences to affect transcription. Therefore, an alternative approach to identify cis-acting regulatory sequences is by gel retardation assays which resolves specific DNA-protein complexes. Using nuclear extracts from wheat endosperm, one or more nuclear factors have been shown to bind in the vicinity of the CACA box. Further work on the identification of other nuclear factor(s) in gliadin gene regulation is underway (Vellanoweth and Okita, personal communications). A similar approach is in progress, to understand the various regulatory factors involved in the expression of the rice glutelin genes.

The storage proteins from both wheat and rice are encoded by complex families of genes. In view of this complexity, it is clear that these genes must be modified for enhanced expression as well as improved nutritional qualities as the introduction of a single modified gene would have little effect. To increase the level of a particular gene product, the promoter region of the storage protein genes can be engineered for elevated expression by a number of ways. For example, an enhanced level of gene expression has been achieved by increasing the copy number of the activating elements of the cauliflower mosaic virus promoter (Kay *et al.*, 1987). Alternatively, a hybrid promoter can be utilized. Ellis et al (1987) has shown that placing the enhancer elements of the octopine synthase gene upstream to the maize Adh1 promoter stimulated transcription. In addition, intron sequences, which may provide greater mRNA stability (Callis *et al.*, 1987), can be inserted in the modified gene to effectively increase transcription (Oards *et al.*, 1989).

Mutagenesis of the coding sequences to encode for a better balance of amino acids is now routine, particularly with the introduction and employment of in vitro mutagenesis and Polymerase Chain Reaction techniques. Although random mutagenesis of the primary sequences of the storage proteins may be undertaken, such an approach may have disasterous consequences on the intracellular transport and packaging of the mutated proteins when introduced back into plants. The storage proteins are synthesized on the rough endoplasmic reticulum, routed and packaged into discrete organelles, the protein bodies, where they aggregate into specific protein-protein complexes. Rational attempts to improve protein quality therefore requires extensive knowledge on the structural conformation of the encoded protein. To date, detailed protein conformation information has been obtained for only the 11S and 7S storage proteins typically accumulated by legumes. Based on a comparison of secondary conformation predictions among different 11S and 7S storage proteins, Argos *et al.* (1985) identified several peptide regions which appear to tolerate large insertions or deletions. Thus such variable peptide regions may be targets for mutagenesis for improved protein quality without disrupting the native conformation of the encoded protein (Argos *et al.*, 1985). Similar variable sites have been identified in the glutelins, the 11S storage protein of rice (Okita *et al.*, 1989a).

II. Plant Transformation

In many cases, transformation of most dicotyledonous plant species can be readily accomplished by infection with Agrobacterium, a soil borne bacterium with its associated disarmed Ti plasmid vector (for details see Hooykaas *et al.*, 1989). Since our laboratory is working mainly on rice and wheat, recalcitrant to Agrobacterium infection, we are attempting to optimize and establish transformation systems for monocots using a direct DNA delivery approach.

Direct DNA delivery into protoplasts

Among the various direct DNA delivery techniques for monocot transformation the delivery of DNA into protoplasts and subsequent growth into intact plants has proven to be the most reliable and reproducible technique (see Davey *et al.*, 1989, Potrykus, 1990). Two basic requirements needed for protoplast-mediated transformation are:

1) A defined protoplast to plant regeneration system and
2) A defined procedure for direct DNA delivery

For monocots we have taken rice as a model system to optimize the cell culture and modes of DNA delivery. In order to obtain good quality protoplasts with the ability to divide and regenerate after transformation, an embryogenic liquid suspension culture is required. In the case of japonica type rice we found that the calli induced from anther tissues was the best source of totipotent embryogenic cells followed by those obtained from immature panicles, young embryos and mature embryos (Aryan and Okita, 1989). Protoplasts can be readily isolated from these highly embryogenic cell suspension lines as described by Thompson *et al.* (1986).

Plasmid DNA harboring the gene(s) of interest and an appropriate selection marker gene can be introduced into protoplasts by either a) electroporation, b) polyethylene glycol (PEG)-mediated DNA introduction, c) liposome DNA-carriers, or d) microinjection into nuclei. A direct comparison of the DNA delivery efficiency into rice protoplasts, as determined by transient expression analysis, showed that electroporation (Aryan *et al.*, 1991) was 4-8 fold more effective than PEG treatment (Maas and Werr, 1989). A similar finding of electroporation being superior to PEG-mediated transformation was also reported by

Yang *et al.* (1988). Although transgenic rice has been obtained by both electroporation (Zhang *et al.*, 1988, Toriyama *et al.*, 1988) and PEG-treatment techniques (Zhang and Wu, 1988, Li *et al.*, 1990) the latter is more popular probably because of its simplicity and less damaging effects on protoplasts (Li *et al.*, 1990). Use of liposome-DNA carriers, prepared with the commercial reagent lipofectin (Felgner *et al.*, 1987), was ineffective in significant DNA uptake into rice protoplasts. In the presence of PEG, however, liposome-DNA carriers resulted in very effective DNA transfection, 1.5-2.0 fold more effective than that obtained alone with PEG-treatment. Thus the combination of PEG and liposome carriers can be an effective way of DNA delivery into fragile monocot plant protoplasts. Other techniques such as microinjection into the nucleus (Crossway *et al.*, 1986) have not been successful in producing transformed monocots (Potrykus, 1990)

The transfected rice protoplasts can be readily cultured in KPR medium (Thompson *et al.*, 1986) at a density ~ $1x10^6$/ml with 0.5% low temperature melting agarose. This method of culture was found to be better than the one without agarose or with agarose beads. The plating efficiency, which can be in the range of 0.1-0.4%, was found to be variable and primarily depended on the cell line and the quality of protoplasts. After ~3 weeks the dividing protocalli can be transferred to the selection medium containing 50 mg/l of G-418 or Hygromycin, depending upon the selection marker gene used. The transformed resistant calli are regenerated, transferred to potting soil, and grown to maturity. Using these optimized protocols for rice transformation we have introduced the promoter constructs of CaMV, rice glutelin and ADPglucose pyrophosphorylase genes fused to the GUS reporter gene into rice plants. A distribution of GUS activity in various tissues during plant development in these transgenic plants will help us understand the temporal and spatial regulation of these genes. Attempts are also being made to apply this cell culture based transformation technology to wheat as well.

Protoplast transformation and regeneration is laborious and cumbersome. The transformed plants, however, originate from a single selected cell and thus is a reliable approach to obtain true transformants. Furthermore, relatively high transformation frequencies, up to 5×10^{-4}, as compared to other techniques can be achieved with this procedure (Li *et al.*, 1990).

Other approaches of direct DNA delivery

In order to circumvent the tedious time consuming steps of establishing embryogenic suspension lines, protoplasting and regeneration, researchers have resorted to alternative approaches of plant transformation. One of the newer techniques is the introduction of DNA into embryogenic calli by particle bombardment. Microprojectiles are coated with DNA and accelerated to high velocity (Klien *et al.*, 1987) which enter the intact cells ultimately delivering the DNA into the nucleus. Stably transformed cells are selected and regenerated into transgenic plants. This technique can be applied to both monocot and dicot plant species and has already resulted in formation of the transgenic soybean (McCabe *et al.*, 1988) and maize (Gordon-Kamm *et al.*, 1990) plants. This method, however, requires expensive instrumentation, suffers from a lack of reproducibility, and entails tedious selection of a relatively low number of transformed cells (Potrykus, 1990).

Another DNA delivery approach is through the pollen tube pathway as described in detail by Lou and Wu (1988). Although this technique is very simple and easy, there is apparently only one successful transformation report so far with rice (Lou and Wu, 1988). Our attempts using this technique with rice and wheat florets gave only transient expression in ~1.0% of the set seeds. Further refinement of this technique, if it does result in stable transformation, are clearly required. Similarly other novel approaches such as dry embryo soaking (Topfer *et al.*, 1989) and electroporation of intact tissues (Dekeyser *et al.*, 1990), which has resulted in only transient expression so far, need further work to assess their potential in producing transgenic plants.

Conclusions

Although gene transfer technology, particularly using the Ti plasmid of Agrobacterium, is routinely being utilized in genetic improvement studies of dicot plants, similar advancements with cereal crops has been slow. With the advent of new DNA delivery approaches, however, one can forsee successful monocot transformation experiments as a routine. To obtain transgenic monocots it appears that DNA delivery into protoplasts and their subsequent regeneration is the most reliable approach although microprojectile gun, with some improvements, appears promising. Thus recombinant DNA technology and plant tissue culture approaches, in collaboration with plant breeding techniques, can play a significant role in future manipulation of agricultural crops. Although this technology has great potential in improving nutritional quality and commercial value of crop plants, every precaution must be taken to study the foreign gene(s) and their possible side effects before introducing them into plants and the environment.

Acknowledgements

This work was supported by a grant from The Rockefeller Foundation. Project 0590, College of Agriculture and Home Economics, Washington State Uni- versity, Pullman, WA 99164. We are thankful to Ms. Kellee Roberti and Kim Stephens for their technical assistance. Travel funds to APA from the Indo-American Hybrid Seeds Co. to attend the Seminars are duly acknowledged.

References

Argos P., Narayana S.V.L. and Nielsen N.C. (1985) Structural similarity between legumin and vicilin storage proteins from legumes. EMBO J. 4: 1111-1117.

Aryan A.P., An G. and Okita T.W. (1991) Structural and functional analysis of promoter from gliadin, an endosperm-specific storage protein gene of Triticum aestivum L. Mol.Gen. Genet. 225: in press.

Aryan A.P. and Okita T.W. (1989) Screening of totipotent rice cell line for efficient transformation. Rice Genet. Newsletter 6: 170.

Callis J., Fromm M. and Walbot V. (1987) Introns increase gene expression in cultured maize cells. Gen. Dev. 1: 1183-1200.

Crossway A., Oakes J.V., Irvine J.M., Ward B., Knauf V.C. and Shewmaker C.K. (1986) Integration of foreign DNA following microinjection of tobacco mesophyll protoplasts. Mol. Gen. Genet. 202: 179-185.

Davey M.R., Rech E.L. and Mulligan B.J. (1989) Direct DNA transfer to plant cells. Plant Mol. Biol. 13: 273-285.

Dekeyser R.A., Claes B., De Rycke R.M.U., Habets M.E., Van Mantagu M.C. and Caplan A.B. (1990) Transient gene expression in intact and organized rice tissues. The Plant Cell 2: 591-602.

Ellis J.G., Llewellyn D.J., Walker J.C., Dennis E.S. and Peacock W.J. (1987) The ocs elements: a 16 base pair palindrome essential for activity of the octopine synthase enhancer. EMBO J. 6: 3203-3208.

Felgner P.L., Gadek T.R., Holm M., Roman R., Chan H.W., Wenz M., Northrop J.P., Ringold G.M. and Danielson M. (1987) Lipofectin: A highly efficient, lipid-mediated DNA-transfection procedure. Proc. Natl. Acad. Sci. USA 84: 7413-7417.

Gasser C.S. and Fraley R.T. (1989) Genetically engineering plants for crop improvement. Science 244: 1293-1299.

Gordon-Kamm W.J., Spencer T.M., Mangano M.L., Adams T.R., Daines R.J., Start W.G., O'Brien J.V., Cambers S A., Adams W.R., Willets N.G., Rice T.B., Mackey C.J., Krueger R.W., Kausch A.P. and Lemaux P.G. (1990) Transformation of maize cells and regeneration of fertile transgenic plants. The Plant Cell 2: 603-618.

Hooykaas P.J.J. (1989) Transformation of plant cells via Agrobacterium. Plant Mol. Biol. 13: 327-336.

Kay R., Chan A., Daly M. and McPherson J. (1987) Duplication of CaMV 35S promoter sequences create a strong enhancer for plant genes. Science 236: 1279-1282.

Klein T.M., Wolf E.D., Wu R., and Sanford J.C. (1987) High velocity microprojectiles for delivering nucleic acids into living cells. Nature 327:70-73.

Larkins B. (1981) Seed storage proteins, In: Conn E.E. ed., The Biochmistry of Plants, 6: 449-487. Academic Press, Inc. USA.

Li Z., Burrow M.D. and Murai N. (1990) High frequency generation of fertile transgenic rice plants after PEG-mediated protoplasts transformation. Plant Mol. Biol. Rep. 8: 276-291.

Luo Z-x. and Wu R. (1988) A simple method for the transformation of rice via pollen tube pathway. Plant Mol. Biol. Rep. 6: 165-174.

Maas C. and Werr W. (1989) Mechanism and optimized conditions for PEG mediated DNA transfection into plant protoplasts. Plant Cell Rep. 8: 148-151.

McCabe D.E., Swain W.F., Martinell B.J. and Christou P. (1988) Stable transformation of soybean (Glycine max) by particle acceleration. Bio/Technology 6: 923-926.

Okita T.W., Hwang Y-S., Hnilo J., Kim W-T., Aryan A.P., Larson R. and Krishnan H.B. (1989a) Structure and expression of the rice glutelin multigene family. J. Biol. Chem. 264: 12573-12581.

Okita T., Aryan A., Reeves C., Kim W-T., Leisy D., Hnilo J. and Morrow D. (1989b) Molecular aspects of storage protein and starch synthesi in wheat and rice seeds. In:

Poulton J.E., Romeo J.T. and Conn E.E. eds.,Plant Nitrogen Metabolism, pp 289-327. Plenum Publishing Corporation.

Oard J.H., Paige D. and Dvorak J. (1989) Chimeric gene expression using maize intron in culture cells of breadwheat. Plant Cell Rep. 8: 156-160.

Potrykus I. (1990) Gene transfer to cereals: An assessment. Bio/Technol. 7: 535-542.

Reeves C.B. and Okita T.W. (1987) Analyses of a/b-type gliadin genes from diploid and hexaploid wheats. Gene 52: 257-266.

Thompson J.A., Abdullah R. and Cocking E.C. (1986) Protoplast culture of rice using media solidified with agarose. Plant Sci. 47: 123-133.

Topfer R., Gronenborn B., Schell J. and Steinbiss H-H. (1989) Uptake and transient expression of chimeric genes in seed-derived embryos. Plant Cell 1: 133-139.

Toriyama K., Arimoto Y., Uchimiya H. and Hinata K. (1988) Transgenic rice plants after direct gene transfer into protoplasts. Bio/Technology 6: 1072-1074.

Yang H., Zhang H.M., Davey M.R., Mulligan B.J. and Cocking E.C. (1988) Production of kanamycin resistant rice tissues following DNA uptake into protoplasts. Plant Cell Rep. 7: 421-425.

Zhang H.M., Yang H., Rech E.L., Golds T.J., Davis A.S., Mulligan B.J., Cocking E.C. and Davey M.R. (1988) Transgenic rice plants produced by electroporation-mediated plasmid uptake into protoplasts. Plant Cell Rep. 7: 379-384.

Zhang W. and Wu R. (1988) Efficient regulation of transgenic plants from rice protoplasts and correctly regulated expression of the foreign gene in the plants. Theor. Appl. Genet. 76: 835-841.

Agrobacterium-mediated Gene Transfer in Citrus reticulata Blanco

GAO FENG[1], HUA XUEJUN[2], FAN YUNLIU[2], ZHANG JINREN[1], AND CHEN SHANCHUN[1]

[1] Citrus Research Institute. Chinese Academy of Agricultural Sciences, Chongqing 630712, China

[2] Biotechnology Research Center, Chinese Academy of Agricultural Sciences, Beijing 100081, China.

Introduction

Agrobacterium-mediated **in vitro** transformation has been applied to a growing number of plant species and used as a new technique to improve their yields and qualities (Fraley et.al. 1986). However, up to now we have not seen any detailed reports about this in citrus fruit trees. In our experiment, leaf disc method developed by Horsch et al. has been simplified and extended for use with explants of leaves, stems and cotyledons of "Hongju" tangerine, to study the possibilities and stabilities of transformation and expression of T-DNA in citrus cells and to provide a new way for citrus improvement by using with genetic engineering.

Materials and methods

Leaves, stems and cotyledons were excised from grown **in vitro** seedlings of "Hongju" tangerine and were cut into sections (about 5 mm) as explants. The explants were briefly inoculated with wild-type A. **tumefaciens** strain B6 S3 grown overnight and diluted 5 times with MS medium, and then blotted dry with sterile filter papers and cultured on MT medium in the dark at 28± 1°C. After 3 days of cocultivation, the explants were transferred to MT medium supplemented with carbenicillin (carb) at 500 mg/1 to remove excess bacteria. The explants which were not inoculated with bacteria were used as controls. All of the explants

J. Prakash and R. L. M. Pierik (eds.), Horticulture – New Technologies and Applications, 123–125.

were cultured with 16h photoperiod at 28±1°C. the results were investigated after 2 weeks. After 1 month, tumour tissues or calli were excised from the original explants respectively and subcultured on the same medium at about 30 days intervals. Octopine synthase activity was assayed according to the method of Xu Yao et al.

Results

After 5 - 10 days cultured on MT medium contained 500 mg/1 carb, the infected explants of leaf, stem and cotyledon initiated tumour tissues. After 2 weeks, many tumour tissues were formed from the infected explants and were visible to the eyes. The results show that the frequency of transformation is biggest in leaves (50.0%), followed by stems (43.5%) and cotyledons (9.5%), but the tumour tissues formed from cotyledons are larges. While no tumour tissue or callus is formed from control leaves and cotyledons, only a small callus is formed from a control stem section. There are some obvious differences between the callus and the tumour tissue in formative sites and appearances. The callus is formed from all of cut surface at the site of end of a stem section and is hemisphere in shape, its surface is rough and loose, while the tumour tissue is formed from a small part of cut surface and is sphere in shape just as a drop of water, its surface is smooth and compact. When subcultured on the hormone-free MT medium, callus grew slow, but tumour tissue grew quite fast.

In order to further identify whether the T-DNA was transferred and expressed in cells of "Hongju" tangerline, octopine synthase activity was assayed in the tumour tissues and callus after subcultured 3 times (about 90 days). The results show that all of the tumour tissues which taken at random exist octopine, but the callus does not. This demonstrated that the T-DNA has been transferred and expressed in the transformed cells of "Hongju" tangerine.

Discussion

There are two main kinds of functional genes in T-DNA of Ti plasmid of A. **tumefaciens**. One is the plant hormone synthase gene, another is the opine synthase gene (Weising, et al ., 1988). Their expression in transformed cell results in that the transformed cell has characteristics of hormone autonomous growth and opine synthesis. Therefore, the two characteristics are usually used as two bases for identifying transformed cells (Xu Yao et al., 1988). The results from our experiment demonstrated that the hormone synthase gene and octopine synthase gene coded by T-DNA were transferred and expressed in the transformed cells of "Hongju" tangerine. This suggests that the Ti plasmid can be used as a suitable gene vector and the explants of leaf, stem and cotyledon can be used as the suitable gene acceptors for citrus improvement. This is the first report of **in** vitro transformation of citrus fruit trees using Agrobacterium.

References

1) Fraley, R.T. et al., 1986. Genetic transformation in higher plants. Rev. Plant Sci. 4:1 - 46.
2) Horsch, R.B. et al., 1985. A simple and general method for transferring genes into plants. Science. 227:1229-1231.
3) Weising, K, et al., 1988. Foreign genes in plants: transfer, structure, expression, and applications. Annu. Rev. Genet. 22:421-477.
4) Xu Yao et al., 1987. A simple and efficient method for the detection of opine synthase activities in plant tissues. Hereditas (Beijing). 9(5): 41-43.
5) Xu Yao et al.,1988,Some characteristics of T-DNA-transformed plant cells. Plant Physiology Communications. (1) :16-21.

Regeneration and genetic transformation studies in watermelon (Citrullus vulgaris L. cv. Melitopolski)

D.K. SRIVASTAVA, V.M. ANDRIANOV* & E.S. PIRUZIAN*
Department of Biotechnoly, Dr. Y.S. Parmar University of Horticulture & Forestry, Solan-173230 India
*Institute of Molecular Genetics, USSR Academy of Sciences, Moscow-123182 USSR.

Key words: Regeneration, Transformation, Neomycin phosphotransferase-II.

Introduction

Tissue culture techniques have opened many new possibilities of crop improvement, since responses at cell level are well defined under controlled conditions. Recent biotechnological developments have enabled investigators to produce transgenic plants that are tolerant of herbicides, insects and resistance to viral infection (Gasser and Fraley, 1989). Agrobacterium tumefaciens, a soil borne bacterium that causes crown gall disease in a wide range of dicotyledonous plants, has been most widely used as a vehicle for gene transfer into plants. The transferred genes are stably integrated into the genomes of the transgenic plants and are transmitted to their progengy as Mendelian factors (DeBlock et al., 1984).

Watermelon is an important food crop grown in a number of countries. The application of genetic engineering in watermelon cultivation is of great value so as to obtain improved or desirable traits like disease resistance. In this paper, we report, the regeneration of watermelon plants from the cotyledon segments and the integration and expression of chimeric NPT-II gene in the transformed calli. This is the first report on regeneration and genetic transformation studies in watermelon.

Methods

Plant Material: Seeds of watermelon were obtained from the National Research Institute of Irrigated Cultivation of Vegetables and Melon, Astrakhan, USSR. The seedlings (7-8 days old) were used for regeneration and transformation experiments.

Bacterial strains and plasmids: Disarmed Agrobacterium tumefaciens strain containing vector pGV 3850 Neo (Zambryski et al., 1983) and Agrobacterium rhizogenes strain containing binary vector pAK 320 Neo+pRi 15834 were used for transfor-

J. Prakash and R. L. M. Pierik (eds.), Horticulture – New Technologies and Applications, 127–130.

mation studies (Piruzian and Andrianov, 1986).
Co-cultivation of agrobacterium tumefaciens and watermelon cotyledon discs: After 2-days of pre-conditioning on appropriate medium, the cotyledon discs were removed from the plates and gently mixed with overnight cultures of *A. tumefaciens* (pGV 3850 Neo, 10^8 cells/ml). The cotyledon discs were then removed from the *Agrobacterium* cells and thoroughly blotted dry between a dried sterile filter papers. The cotyledon discs were then returned to the same plates. After two days of co-cultivation period, the cotyledon discs were transferred to the selective medium with appropriate hormones. Cotyledon discs with developing callus were transferred every 3-4 weeks to the fresh selective medium for tissue amplification and regeneration.
Agrobacterium inoculation and selection of transformed tissue from hypocotyl segments: Hypocotyl segments (1-1.5 cms in length) were cut from aseptically grown 7-8 days old seedlings. The inoculum (*A. rhizogenes* containing binary vector pAK 320 Neo + pRi 15834) was collected on a sterile bacteriological loop and smeared gently on the cut surface of the inverted hypocotyl segments placed vertically in hormone free MS medium. The petriplates were then sealed and incubated in the culture room. One week after inoculation, the hypocotyl segments were cut above the agar surface and transferred into hormone free MS medium supplemented with kanamycin (50 mg/l) and cefotaxime (100 mg/l). The roots, produced on the inoculated surfaces were excised and placed on MS medium containing different combinations of 2,4-D, NAA and BAP. The roots grew well on the selective medium and formed callus. Subcultures were made every 3-4 weeks to the fresh selective medium for tissue amplification and regeneration.
Neomycin phosphotransferase-II assay: Frequency of transformation by vector DNA was assessed in the transformed calli using an assay for NPT-II enzyme (Reiss et al., 1984).

Results

The presence of BAP in the nutrient medium alone, promoted the regeneration of shoots from the cotyledon segments. Out of the various concentrations of BAP, 4.5 µM was highly effective for the regeneration of shoots followed by callus formation (Fig.1A). Root regeneration took place from the regenerated shoot when transferred to the medium supplemented with 0.5 µM NAA (Fig.1B).

The co-cultivated cotyledon discs were transferred to the fresh selection plates. Callus could be seen after 2-weeks at the cut edges of the cotyledon and also at the wound sites where the tissue was damaged during inoculation process. The calli were maintained on the selective medium for further proliferation and regeneration. The calli which were positive to NPT-II (Fig.1F) were used for plant regeneration.

The *Agrobacterium* strain containing the vector pAK 320 Neo + pRi 15834, was infectious on watermelon hypocotyl as indicated by a dense green colour callus that appeared on the inoculated surfaces 8-10 days after inoculation. 5-7 days later this callus produced roots (Fig.1C). There was no root formation in the uninoculated hypocotyl segments. The roots harvested from the inoculated hypocotyl segments were cultured on the selective MS medium containing different combinations of 2,4-D, NAA and BAP. Out of the various combinations of hormones tried, 5 μM 2,4-D, 1 μM NAA and 1 μM BAP gave comparatively better growth of callus (Fig.1D). The calli were maintained on the selective medium (Fig.1E) and tested for NPT-II assay (Fig.1F). The calli which were positive to NPT-II were used for plant regeneration.

Discussion

The watermelon plants were regenerated from the cotyledon discs. Phenotypic and enzymatic data indicate that the chimeric kanamycin resistance gene was expressed in watermelon cells using both cointegrating and binary vector agrobacterial systems. Watermelon transformation technology in combination with advances in watermelon regeneration (Srivastava et al., 1988, Srivastava et al., 1989), may eventually lead to the recovery of a transformed watermelon plant.

References

DeBlock M. Herrera-Estrella L. Van Montagu M. Schell J. and Zambryski P. (1984) Expression of foreign genes in regenerated plants and their progeny. EMBOJ.3:1681-1684.

Gasser C.S. and Fraley R.T. (1989) Genetically engineering plants for crop improvement. Science 244:1293-1299.

Piruzian E.S. and Andrianov V.M. (1986) Cloning and the analysis of a replication region of the *Agrobacterium tumefaciens* C58 nopaline Ti-plasmid and its application for foreign gene transfer into plants. Genetika (USSR) 22: 2674-2683.

Reiss B. Sprengel R. Will H. and Schaller H. (1984) A new sensitive method for qualitative and quantitative assay of neomycin phosphotransferase in crude cell extracts. Gene 30: 211-218.

Srivastava D.K. Andrianov V.M. and E.S. Piruzian (1988) Organogenesis and regeneration in watermelon tissue culture. Soviet Jr. Plant Physoil. 35. 1243-1247.

Srivastava D.K. Andrianov V.M. and E.S. Piruzian (1989) Tissue culture and plant regeneration of watermelon (*Citrullus vulgaris* L. cv. Melitopolski). Plant cell Rep. 8: 300-302.

Zambryski P. Joos H. Genetello C. Leemans J. Van Montagu M. and Schell J. (1983) Ti-plasmid vector for the introduction of DNA into plant cells without alteration of their normal regeneration capacity. EMBOJ. 2:2143-2150.

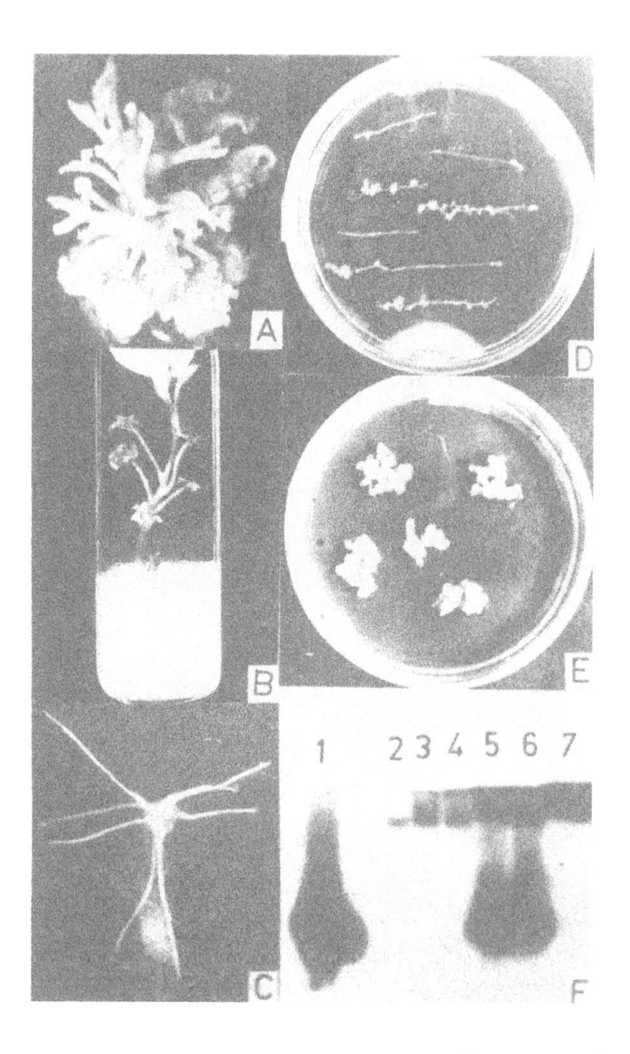

Fig.1 A-F Regeneration and transformation studies in Watermelon (Citrullus vulgaris L.)

A. Regeneration of shoots from cotyledon explants.
B. Regenerated shoot with root growth.
C. Development of hairy roots from the inoculated hypocotyl with the Agrobacterium rhizogenes strain containing vector pAK 320 Neo + pRi 15834.
D. Initiation of callus from the transformed roots.
E. Transformed calli growing on the selective medium.
F. NPT-II assay on transformed calli obtained from cotyledon disc and transformed hairy roots from hypocotyl. No.1 and 2 are positive and negative controls respectively whereas 3-7 are samples of transformed calli.

Molecular aspects of Cytokinin's stymied action on Auxin mediated new root formation in the hypocotyls of Phaseolus vulgaris L.

G. R. KANTHARAJ* and G. PADMANABHAN**
* Dept. of Botany, The National College, Bangalore - 560 004, India. (Correspondence address)
** Dept. of Biochemistry, Indian Institute of Science, Bangalore - 560 012, India

INTRODUCTION

Elucidation of molecular events that regulate organogenesis either in an *in vitro* or in an *in vivo* system is a daunting task fraught with many imponderables. It is deja vu that a set of phytohormones control the growth and development of plant body. It is also fiat accompli that any two growth hormones such as auxin or cytokinin, though having their own specific effects on different target tissues, yet their combined action on the same tissue varies and it is concentration dependent (Miller, 1961). The interplay of different hormones at molecular level in inducing or inhibiting one or the other morphogenic structures such as root or shoot or redifferentiation of differentiated tissues into new organs is both befuddling and contentious.

Among many factors which regulate adventitious root formation, auxin is of prime importance. During auxin induced new root formation either in epicotyls or hypocotyls and other plant tissues, several biochemical changes including increase in DNA, RNA, protein and even enzyme levels have been reported (Davies and Larkins, 1973; Travis et al., 1973; Travis and Key, 1976). Whether or not the differential gene activity leading root formation involves "Homeotic" kind of gene complex and compartmentalisation, as found in Drosophila species, also exists in plant systems cannot be averred for the lack of specific mutants (Scott, 1985).

In earlier investigations, we have preferred excised segments of 3 - 4 cm length of *Phaseolus vulgaris, Linn.*, to *in vitro* tissue culture methods, for raison d' etre that the segments in response to 5×10^{-5} M concentration of Indole Butyric Acid (IBA) produce 30 - 40 new roots in about 48 hrs (Kantharaj et al., 1979). During IBA mediated root initiation, early enhancement (30 mins) in the rate of protein synthesis was due to auxin activation of protein synthesising machinery and not due to early increased synthesis of total RNA including poly(A)$^{+}$RNA. The increased protein synthesis at later stages was augmented by the enhanced RNA including poly(A)$^{+}$RNA

J. Prakash and R. L. M. Pierik (eds.), Horticulture – New Technologies and Applications, 131–139.

synthesis. Among many changes, in IBA enhanced or induced proteins, the 55 - 58 Kd proteins were found in higher quantities, and the same have been identified as Tubulins (α & β units) by their mobility on the gel and also immunological methods (Kantharaj et al., 1985). The present study is actuated to understand molecular events that regulate the inhibition of IBA mediated new root initiation by Benzyl Amino Purine (BAP) - a synthetic cytokinin.

METHODS

Materials French bean seeds were purchased from a local seed nursery. Hypocotyl segments of 3 - 4 cm length from the dark grown seedlings were cut. Then they were abated of auxin by washing in distilled water for 30 mins and made certain that none of the segments produced any new roots on their own. Such segments were incubated for 30 mins in half strength Hoagland's medium (Arnold, 1968) containing 5×10^{-5} M IBA or 4.5×10^{-6} M BAP or both. Then the segments were surface washed and further incubated in nutrient medium for the required durations. Aseptic condition was maintained strictly in all experiments.

^{14}C - labelled Chlorella hydrolysate (270/m atom C) and ^{32}P - carrier free orthophosphate were purchased from Bhaba Atomic Research Centre, Bombay. ^{3}H - Leucine (105 Ci/mM) was purchased from Radiochemical Centre, Amarsham, U.K., Oligo dT-cellulose and Poly(U) - sepharose were purchased from Colloborative Research Inc., Waltham, U.S.A. All other biochemicals were purchased from Sigma, U.S.A.

Rate of protein synthesis and RNA synthesis Hormone pre-treated and control segments at different stages of development were exposed to ^{14}C - Chlorella hydrolysate or ^{3}H - UTP or ^{32}P - orthophosphate for 30 mins; then they were surface washed and processed for determining the protein synthesis by the methods of Rosen, 1957 and Lowery, 1957, and RNA synthesis by the methods of Pennman, 1966 and Padmanabhan et al., 1975.

Quantification of mRNA activity and in vitro protein synthesis Isolated total RNA or poly(A)$^{+}$RNA were translated in a cell free system derived from wheatgerm using ^{3}H - leucine as the label by Roberts and Patterson method, 1973.

Protein analysis Membrane proteins and in vitro radiolabelled poly(A)$^{+}$RNA translated proteins were subjected to Sodium dodisulphate - polyacrylamide gel electrophoresis (SDS-PAGE) by the methods of Laemli, 1970. For measuring radioactivity of proteins in the gels, the gels were first stained with Coomassie brilliant blue and then the required gels were sliced to 1 mm pieces, and digested, and counted for radioactivity.

RESULTS

Anatomical and cytological changes: The pericyclic cells found in-between the split exarch xylem elements in the hypocotyls treated with auxin were found in transformed state at 12 hrs and some of them were mitotically active with periclinal orientation. At 24 hrs, a distinct meristematic dome for new roots was found in the same locus. On the contrary, in the hypocotyls treated with both IBA and BAP, the same pericyclic cells were in a state of proliferation and no root initials were found. However, the hypocotyls exposed to BAP alone did not show any changes in pericycle region even after 48 hrs, except for the enlarged state of cells (figures not shown).

Rate of protein synthesis The rate of protein synthesis, as measured by the above mentioned method (Table 1) show increased levels in all those segments treated with IBA, BAP or both. Such increased levels of proteins at later stages were found in all hormone treated tissues.

Table 1: Rate of protein synthesis: Hypocotyl segments of different stages of development were labelled with ^{14}C - Chlorella hydrolysate for 30 mins. From the homogenate, specific activity of proteins (cpm / µg protein = A) and specific activity of total aminoacid poll (cpm / µg protein = B) were determined. A / B gives the measure of the rate of protein synthesis.

TREATMENT	30 mins	3 hrs	6 hrs	12 hrs	24 hrs
Control	0.07	0.08	0.07	0.08	0.11
IBA	0.15	0.19	0.20	0.29	0.46
BAP	0.13	0.14	0.31	0.25	0.40
IBA + BAP	0.11	0.12	0.14	0.30	0.47

Rate of RNA synthesis The rate of RNA synthesis measured in terms of incorporated radioactive ^{32}P - orthophosphate (carrier free) as shown in Table 2, indicates that increased levels of total RNA as well as poly(A)$^+$RNA (data for poly(A)$^+$RNA specific activity not shown) were detected at 2 hrs in the segments treated with IBA alone. Such increase in RNA synthesis was found in BAP and IBA + BAP treated segments at 6 hrs. At later stages, the RNA content including poly(A)$^+$RNA steadily increases upto 48 hrs. Maximum increase was found in IBA + BAP treated tissues than in IBA treated segments. Cytokinin treated segments showed significant increase but nowhere to the level of IBA treated tissues. Quantification of mRNA activity in total RNA determined by in vitro translation of total RNA in a cell free system derived from wheatgerm also shows increased activity of template RNA at 12 and 24 hrs.

Table 2. **Rate of** RNA synthesis: Hormone treated and control hypocotyls were labelled for 30 mins with ^{32}P - Phosphate (carrier free). Total RNA was precipitated. Specific activity of total RNA i.e., A, in cpm/A 260 and specific activity of total nucleotide pool i.e., B, in cpm/A 260 were determined. The rate of RNA synthesis is given by A / B.

TREATMENT	20 mins	2 hrs	6 hrs	12 hrs	24 hrs
Control	0.13	0.12	0.13	0.18	0.18
IBA	0.12	0.20	0.24	0.29	0.33
BAP	0.11	0.12	0.15	0.27	0.28
IBA - BAP	0.12	0.13	0.19	0.30	0.41

Table 3 Quantification of poly(A)$^{+}$RNA: Total RNA was isolated from different hypocotyl segments and then 20 µg of RNA was translated in a cell free system derived from wheat-germ containing ^{3}H - leucine as label. Three seperate experiments were conducted and three aliquots for each experiment were counted, and the average cpm for each experiment is presented as a measure of template activity.

TREATMENT	12 hrs			24 hrs		
	Expt I	Expt II	Expt III	Expt I	Expt II	Expt III
Control	46640	49500	48000	45750	54300	52300
IBA	75780	80120	78650	136170	142000	138200
BAP	72500	70150	70100	160580	150000	155300
IBA - BAP	88670	89720	87680	164250	168000	167000

Analysis of In Vivo and In Vitro proteins The analysis of soluble proteins by SDS-PAGE methods do not show any discernable changes in the overall protein profiles obtained from different tissues. On the contrary, **membrane proteins** show certain distinct changes in the overall pattern, but the others though show subtle changes however are not clearly discernable. In IBA treated tissues, among many changes, the 55 - 58 Kd proteins were distinctly synthesised in higher quantities than in control. A similar enhanced level of same proteins are also found in BAP and IBA + BAP treated segments. However, a distinct band of 135 Kd protein found only in IBA treated tissue was distinctly absent in other tissues. The said protein profiles were more or less identical in both membrane and in vitro translated proteins (Fig. 1 **and** Fig. 2).

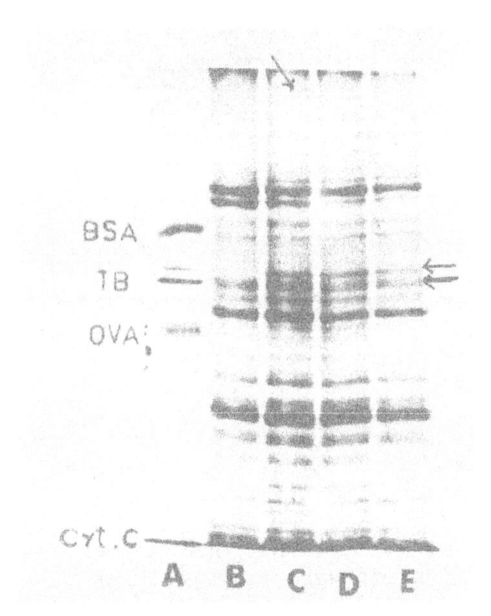

Fig. 1 SDS-polyacrylamide slab gel electrophoresis of membranes obtained from control and hormonal treated hypocotyls. Total membrane proteins were isolated from hypocotyls. The solubilised proteins were subjected to SDS-PAGE and stained with Coomassie brilliant blue. (A) Marker proteins - BSA(Bovine Serum Albumin), TB(Tubulins), OVA(Ovalbumin), Cyt.C(Cytochrome C); (B) Control; (C) IBA treated; (D) BAP treated; (E) IBA - BAP treated. All hypocotyls were exposed for 12 hrs. Note the arrow in (C) shows a new protein band not found in others. Arrows against TB shows increased amount only in hormone treated segments.

DISCUSSION

Differential gene activation or inactivation in the epicotyls or hypocotyls of pea plants and the hypocotyls of soybean in response to the combined action of auxins and cytokinins or individually, as short and long time responses, has been demonstrated by the methods of differential screening of C-DNA clones and also by the analysis of in vitro translated radioactive proteins by 2-Dimensional gel electrophoresis (Theologis, 1982; Keinschmidt, 1984; Theologis, 1986; Kuhlemeir, 1987). It is fiat accompli that redifferentiation of differentiated tissues in response to individual or the combination of hormones into specific organs encompasses differential gene activity in the target tissues entailing cell transformation into embryonic kind of meristems. Further differentiation is imposed by cell polarity fixation. The role of cytoskeletal elements including tubulins and its associated proteins such maps, tau and other proteins in cell transformation and cell polarity fixation for directional cell division for further differentiation has been implied and such work has been extensively reviewed (Hopler, 1974; Quatrano, 1978; Lloyd, 1987).

The excised hypocotyl segments of French beans have been used as beau ideal system for understanding the role and interplay of auxins and cytokinins in the induction or repression of new root formation. The IBA at the concentration of 5×10^{-5} M or more induces maximal number of roots, but BAP at the concentration of 4.5×10^{-6} M or more totally inhibits IBA induced root initiation. However, the BAP at a concentration of 1.5×10^{-6} M or less, with IBA enhances auxin mediated root formation which is not significant (data not given) the observation of which adduces that adventitious root formation or its inhibition is dependent on the balance between the auxin and cytokinin concentration ratio. In all our experiments, inhibitory concentration of BAP (4.5×10^{-6} M) is used and the ratio between them is consistently maintained.

Anatomical and cytological studies luculently affirm that the pericycle cells located inbetween split exarch xylem elements act as target cells for IBA's potentation into new roots. But cytokinins at higher concentrations inhibit the transformation of IBA's target tissue into new root initials, instead the target tissue and its adjunct cells get activated into undifferentiated proliferating kind of cells. The quintessential feature of cytokinin's stymied action is that it is effective only if auxin treated segments are exposed to BAP well before 12 hrs, but at later time, it fails to abolish auxin induced root formation. This avowedly proffers, whatever molecular events actuated in the form of gene activation or repression, within the said time-frame are critical and as well as potentating events for root initiation or inhibition.

To adduce differential gene activity in response to hormones, studies on changes in the pattern of protein synthesis pari passu RNA synthesis in time-frame starting from 30 mins to 24 hrs, obviously imply and infer that both IBA and BAP independently and in combination enhance the rate of protein synthesis very early (30 mins) without augmenting or activating transcriptional activity. So the hormonal activation of increased protein synthesis can be adduced to the activation of protein synthesising machinery. The auxin alone, however, potentates enhanced transcriptional activity entailing the production of higher amounts of RNA pari passu poly(A)$^+$RNA at 2 hrs and onwards, but BAP alone, inspite of having no effect on tissue into any apparent morphological manifestations, enhances transcriptional activity only after 6 hrs of treatment. On the contrary, BAP represses IBA activated early (at 2 hrs) transcriptional activity and maintains its repression till 6 hrs, which proffers that BAP's repression is transitory but regulatory. Notwithstanding, the enhanced protein synthesis at later periods i.e., 12 to 24 hrs in hormone treated segments is bolstered by increased rate of transcriptional activity. Determination of the rate of total RNA synthesis (Table 2), specific activity of poly(A)$^+$RNA synthesis (data not given) and quantification of template activity of total RNA (Table 3) by the way of in vitro translation, affirm that mRNA content increases in all hormone treated segments over 24 hrs of time, by two to four fold over control tissues.

Analysis of both in vivo and in vitro synthesised proteins derived from 24 hrs stage of segments by SDS-PAGE methods have been used to determine and differentiate cytokinin's stymied action on IBA induced gene activited or its repressed products. Protein profiles of soluble proteins (10^5 x G supernatant fraction) don't show any distinctly identifiable changes in the overall pattern. Among the melange of protein profiles obtained from in vivo and in vitro methods, inspite of many subtle and not so clearly discernable changes found in the overall pattern, the most distinct and delineative changes among the pot pouri radioactive protein profiles obtained from in vitro translated poly(A)$^+$RNA products (Fig. 2) which matched exactly with the in vivo membrane protein profiles (Fig. 1) are identified. And their molecular weights have been determined as 55 - 58 Kd and 135 Kd by using standard proteins as markers (on a semi-log graph).

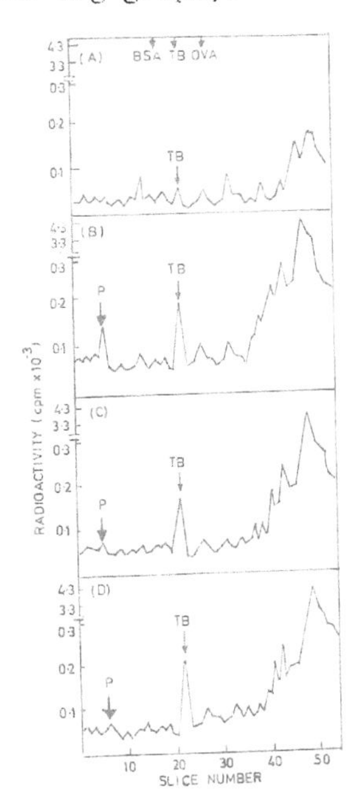

Fig. 2 Radioactive profile of in vitro translated poly(A)$^+$RNA products obtained from control and hormone treated hypocotyls at 12 hrs. The translated products were separated by SDS-PAGE method. The gels were sliced into 1 mm pieces, digested with H_2O_2 and the radioactivity was measured. (A) Radioactive protein profiles from control segments; (B) IBA treated segments; (C) BAP treated segments; (D) IBA + BAP treated segments. BSA(Bovine Serum Albumin), TB(Tubulins), OVA(Ovalbumin), Cyt.C (Cytochrome C) and P(Putative protein). Note the newly induced putative "P" protein in (B) and its absence in (C) and (D), and also note increased amounts in TB only in hormone treated segments and not in control.

Substantial increase in 55-58 Kd proteins only in those segments treated with different hormones proffers that both IBA and BAP independently or in combination, potentate and enhance the transcriptional activity of 55-58 Kd protein encoding genes. In earlier studies (Kantharaj et al., 1985) the 55-58 Kd proteins have been identified as α and β tubulins by immunological, and other methods. Though the requirement of tubulins in higher quantities can be accounted for IBA and IBA + BAP treated tissues for their root initiating and cell proliferating activities, their requirement in BAP tissues is shrouded with enigma for the tissue does not manifest in any anatomical or cytological changes.

In contrast to tubulins, the novel & contre temp 135 Kd protein band distinctly found only in IBA treated segments, is unequivocally marked by its total absence in other tissues. This affirms that the putative protein is an induced gene product ostensibly activated by IBA. The repression of the said protein by cytokinin suggests that cytokinin plays a regulatory role on auxin activated gene. However, the 135 Kd protein's role and character could not be ascertained at this point of time except for the fact that both 55 - 58 Kd and 135 Kd proteins are associated with membranes. Tubulins and their associated high molecular weight proteins have been implicated to act as nucleating centres in the membranes and are also involved in the orientation of mitotic spindle fibres in mitotically transformed cells (Hopler, 1974; Quatrano, 1978). Whether or not the IBA induced gene products have any such role in root initiation and the cytokinins play a regulatory role in inhibiting it, is still a profoundly esoteric phenomenon to be expedited.

In an avowed effort to delineate molecular events that regulate IBA induced root initiation and its inhibition by the regulatory activity of cytokinin, C-DNA clones have been prepared against different mRNA populations obtained from different tissues and the screening process is in progress.

ACKNOWLEDGEMENT

G. R. Kantharaj sincerely acknowledges with a deep sense of gratitude Prof. G. Padmanabhan, for being a lodestar and a friend in his research activities.

REFERENCES

1. Arnold, D. and Joseph, A (1968) in Experimental Physiol, (Holt - Rinehart & Winston, London., ed.), pp. 265-268.
2. Davies, E. and Larkins, B. (1973), Plant Physiol., 52, 339-343.
3. Hagen, G et al., (1984), Planta, 162, 147-153.
4. Hopler, P.H & Palevitz, S.A (1974), Ann. Rev. Plant Physiol., 25, 309-316.
5. Kantharaj, G.R et al., (1979), Phytochemistry, 18, 383-387.
6. Kantharaj, G.R et al., (1985), Phytochemistry, 24, 23-27.
7. Kuhlemeier, C et al., (1987), Ann. Rev. Plant Physiol.,38, 221-257.
8. Laemli, U.K (1970), Nature (London), 227, 680-682.
9. Lloyd, C.W (1987), Ann. Rev. Plant Physiol., 38, 119-139.
10. Lowry, O.H et al., (1957), J. Biol. Chem., 193, 265-270.
11. Miller, C.O (1961), Ann. Rev. Plant Physiol., 12, 259-267.
12. Padmanabhan, G et al., (1975), Proc. Nat. Acad. Sci., USA., 72, 4293-4297.
13. Penman, S (1966), J. Mol. Biol., 17, 117-122.

14. Quatrano, R.S (1978), Ann. Rev. Plant Physiol., 29, 487-494.
15. Roberts, B.E & Patterson, B.M (1973), Proc. Nat. Acad. Sci., USA., 70, 233-237.
16. Rosen, H (1957), Arch. Biochem. Biophys., 67, 10-14.
17. Scott, M.P (1985), Trends in Genetics, 1, 74-80.
18. Theologis, A & Ray, P.M (1982), Proc. Nat. Acad. Sci., USA, 79, 418-421.
19. Theologis, A (1986), Ann. Rev. Plant Physiol., 37, 407-438.
20. Travis, R.L et al., (1973), Plant Physiol., 52, 608-614.
21. Travis, R.L & Key, J.L (1976), Plant Physiol., 57, 936-940.
22. Trewavas, A.J (1976), in Molecular aspects of Gene expression in plants, (Bryant J.A., ed.), pp. 249-259, Academic Press, NY, USA.

COMMERCIAL ASPECTS OF MICROPROPAGATION

R.L.M. Pierik
Department of Horticulture
Wageningen Agricultural University
P.O.Box 30
6700 AA Wageningen
The Netherlands
Publication 581

Abstract

Since detailed information was available for micropropagation in the Netherlands for the period 1980-1989, special attention has been paid to this country which accounts for 29% of the West European production. The Netherlands has 76 commercial laboratories and had a production of 80 million plants in 1989. The three most common produced plants in this country are Nephrolepis (14.4 million), gerbera (17.1 million), and lily (16.3 million).

An analysis has been made of commercial micropropagation in 15 West European countries. In 1988 Western Europe had a total of 248 commercial tissue culture laboratories of which 37 each produced more than one million plants per year. The total West European production in 1988 was 212.5 million plants.

Although there are few accurate statistics on micropropagation in other countries in the world, a global survey of the world production will be presented.

Special attention will be paid to the demands of the Dutch customers of micropropagated plants and to marketing strategy when exporting micropropagated plants e.g to West European countries.

Commercial cloning in vivo and in vitro of adult or mature woody plants is adversely affected by characteristics accompanying maturation, such as reduced growth rate and reduced or total lack of rooting. For that reason attention will be paid to the complex phenomena of maturation and rejuvenation. Possibilities to rejuvenate will enable a wider application of tissue culture technology among woody species.

A comparison will be made of the various micropropagation methods used: single-node culture, axillary branching, adventitious organ formation, and callus systems. The chances of obtaining genetic variation and mutations when using various methods will be analyzed.

In recent years several inert synthetic supports for micropropagated shoots have been developed. The cost price of these supports has often been the limiting factor for application on a large scale. Since 1988 a cheap microplug, the so-called 'Rockwool' plug, has been introduced which is especially developed for rooting of micropropagated shoots under semi-sterile conditions. Rockwool or mineral wool, already well known for many horticultural crops as a growing medium without soil in artificial substrates, is used in the Netherlands, especially to reduce labour costs and to increase the quality of the micropropagated plants.

A new automatization system for micropropagation has been developed in the Netherlands, involving preparation and sterilization of culture media, sterile filling of membrane capsules as 'containers', and sterile automized inoculation.

J. Prakash and R. L. M. Pierik (eds.), Horticulture – New Technologies and Applications, 141–153.

One of the most complicated and complex factors influencing micropropagation is the solidifying agent agar, produced from seaweed. A survey is given of agar analysis together with the main effects of using various agar brands.

1. Introduction

Facts and figures (Table 1) about the Dutch horticultural industry show that the Netherlands is a leading country particularly in the world of ornamental plants. For that reason micropropagation plays an important role in our ornamental industry.

Table 1. The Dutch horticultural industry in 1989.

Total production area under glass in ha	9 551
Total production area under glass, ornamental plants in ha	4 727
Turnover of the Dutch flower auctions in Dfl.	4 264 000 000
Flower and plant export in Dfl.	4 900 000 000
Vegetable and fruit export in Dfl.	4 420 000 000
Flower bulb export in Dfl.	1 200 000 000
Nursery stock export in Dfl.	495 000 000
% of the world export of cut flowers	68
% of the world export of pot plants	51

Dfl. 1.00 = U.S. $ 0.6

2. Commercial micropropagation in the Netherlands

Using detailed information available (Pierik, 1990a,c), special attention has been paid to this country, which accounted for 29% of micropropagation in Western Europe in 1988. A recent study (Table 2) in 1989 (Pierik, 1990a) showed that the Netherlands has a total of 76 commercial tissue culture laboratories with a production of 80 million plants. The majority of these are situated in the Western part of the country, where the ornamental industry is concentrated. The number has increased from 28 in 1980 to 76 in 1989. Table 2 shows a large number of relatively small-sized laboratories found in the Netherlands, many being micropropagation laboratories within plant breeding companies.

Table 2. Micropropagation of ornamentals in the Netherlands in 1989 (Pierik 1990a). Numbers of plants are given in millions.

Plants produced per laboratory	Commercial laboratories
Less than 0.1	39
0.1-1.0	21
1.0-5.0	10
More than 5.0	6
Total	76

The most important pot plants	
Nephrolepis	14.4
Saintpaulia	5.1
Spathiphyllum	4.7
Syngonium	3.3
Ficus	2.8
Anthurium scherzerianun	2.5
Platycerium	0.7
Cordyline	0.6

The most important cut flowers	
Gerbera	17.1
Rosa	1.1
Anthurium andreanum	0.9

Other important ornamentals	
Lilium	16.3
Cymbidium	1.5
Other orchids	1.5

3. Commercial micropropagation in Western Europe

An analysis has recently been made of commercial micropropagation in 15 West European countries (Pierik, 1990b,c). In 1988 Western Europe had a total of 248 commercial tissue culture laboratories of which at least 37 each produced more than one million plants per year. The total production for 1988, in the various categories are summarized in Table 3, which shows that ornamental plant species (157 million or 74% of the total propagated) dominate micropropagation in Western Europe. The most frequently cloned ornamental plants are: Ficus, Syngonium, Spathiphyllum, Gerbera, Rosa, Philodendron, Saintpaulia, Nephrolepis, Cordyline, Anthurium, Calathea, Cymbidium, Dieffenbachia, and Rhododendron. The leading country in almost all categories of ornamental plants is the Netherlands, which in 1988 had 67 commercial tissue culture laboratories, followed by Italy (35), Spain (27), France (22), West Germany (21), United Kingdom (18), Belgium (16), Denmark (9), Ireland (8), Greece (6), Portugal (5), Switzerland (4), Finland (4), Sweden (4), and Norway (2). No data were collected from Luxembourg and Austria.

Table 3. Micropropagation in Western Europe in 1988. Numbers of plants are given in millions.

Categories		Categories	
1. Pot plants	9 234	8. Perennial garden plants	298
2. Cut flowers	3 784	9. Agricultural crops	242
3. Fruit trees	1 943	10. Miscellaneous ornamentals	194
4. Ornamental bulbs and corms	1 316	11. Vegetables	138
5. Small fruits	935	12. Trees (forestry)	129
6. Orchids	529	13. Herbs	003
7. Ornamental trees/shrubs	389	14. Not specified	2 113
		Total	21 246

4. Commercial micropropagation in the whole world

Several authors in the new handbook on micropropagation (Debergh and Zimmerman, in press) gave a description of micropropagation in the whole world. The chapters giving the production figures from all parts of the world are summarized in Table 4.

Table 4. Commercial micropropagation in the whole world in 1989 (estimated by R.L.M. Pierik).

	Millions of plants produced
The Netherlands	80
France	46
Italy	34
Other West European countries	60
East European countries	34
Israel	6
Africa	?
Middle and South America	?
Australia and New Zealand	46
Asia	92
United States of America	115
Total	513

5. Demands of Dutch customers, problems in the Dutch tissue culture laboratories and avoiding peaks in the demand of labour

Major demands of the Dutch customer where micropropagated plants are concerned are:

1. Delivery on time.
2. Reasonable and competitive prices.
3. Consistantly high quality plantlets.
4. Periodical delivery of large quantities necessary.
5. Good plant form and not too bushy.
6. Homogeneity and uniformity of the plants.
7. No mutations.
8. Easy to acclimatize and to bring into production.
9. No intervention by third parties which increases the cost price.
10. Culture and acclimatization advice necessary.
11. Acquisition from more than one source to spread risks.
12. A low percentage of infections when transferred to soil.
13. Realiability.

Problems encountered in micropropagation in the Netherlands can be summarized as follows:

Economic

1. High research costs.
2. High overhead costs.
3. High labour costs.
4. High fixed charges.
5. Financing and setting up a laboratory.
6. Profits do not come from the standard items.
7. Difficulties in lowering the cost price.
8. Difficulties in increasing efficiency.

Marketing

1. Strong competition from countries with low labour costs.
2. Overproduction of a number of classical crops.
3. Summer dumping.
4. Difficulties in finding new products and markets.
5. Lack of marketing expertise.
6. Not familiar with the market of a certain country.
7. No entrance into a market.
8. Language and communication problems.

Technical

1. Lack of acclimatization in vitro and ex vitro.
2. External and especially internal infections.
3. Induction of dormancy in bulbous crops difficult.
4. Insufficient automatization.
5. Genetic instability in callus systems.
6. Technical problems with certain crops.
7. Sometimes unwanted after-effects (e.g. branching)
8. Each cultivar of a species requires different media, etc.
9. Rejuvenation not attainable in shrubs and trees.
10. Excretion of toxic substances unavoidable.
11. Problems with scaling up.

12. Problems in production management.
13. No diversification of product.
14. Some companies have poor guidance.

As far as possible peaks in labour demand should be avoided by use of the following:

1. Acquisition of temporary labour.
2. Contracting out during production peaks.
3. Organization of labour shifts.
4. Staggering of labour throughout the year.
5. Choosing complementary crops.
6. Cold storage.
7. Automatization of the production processes.

6. Rejuvenation

Maturation is the major problem preventing a wider application of tissue culture technology among woody species, especially trees. Successes with a small number of trees has been achieved mainly with special starting material, certain pre-treatments and manipulations (Pierik, 1990d). The transition from juvenile to adult, which is localized in the meristems, the characteristics of mature woody plants and the possibilities to eventually micropropagate adult trees and shrubs are summarized below.

Mature woody plants show:

1. A reduced growth rate.
2. A reduced or total lack of rooting ability.
3. Plagiotropy.
4. Ability to flower.

Ways to micropropagate adult shrubs and/or trees:

1. Using juvenile buds and/or epicormic buds (stump sprouts) at the base of the tree.
2. Etiolation, ring barking or girdling.
3. Forcing spaeroblast.
4. Use of root suckers.
5. Isolation of shoots from lignotubers, water shoots, or reiterated shoots.
6. Special (pre-)treatments to rejuvenate or to reinvigorate:
   - Hormone treatment, especially cytokinins (anti-ageing substances).
   - Selection of vigorously growing buds.
   - Isolation of meristems or shoots.
   - Severe hedging, pruning, cutting back, stool beds or stool layering.
7. Repeated subculture in vivo or in vitro of (micro-)cuttings or shoots.
8. (Micro)grafting of adult meristems/shoots on juvenile rootstocks.
9. Adventitious shoot and/or somatic embryo formation.
10. Sexual reproduction of plants which show apomictic or nucellar embryony.
11. Co-culture of juvenile and adult shoots in one container.
12. Meristem culture.
13. Use of nucellar tissues as centres (islands) of juvenility.

7. Micropropagation systems

The methods summarized below have been developed to micropropagate plants, although they vary in reliability and convenience of application (Pierik, 1987).

Single-node culture

This method is the simplest, most natural and safe method (no problem with mutations) with plants which elongate (e.g. potato, tomato, lilac and grape), forming a stem with leaves with buds in their axils, but is difficult with rosette plants. The rate of propagation is strongly dependent on the number of nodes formed within a particular time interval. When cloning shrubs and trees, serious problems can arise with this method (dormancy of buds and failure to elongate the stem).

Axillary branching

Axillary buds have their dormancy broken by breaking apical dominance with cytokinin. This method has become the most important propagation method being simple and quite safe. Another advantage is that the propagation rate is relatively fast and the genetic stability is usually preserved. However, mutations can occur when adventitious buds are formed as a result of high cytokinin levels.

Regeneration of adventitious buds/shoots

This method includes the formation of adventitious buds/shoots on explants from leaves, petioles, stems, scales and floral stems. However, the percentage of plant species that can regenerate adventitious buds is relatively small and is often restricted to herbaceous plants. The chances of obtaining mutations is much higher with this method than with the first two methods described, particularly with so-called chimaeric plants. However, it is successfully applied for Saintpaulia ionantha, lily, hyacinth, Achimenes, and Streptocarpus (Pierik, 1988).

Regeneration of plants from callus, cells, and protoplasts

Despite claims to the contrary, cloning of higher plants through callus, cells, and protoplasts has been found to have many disadvantages. The greatest difficulty with callus cultures is their genetic instability. Only a few plant species such as Anthurium andreanum (Pierik, 1988), are cloned (partially) through a callus phase in the Netherlands.

8. Ex vitro rooting in Rockwool microplugs

In recent years inert synthetic supports for micropropagated shoots have been developed e.g. by Baumgartner Papier in Lausanne (Sorbarods) and by Milcap France (substrates 'Milcap', lumps and plugs). These supports are useful (Roberts and Smith, 1990), but the cost price is relatively high. However, the cost price of Rockwool supports is much lower (U.S. $ 0.03).

In 1988 Grodan-Rockwool in cooperation with the Department of Horticulture, Wageningen Agricultural University in the Netherlands, introduced Rockwool microplugs which are especially developed for rooting of micropropagated shoots in vivo. Rockwool, already well known as a growing medium, in artificial substrates without soil, for many horticultural crops (3000 ha in the Netherlands), is at present mainly used for Gerbera and Anthurium andreanum microcuttings. The microplug of mineral wool has now also been introduced in the Netherlands, especially to reduce labour and material costs and to increase the quality of the micropropagated plants.

The following are some of the unique properties of Rockwool:

1. It replaces soil so that export of plants becomes possible to those countries whorefuse

soil as part of their plant protection regulations.
2. It has the capacity to achieve a homogeneous water/air balance, including an optimal distribution of water, nutrients and air in the root environment.
3. Roots can penetrate the well drained substrate very easily and shoots can easily be transferred to other Rockwool blocks of larger size or to soil to continue further growth. Rockwool enables transport without damaging the root systems and quick regrowth after transplantation.
4. It is sterile, eliminating the need for soil steaming or chemical sterilization.
5. It allows the regeneration of normal roots in comparison to abnormal root formation in agar media. The plugs have a special fibre structure to promote rooting.
6. It is almost chemically inert, containing no plant nutrients, or contaminants of organic origin normally present in agar.
7. Tray systems for microplugs facilitate transport.

However, there are a number of special requirements to be kept in mind:
1. Plugs should be fully saturated with a weak nutrient solution (Electrical conductivity 0.5-1.0 µS/cm) before shoot cuttings are inserted. Plugs should be allowed to leach out before inserting the cuttings.
2. Auxin requirement for cuttings is strongly dependent on the plant species and cultivar, but also on the hormonal composition of the culture medium just before shoot cuttings are taken.
3. Shoots should be washed to remove sugar from the culture medium, otherwise infections easily occur.
4. Since Rockwool has initially a slight basic reaction, the nutrient solution (dependent on the plant species) should be used at a pH of 5.2.
5. Relative humidity (90-95%) should be carefully controlled, especially during the first 1-2 weeks after inserting the shoot cuttings. Hardening off should take place by gradually decreasing the relative humidity.
6. Regular application of water and nutrients is required, preferentially by hand, depending on water loss and location in the greenhouse.
7. The quality and size of the minishoots is more critical in Rockwool plugs than in agar media.
8. The greenhouse climate (temperature and relative humidity) needs to be carefully controlled during rooting and acclimatization in Rockwool. This has been shown particularly for Gerberas.
9. When transplanting rooted plants in plugs to soil, care must be taken that plants do not dry out in the first week, since soil easily removes water from the relatively small plugs.

9. A new automatization system with membrane capsules

Quite recently a Dutch company in Wageningen, called PermX, has brought onto the market a new revolutionary production technique which enables automatization of micropropagation. The most important principle of the PermX machine is that presterilized membrane capsules (see section 10) are automatically filled with sterilized nutrient media. The PermX technology reduces costs because many activities applied in conventional tissue culture laboratories are now automated. The inoculation of e.g. scale explants from lily and sealing (closing) of the membrane capsules is completely automated. The cutting of explants is still done by hand. The advantages of the machine are summarized below.

1. Reduces costs.
2. Many activities are now automated.

3. Use of ecologically sound materials.
4. The system approaches natural circumstances.
5. More efficient production process.
6. Better plant quality and viability.
7. Flexibility for both producer and consumer.
8. Application of membrane capsules.
9. Optimizes quantity and composition of medium.
10. Individualized cultivation method.

10. Micropropagation in membrane capsules

Culture of shoots and explants in the new automatization system described in section 9 is carried out in membrane capsules. Shoots or explants are cultured on agar media in a modified atmosphere through membrane diffusion in polyethylene bags. The new container consists of a membrane capsule, allowing a maximum of gas diffusion (exchange) with the environment (Gerrits, 1991). The advantages of membrane capsules for plant tissue culture and micropropagation are summarized below.

1. Higher multiplication rates can be obtained.
2. The quality of the microplants increases.
3. Better plant vigour.
4. No competition between plants.
5. Absence of risk of infection by microorganisms.
6. Membranes can be chosen with different gas permeability.
7. Possibilities of extended plant development.
8. Aseptic rooting in Rockwool plugs is possible.
9. Early acclimatization takes place.
10. Membranes are impermeable to water.
11. Membranes are an absolute barrier for microorganisms penetrating into other bags.
12. The diffusion capacity of the membrane is controllable.
13. Storage density in the culture rooms increases considerably.
14. The production capacity can be increased.

11. Agar quality

Agar is a hydrophilic colloid extracted from seaweeds, belonging to the Rhodophyceae. It consists of 2 fractions (agarose and agaropectin) and is the classical gelling agent used in plant tissue culture (Armisen and Galatas, 1987). Although a natural product, it is in principle purified by the manufacturers and should contain no toxic compounds. The ash content (inorganic salts) of agar varies from 2.5-5.0% (w/w). The ash content of agarose is much lower than that of agar due to the absence of ionic groups.

Analyses of agars have shown that they contain many organic and inorganic impurities (Kohlenbach and Wernicke, 1978; Debergh, 1983; Pierik, 1987; Scherer et al., 1988; Kordan, 1988). A recent analysis (R.L.M. Pierik and H.J. Scholten, unpublished) is shown in Table 5. It is not known which undesired contaminants should be removed to obtain a qualified agar or even which ash content is acceptable and which ranges of the various elements in the agar can be tolerated. Toxicity is certainly dependent on the element under consideration. Debergh (1983) demonstrated that inorganic impurities introduced with the agar are responsible for significant differences in the concentration of several ions in addition to the normal ions added with the mineral nutrition. Nothing about the nature of organic contaminants can be found in literature.

Table 5. Range of concentration of various elements found in 20 agar brands.

Determined by spectrophotometry:

	mmol		mmol
N	total: 10-126	Ca	total: 0-374
P	total: 1-121	Mg	total: 1-45
Na	total: 16-700	Mn	total: 1-57
K	total: 1-24	Zn	total: 1-19

Determined by instrumental neutron activation (atom spectroscopy):

	ppm		ppm
Fe	total: 18-417	Cl	total: 420-6960
Co	total: 0,013-0,187	I	total: 2-234
V	total: 0,02-2,13	Br	total: 3,2-151,0
Cr	total: 0,41-2,42		
Sr	total: 2,2-24,0		
Th	total: 0,01-0,05		

Moisture content can vary strongly from 3.6-20.2%, and the colour varies strongly from bad (brown/yellow-brown) to good (pure white). Acidity (pH) after dissolving agar in water (without minerals, etc.) varies from 4.91-7.18. The EC (Electric conductivity in μScm^{-1}) varies from 76-576, and gives an indication of the total inorganic salt content. Purification of agar (agarose) results in repression of growth and development when compared with non-purified agar (Figure 1). The world consumption of agar for micropropagation in 1990 was recently estimated (R.L.M. Pierik, unpublished). The results of this are shown in Table 6.

Figure 1. The effect of type of agar and agar brand on shoot length and number of leaves formed by exised shoot tips of rootstock A2 and A3 of Syringa vulgaris L. (R.L.M. Pierik and P.A. Sprenkels (unpublished).

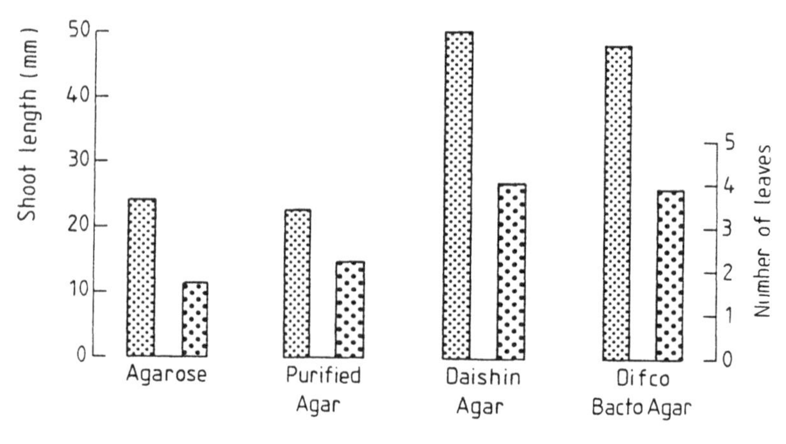

Table 6. Agar consumption for micropropagation in the world in 1990 (estimation R.L.M. Pierik, based on Dutch production figures).

Geographical location	kg
The Netherlands	5 500
Other Western European countries	9 625
Israel, Africa, Middle and Far East	2 750
Middle and South America	2 060
Australia and New Zealand	3 160
Asia	6 305
United States of America	7 900
Canada	700
Total	38 000

Remarkable effects of various agar brands have been described in the literature. The following processes were strongly influenced by the agar brand: pollen germination and pollen tube growth (Kordan, 1988), differentiation of tracheary elements (Roberts et al., 1984), anther regeneration (Kohlenbach and Wernicke, 1978), regeneration of sugar cane (Anders et al., 1988) and Kalanchoë (Hauser et al., 1988), and shoot proliferation of apple and pear (Singha, 1984).

To find a correlation between agar quality (physical and chemical properties) and growth response of 3 plant species on 20 different agar brands the following analyses and bioassays were carried out:

1. A number of elements in the agars (N, P, K, Ca, Mg, Na, Cl, Br, I, Fe, Mn, Zn, Co, Cr, Th, Sr, V and several others) were determined spectrophotometrically or by means of atomic absorption spectroscopy.
2. Electrical conductivity, moisture content, colour, and pH before and after autoclaving of the agars were determined.
3. Shoot elongation of the rose cv. 'Motrea' was examined on a basic culture medium with 4.5% (w/v) sucrose as suggested by Marcelis-van Acker (1990).
4. Shoot elongation of the lilac (Syringa vulgaris) rootstocks A2 and A3 were followed under the conditions previously described by Pierik et al. (1988).
5. Shoot fresh weight and shoot multiplication of the Gerbera cv. 'Joyce' were determined using the shoot multiplication system described by Pierik et al. (1982) with 2.5 mg/l kinetin.

From this work the following conclusions can be drawn:

1. The brand of agar strongly influences the shoot length of rose and the fresh weight of Gerbera (Figure 1). However, lilac is very irregularly affected, indicating that impurities in the agar are probably of little importance for the growth and elongation of shoots of this species. This can be explained by its toleration of a medium with a very high inorganic salt content (Pierik et al., 1988).
2. There was a large difference in shoot elongation in rose between the worst (No. 9) and the best agar (No. 18). Fresh weight of Gerbera shoots also showed large differences on the worst agar (No. 8) and the best agar (No. 13).
3. When the growth of rose and Gerbera are compared (Figure 1), a similar pattern of growth is seen, although there are exceptions to this rule, e.g. agar No. 17.

4. A strict correlation between the growth response of rose or Gerbera and moisture content, pH before or after autoclaving, colour, or EC was not found. A high pH before autoclaving generally coincided with a high EC.
5. Agar No. 9, on which growth of rose as well as Gerbera was poor, can be characterized by a very high EC (415 µS/cm), a relatively high pH before autoclaving (7.0), a 'bad' colour (grey-brown), and a high concentration of N, Ca, Na, Mn, Cl, Br, I, V, Cr and Sr.
6. Agar No. 13 the best agar for Gerbera and a good agar for rose, has a 'good' colour (white), a low EC (122 µS/cm), a low pH before autoclaving (6.2), and a low concentration of N, Na, Br, Cl, I and V, but a relative high concentration of Ca, Mn, Mg, Co, and Fe.
7. Agars bad for both rose and Gerbera can be characterized by high levels of several elements (Ca, Na, Mg, I, Br, V and Cr), a relative high EC, a higher pH before autoclaving, and a grey-brown colour.

Figure 2. Effect of the agar brand on shoot length (mm) of the rose cv. 'Motrea', shoot length (cm) of Syringa vulgaris L. and shoot weight per explant of the Gerbera cv. 'Joyce' (Pierik, 1991).

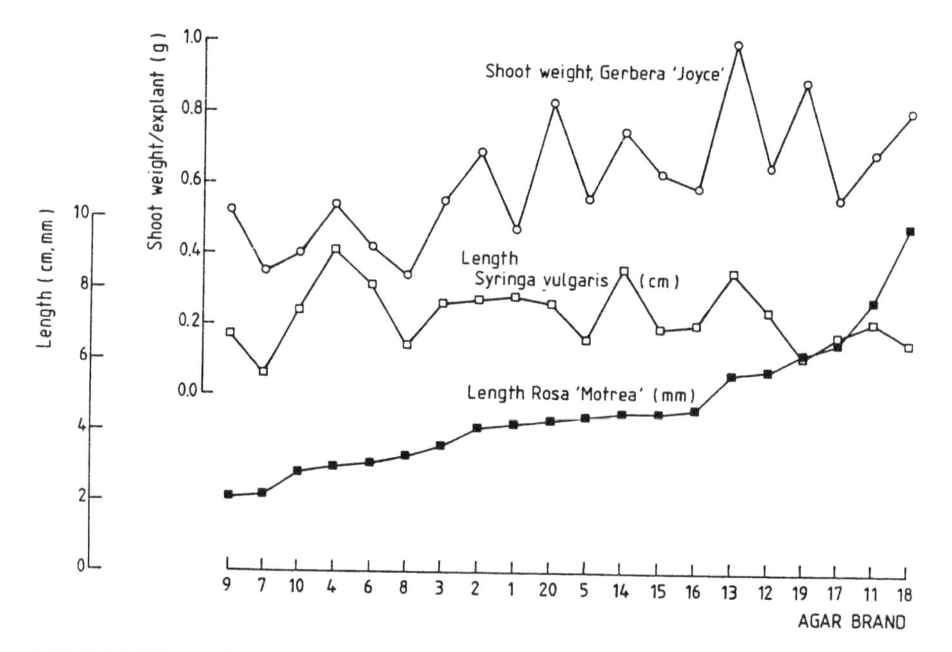

ACKNOWLEDGEMENT

I am indebted to Mrs. C.J. Kendrick for the English translation and corrections.

REFERENCES

Anders, J., Larrabee, P.L., and Fahey, J.W., 1988. Evaluation of gelrite and numerous agar sources for in vitro regeneration of sugar cane. HortSci. 23(3-2):755.

Armisen, R. and Galatas, F., 1987. Production, properties and uses of agar. In: McHugh, D.J. (editor), 1987, Production, and utilization of products of commercial seaweeds, FAO Workshop, China:1-47.

Debergh, P., 1983. Effects of agar brand and concentration on the tissue culture medium. Physiol. Plant. 59:270-276.

Debergh, P. and Zimmerman, R.H. (editors), 1991. Micropropagation. Kluwer Academic Publishers, Dordrecht, the Netherlands: 1-484.

Gerrits, W., 1991. Plant tissue culture in a modified atmosphere through membrane diffusion. Proceedings Eucarpia Meeting, Wageningen, The Nederlands. Pudoc, Wageningen, The Netherlands (in press).

Hauser, B., Geiger, E. and Horn, W., 1988. Eignung von Gellan und verschiedenen Agarqualitäten für Gewebekulturen von Kalanchoë-Hybriden. Gartenbauwiss. 53:166-169.

Kohlenbach, H.W. and Wernicke, W., 1978. Investigation on the inhibitory effect of agar and the function of active carbon in anther culture. Z. Pflanzenphysiol. 86:463-472.

Kordan, H.A., 1988. Inorganic ions present in commercial agars. Biochem. Physiol. Pflanz. 183:355-359.

Marcelis-van Acker, C.A.M. 1990. Plant tissue culture as a tool to determine growth potential of axillary rose buds. Abstr. VIIth Intern. Congress Plant Tissue and Cell Culture, Amsterdam A3-128:114.

Pierik, R.L.M., 1987. In vitro culture of higher plants. Kluwer Acad. Publ., Dordrecht, the Netherlands: 1-344.

Pierik, R.L.M., 1988. In vitro culture of higher plants as a tool in the propagation of horticultural crops. Acta Hortic. 226:25-40.

Pierik, R.L.M., 1990a. Vegetatieve vermeerdering in kweekbuizen in Nederland. Grotere bedrijven en een stijgend aantal planten. Vakblad Bloemisterij 45(23):32-33.

Pierik, R.L.M., 1990b. Vegetatieve vermeerdering in vitro in West Europa en Israel. Nederland koploper is siergewassen. Vakblad Bloemisterij 45(23):34.

Pierik, R.L.M., 1990c. Commercial micropropagation in Western Europe and Israel. In: Debergh, P. and Zimmerman, R.H. (editors), 1991, Micropropagation. Kluwer Acad. Publ., Dordrecht, the Netherlands: 155-165.

Pierik, R.L.M., 1990d. Rejuvenation and micropropagation. In: Nijkamp, H.J.J. et al. (editors), Progress in plant cellular and molecular biology. Kluwer Acad. Publ., Dordrecht, the Netherlands: 91-101.

Pierik, R.L.M., 1991. Micropropagation of ornamental plants. Acta Hortic. (in press)

Pierik, R.L.M., Steegmans, H.H.M., Verhaegh, J.A.M. and Wouters, A.N., 1982. Effect of cytokinin and cultivar on shoot formation of Gerbera jamesonii. Neth. J. Agric. Sci. 30:341-346.

Pierik, R.L.M., Steegmans, H.H.M, Elias, A.A., Stiekema, O.T.J., Velde, van der, A.J., 1988. Vegetative propagation of Syringa vulgaris L. in vitro. Acta Hortic. 226:195-204.

Roberts, A.V. and Smith, E.F., 1990. The preparation in vitro of chrysanthemum for transplantation to soil. I. Protection of roots by cellulose plugs. Plant Cell Tissue Organ Culture 21:129-132.

Roberts, L.W., Stiff, C.M. and Baba, S., 1984. Effect of six different agars on tracheary element differentiation in explants of Lactuca. Plant Tissue Culture Letters 1:22-24.

Scherer, P.A., Müller, E., Lippert, H. and Wolff, G., 1988. Multielement analysis of agar and gelrite impurities investigated by inductively coupled plasma emission spectrometry as well as physical properties of tissue culture media prepared with agar or the gellan gum gelrite. Acta Hortic. 226:655-658.

Singha, S., 1984. Influence of two commercial agars on in vitro shoot proliferation of 'Almey' crabapple and 'Seckel' pear. HortSci. 19:227-228.

Large Scale Micropropagation: It's Strategic Use in Horticulture

K.L. GILES
Department of Horticulture Science
University of Saskatchewan
Saskatoon, SK S7N 0W0 CANADA

Key Words: micropropagation; climactic change; new variety development; strategic plans

Introduction

In considering the problems and opportunities of large scale micropropagation at the commercial level, it has become increasingly apparent to me that the technology holds long term strategic opportunities, as well as the shorter term tactical uses it may be put to commercially. These longer term applications are likely to be forced upon us by a number of pressing environmental and climatic factors, which will increase the need for new variety introduction in the face of changing growing conditions. These challenges will force all of Agriculture to formulate very precise goals and develop practical plans to achieve them. At any time when resources are limited, clear cut programs of research and development are required in order that effective use of those resources can be established. Micropropagation provides one such technology for the production of a framework to establishment a targeted and flexible Horticultural program.

Background Information

Our planet is currently under stress. It has a rising temperature, an increasing population and a worsening case of atmospheric and water pollution. These problems are already beginning to influence climatic patterns and as such will influence Horticultural production patterns world wide. In the last 40 years the world population has doubled. In 1950 the population was approximately 2.5 billion and by the year 1990 it is slightly over 5 billion. If nothing can be done to reduce this population increase, by the year 2030 we can anticipate having approximately 10 billion people on this planet. Looking at figure 1, which describes the percentage change in food production compared with the percentage change in population from

J. Prakash and R. L. M. Pierik (eds.), Horticulture – New Technologies and Applications, 155–160.

1962 to 1986, it becomes clear that although population growth over that period has slowed slightly. When expressed as a percentage the increase in the production of food has dropped dramatically. The world's food stores are now lower than they were 30 years ago. These figures are critical when taken together with the growing amount of evidence suggesting that global warming is occurring at a significant rate, as a result, in part at least, from the pollution produced by the massive increases in population.

A quick review of the contents of the magazine "Science" over the past few years indicates the amount of work and evidence that is building relating to the overall problems of global warming. For instance, December 1988, "Origins of the 1988 North American Drought", Trenberth, et. al.; March, 1989, and in March, 1989, "Precise Monitoring of Global Temperature Trends from Satellites",Spencer and Christy. There is clearly growing evidence and literature relating to the trends that are currently observable by careful monitoring of world temperature and pollutant levels.

If we seek evidence that our environment is changing it is also readily available from observations on population growth and the changes in human activity worldwide. There is a tendency for rural populations to move to the city to become more involved in industrial processes rather than traditional Agricultural or Horticultural. Satellite imagery has allowed detailed studies of land cover changes, both using infrared photography and image analysis programs. There is very significant loss of forest and wooded land worldwide, both in the Amazon basin and more temperate regions. This is causing distinct changes in global heat transfer patterns. The greenhouse gases, as they have become labelled, carbon dioxide, methane and the chlorofluorocarbons as well as nitrogen oxides, are increasing at alarming rates. These gases help trap solar heat in and around the planet causing temperature increases. The somewhat alarming discoveries of Antarctic and now possibly Arctic holes in the ozone layers, protecting the earth from excessive ultra-violet radiation, are again probably the result of greenhouse gases, particularly the fluorocarbons. Here, however, the precise reasons for fluctuation in the size of the holes is still uncertain. Worldwide climatic variation has been significant over the past 25 years and climatic prediction has become significantly more difficult with the almost volatile weather patterns that have developed in some regions of the world. Specific pollution problems, such as erosion, acid rain, toxic wastes, etc. are obvious in almost all developed and developing parts of the world. These too can have dramatic local effects on productivity and food safety.

Table 1 outlines the approximate levels of greenhouse gases in parts per billion. All of them show significant increases over the last 100 years and indications are that over the next 40 years significant increases in the concentrations will occur in most of these gases. For instance, carbon dioxide currently at about 350 x 10^3 part per billion is liable to rise to 450 parts per billion, which will be approximately 1.5 times as high as it was only 100 years ago. Methane is rising rapidly and appears likely to continue to do so. It's atmospheric residence time is short, but it is still rising quickly. This is due mainly to the large scale production of rice and cattle. Some of the pollutants such as the CFC's (chlorofluorocarbons) were absent completely 100 years ago but show now some significant increases, though strenuous efforts are now underway to curtail any further increases. These are significant changes in our atmospheric environment and are almost certainly associated with the worldwide increase in atmospheric temperatures noted over the last 50 to 100 years. These changes will have profound effects on the distribution and optimal production levels of a wide range of crop species.

These challenges represent significant opportunities for Agricultural research and development and specific opportunities in the area of Horticulture. Worldwide there must be an increased use of soil conservation methods and global concerns about alternative strategies of Agriculture. The advances in molecular biology have created important tools in pure and applied plant biology, and plant defense mechanisms against predators, diseases and stress will not be long in coming to the market place. The excessive use of herbicides and pesticides is likely to be curtailed with the increasing use of non-chemical means of insect control. Legislation and societal pressures aimed at reducing these chemicals and other forms of pollution are imminent. Clearly animal and poultry scientists are now ready to use biotechnology and genetic engineering to develop new characteristics in animals to meet environmental and production needs. Horticulture has also set to take on the challenges needed to develop economic stability and growth. Diversification and the introduction of newly improved lines of horticultural crops is likely in the next decade.

Micropropagation provides a rapid, reliable system for the production of large numbers of genetically uniform plantlets. These can be of many different crop types ranging through vegetable, ornamental, fruit and forestry crops. The products of the micropropagation process may be used for product, breeding program work or seed production purposes. Where direct product is the final goal very heterozygous material can be selected and used directly, after propagation, avoiding lengthy breeding programs prior to the introduction of the new variety. This increased speed of introduction could be critical in the changing environmental picture of today. The tactical use of micropropagation in breeding programs for parental line and in-bred line maintenance can increase the efficiency and productiveness of breeding programs. The timely introduction of improved seed varieties is an important consideration under our present circumstances.

Micropropagation is only one tool in the machinery available for us to react to the challenges that lie ahead. It must be integrated into the family of plant improvement and propagation techniques already in use. Molecular biology offers the hope of directed gene modification for disease, pest and stress protection without the costly and injurious application of pesticides and fungicides, or the threat of salt build-up from irrigation excesses. Micropropagation has application in the propagation and trialling of transgenic plants and then bulking up for breeding and large scale introductions. Mechanisms must be put in place for dialogue and interfacing of grower, breeder and propagation groups in order to facilitate their interaction.

These developments in technology will require that clear and specific priorities are set for its introduction into Horticulture, especially in countries such as India with its very varied climate and spectrum of products. These priorities will have to be set on both a scientific and economic basis. The commitments must reach deep into the scientific infrastructure, they will almost certainly affect educational programs both at schools and universities in the areas of economic sociology as well as science. Centres of excellence should be developed committed to specific goal oriented research that can allow the horrendous problems of climatic change to be integrated into a well coordinated development and research program. These commitments will have to be in the areas of cultural techniques, plant breeding, cell culture, molecular biology, environmental studies and ,of course, the ancillaries of sociological and economic impacts that may result from changes.

Evidence of the challenges of change are here today. It is up to the scientific community in India and all other countries to respond by organizing and equipping scientific research to provide answers to some of these problems as soon as possible.

References

Spencer, R.W. and Christy, J.R., 1990. Precise Monitoring of Global Temperature Trends from Satellite. Science 347: 1558-1562.

Trenberth, K.E., Branstator, G.W. and Arkin, P.A. Origins of the 1988 North American drought. Science 242: 1640-1645.

The global warming is real. News and Comment. 1989, 243: 603.

Table 1 "Greenhouse Gases" in the Atmosphere.

GAS	ATM. RES. TIME	CONC - 100 Y	CONC NOW	CONC 2030
CO	MONTHS	40-80	100-200	>200
CO_2	100 YEARS	290,000	350,000	450,000
CH_4	10 YEARS	900	1,700	2,200
NO_X	170 YEARS	285	310	340
SO_2	DAYS	0.03	0.03-0.5	0.03-0.5
CFC	100 YEARS	0	3	3

% CHANGE IN FOOD PRODUCTION (SOLID BARS)
VERSUS % CHANGE IN POPULATION (STRIPED)

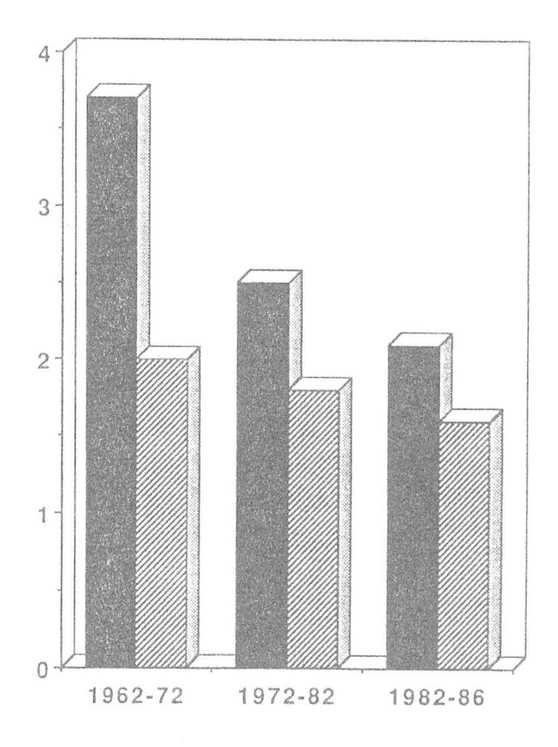

FIGURE 1

Micropropagation of Bulb Crops

PETER C HARPER
Microplants Ltd.
Longnor. Derbyshire. United Kingdom

Keywords: Micropropagation, bulb crops, flower bulbs, selling prices, production costs, economic criteria

1. Introduction

In vitro culture of the tissues of most higher plants has been accomplished successfully, and in almost all species it has been found possible to regenerate whole, free-living plants. The technique has been in commercial use for twenty five years, becoming an established, almost conventional system for plant propagation alongside seed germination and the various methods of vegative propagation. Its value for the very rapid production of large numbers of healthy starter plants is well recognised.
Such information as is now available shows that, although propagation systems have been developed for all the important categories of cultivated plants, in practice only a narrow range of plants has entered commercial production.
World wide annual production of micropropagated plants is estimated to be a least 500 million units, most of which are known to be of ornamentals, perhaps as high as 90%. This figure is well illustrated in the most accurate of all the data so far collected, that from Western Europe particularly the Netherlands, collected over the last decade by Professor R.Pierik of Wageningen.
Of Europe's 212 million microplants close to 170 million are of foliage pot plants, cut flowers, fruit trees and ornamental bulbs and corms. Within each category the restricted nature of the product becomes even more evident. Of the 35 genera of foliage plants 5 species claim 80% of the production. In the cut flower crop Gerbera is the leading product with 90% and in the rootstocks, peaches form 70% and citrus 20% It is the same with the flower bulbs. 97% of tissue cultured flower bulbs are lilies but they are no means the most numerous in cultivation.
Holland accounts for over 90% of dry bulb production in Europe so statistics from that country represent the Western production accurately.

J. Prakash and R. L. M. Pierik (eds.), Horticulture – New Technologies and Applications, 161–168.

Table 1. Micropropagation in 15 countries of Western Europe in 1988

CATEGORIES	Millions of Plants
1.Pot Plants	92,337,417
2.Cut Flowers	37,840,557
3.Fruit Trees	19,427,600
4.Ornamental bulbs/corms	13,161,960
5.Small fruits	9,351,525
6.Orchids	5,287,990
7.Ornamental trees/shrubs	3,889,290
8.Perennial garden plants	2,976,050
10.Miscellaneous ornamentals	1,943,528
11.Vegetables.	1,377,825
12.Trees(forestry)	1,289,500
13.Herbs.	30,000
14.Not specified	21,132,000
Total	212,460,742

Source Pierik 1990

By 1988 the lily crop occupied the second largest area in the Netherlands (Table 2), well behind tulips, equal to gladiolus but still leaving hyacinth, iris, narcissus and crocus with significant production. The figures for bulbs exported, which constitute nearly 70% of Dutch output show that lily comes fifth in terms of numbers of bulbs (Table3). In terms of productivity each of the nine bulbs in Table 3 constitutes a sizeable market for starter plants yet lily alone is produced by commercial micropropagation. This article examines a number of possible reasons for the restriction.

Table 2. Bulb production in the Netherlands
Total number of bulbs produced=8,500,000,000 with a value of £483,000,000. Of which 68% is exported half as dry bulbs and half to flower growers.Of the bulbs retained in Holland most go to flower growers.

Production in hectares.	1980	1988
Tulip.	6086	7389
Lily.	1119	2192
Gladiolus.	1902	2114
Narcissus.	1523	1658
Hyacinth.	827	1057
Iris.	952	916
Crocus.	398	491
Others.	535	611
Total	13,342	16,428

Source P.V.S.

Table 3. Number of flower bulbs exported from the Netherlands in millions of units

Kind	1984/5	85/6	86/7	87/8	88/9
Tulip	1606	1619	1652	1762	1699
Gladiolus	1493	1390	1485	1424	1420
Iris	488	483	417	420	469
Crocus	400	428	388	374	375
Lily	213	272	303	301	356
Narcissus	306	300	313	313	306
Anemone	209	220	277	258	254
Hyacinth	155	161	159	156	159
Freesia	110	122	126	126	138
All other(43)	566	614	677	673	674
Total	5546	5609	5851	5807	5850

Source P.V.S.

2. Criteria used in selecting subjects for micropropagation

When considering the various methods by which plants are to be propagated, micropropagation should never be the first choice. By far the cheapest and most convenient method is from seed and where possible this should be the preferred system. Vegetative means of propagation are usually simple and widely known. If the multiplication rate is slow there are many means by which it can be enhanced without recourse to high technology solutions. Successive cropping of stem cuttings from potatoes and chipping of bulbs are well known examples.

Micropropagation is an expensive and complicated method which requires the coming together of several kinds of skills. The great advantage is its capacity to generate many thousands of plant propagules in a few months, at any time of the year. In selecting a suitable propagation system the expense and technology of this sterile culture technique must be justified by reference to sound commercial criteria.

The list given below is not exhaustive but it provides sufficient factual evidence on which to base an informed decision. There should be a large and preferably expanding market for the product.

Production systems should be straightforward and lack unusual features. There must be a significant advantage over existing methods. Large contracts should be available each year.

Prices for the final product should be high. Let us use the bulb crop to discuss these criteria.

2.1 *Available Market*

Most perennial flowering plants which pass the dormant season as bulbs or corms have large brightly coloured flowers, often with attractive perfumes. Breeding and selection over some 4 centuries has provided varieties suitable for cut flowers, for pot plants, for bedding displays outdoors or under protection, and for more permanent garden or landscape planting.
Over 50 genera are used as ornamentals, with most being small bulbs such as the musaris and pushkinias. Really large volume bulbs belong to a few genera with big bright flowers having several uses(see Table 3) Tulip and gladiolus are the leaders in terms of numbers, weight, and value. Well behind come iris, crocus, lily and narcissus followed by hyacinth and freesia.
This group comprises more than 80% of world trade in bulbs or bulb flowers. Statistics for the total world production of ornamental bulbs and corms are difficult to interpret but an indication of the size and importance of the crop can be gauged from the trade figures released by the International Horticultural Statistics for 1989.
The countries of the European Community export to each other and to the rest of the world 160.000 tons of flower bulbs worth some 750 million Swiss francs. In the same year EC countries imported 113,000 tons with the value of 500 million Swiss francs. Some 92% of this trade is in bulbs from the Netherlands which has probably the greatest concentration of growers in the world. Holland produces 8,500,000,000 individual bulbs with a value to the growers of £483,000,000 (16,422,000,000 rupees). Almost 70% of this is exported and most of the crop goes to flower growers who further add value to the product.

2.2 *Micropropagation and other propagation systems*

All the major bulbs may be successfully propagated by conventional vegetative means. Commercially important cultivars cannot be derived from seed. This natural rate of propagation varies from the easy hyacinth to the slower tulips. Gladiolus readily produces 20 to 30 small bulbils at the base of the old corm, as does the crocus. Hyacinth is induced to behave similarly by cutting into the basal plate tissue. These rates of increase are sufficient to provide for adequate replenishment of stocks of the main varieties in use. Tulip, narcissus and iris are naturally slow. For narcissus, natural increase is around *2.7 times per year, but once a basic stock has been acquired this is quite sufficient to provide an economic yield and a maintenance stock for replanting in subsequent seasons. Much the same applies to tulips.

There are techniques for enhancing multiplication of most bulbs, the newest of which consists of "chipping" or slicing bulbs into slivers of scales or dividing into individual scales which will provide multiplication rates of between twenty and thirty times. Narcissus, iris and lily are all propagated by these methods.
At present there is no micropropagation system for tulips which is sufficiently simple and straight forward that it can be a basis for a commercial production system. Such systems do exist for the other genera. Initiation and multiplication are fairly easy but the root development and bulb formation phases can be prolonged and the balance between leaf formation and bulb expansion may depend upon sensitive environmental control, particularly in crocus and iris.

2.3 *Repeat business*

A micropropagation laboratory is expensive in equipment and in the training of those who work in it. To be successful there must be a guarantee of continuing business for years ahead. The research, development and "fine tuning" which is involved in setting up any production line and in implementing high standards of quality control can only be undertaken with the assurance of long runs for several years ahead. This is wholly determined by the life cycle of the plant and the cropping system adopted by the grower. For strawberry, which has a three year cycle, one third of the area is re-planted each year, but many fruit crops, especially those on shrubs or trees may be on an eight, sixteen or a thirty year rotation. Although forest trees may be requested in millions the eighty to one hundred and fifty year life span reduces their potential for continuing business.
In the bulb crop most producers will build up a stock of sufficient size to give both crop and replanting material for the next cycle. In tulips, narcissus, gladioli, iris and crocus where old varieties are replaced by new only slowly, the stocks may last 15 to 20 years before they need to be replaced due to the accumulation of pest and disease. Generally the grower will buy in stocks of a few thousand disease-indexed starter bulbs of a new variety some years before they are needed and will build the stock gradually, all the while acquiring a knowledge of its particular requirements.
Only the lily, with its rapid succession of new varieties offers a prospect of high volume demand for starter plants. The bulbs are grown by specialist producers who sell on to flower growers for the cut flower trade. Propagation until now has been by induction of small shoots from single bulb scales. Micrpropagation which generates small bulblets does not impose any great change in the existing system.

The lily,amaryllis, and some of the hyacinth crop are the ones in which annual repeat orders will be available.

2.4 *Price structure*

As its name implies, a starter plant is the beginning of a production chain which includes growing on the plantlet, and perhaps selling on to a flower grower or through a wholesaler to retail outlets. Distributors and transporters play their part. Always there are steps and stages between the micropropagator and the final consumer, and each step brings a price increase. Therefore the end-user-price must be high enough to include an acceptable price for all the participants.
Bulb flowers are amongst the least expensive cut flowers available in commerce. Dry bulbs are extremely cheap in the developed countries. Table 4 gives examples of the selling prices of a range of bulbs while Table 5 gives the range of prices obtained by Dutch growers at Auction during 1989 for flowers.
Apart from lily and amaryllis the unit prices are extremely low.

A full understanding can be gained by comparing these producer and end-user prices with figures for production costs incurred in a commercially run micropropagation factory.

Table 4

Retail prices for dry bulbs in Holland.
1987/89

		Unit price	
		pence	rupees
Tulip	£1 per 10	10	3
Narcissus	£1 per 10	10	3
Hyacinth	30p each	30	10
Iris	£0.75 for 10	7.5	2.5
Gladiolis	£3.50 per 100	3.4	1
Lily		125	45

Dutch Ministry of Agriculture and Laboratory for bulb culture

Table 5. Range of prices obtained by growers for flowers at Auction 1987-89

Single flower stems.	£'s	rupees
Tulip	0.02 - 0.10	1 - 3
Narcissus	0.01 - 0.08	0.3 - 3
Iris	0.05 - 0.12	2 4
Gladiolis	0.03 - 0.10	1 - 3
Lily (Stargazer)	0.03 - 0.96	10 - 32
Lily(Enchantment)	0.10 - 0.32	3 - 10

Dutch Ministry of Agriculture and Laboratory for bulb culture.

Micropropagation is more labour intensive than any other modern industry. The high wages of Western Europe ensure that labour contributes 65%-75% of the production costs. Given that expense, efficiency must be maintained and operator skill and speed is vital to success. The limits of production are determined by the limitations of the operators.
With a working year of 240 days and 7 effective working hours per day of a 5 day week the operator can handle close to 300.000 cultures per year.
The normal ratio of number of cultures handled to completed plantlets sold off the factory is at best 3:1. Often losses from a variety of causes will increase this figure to unacceptable levels.
Low multiplication rates at Stage II and poor establishment at Stage lV are the most damaging and most difficult to rectify. Death of cultures through contamination by bacteria, fungi or mites is almost as dangerous. These are ever present factors which increase costs and much of the difficulty of tissue culture lies in constantly combating their influence.
Given multiplication rates above *2.5 and contamination rates below 2% and acclimatisation and establishment figures of at least 95% one operator may be expected to produce some 80,000 to 100,000 plantlets ready for sale in a year. Using current wage rates in Western countries, to which must be added various State provisions for insurance, pensions and health the cost of handling a culture once is 3 pence equivalent to one rupee.
Thus the cost of all in vitro phases will be 9 pence(3 rupees).
To this must be added incubation room costs, the weaning house operating and administration and despatch which brings a total of 12 pence(4 rupees). This is a very basic calculation because a functional laboratory capable of producing several million plants per year carries a range of management, administration and ancillary staff and

accompanying facilities which add another 3-4p(1-1.5 rupees) to the costs.
Without proceeding to a full cost analysis, a comparison of these basic figures with unit price per bulb in Tables 4 & 5 shows clearly that micropropagation is not likely to be profitable for most genera.

3. Conclusion

The trade in ornamental bulbs and the flowers from them is huge involving billions of units per year.It is highly organised with a massive international export and import activity. On the surface there is a strong demand for starter plants which ought to provide opportunities for micropropagation laboratories.
That this is restricted to one genus the lily, is due to the failure of all other bulbs to meet some of the commercial criteria set out above. The main unfavourable circumstance is price which excludes tulips, gladiolus, narcissus, iris and crocus. Tulips and crocus are also excluded by the absence of good production systems. Tulips, narcissus, gladiolus and iris do not offer annual repeat business. Gladiolus and hyacinth may be propagated abundantly by simple ordinary methods.
These findings are applicable to all potential subjects for micropropagation. It was the neglect of a rigorous commercial appraisal which has been the main cause of failure for so many micropropagation enterprises.

4.References

Handbook of bulb cultivation, season 1989-90
Hobaho, Lisse. Netherlands 1989

Kwantitatv. Informatien voor de Bloembollen en Bolbloementeelt. Laboratarium voor Bloembollemonderzoek, Lisse. Netherland 1990

Pierik. R.L.M. 1990 Commercial micropropagation in Western Europe and Israel in P.C. Debergh and R.H.Zimmerman eds : Micropropagation of Horticultural crops. Kluwer Academic Publishers,Dordrecht,Netherlands.

Yearbook of International Horticultural Statistics 1989. Non-edible horticultural products, Vol.37. International Association of Horticultural Producers. Hanover 1989

Export bloemenbollen assortiment per land 1984-89 Report No 254. Produktschap voor Siergewassen (P.V.S.) Den Haag, Netherlands 1989.

Current Status of Forest Biotechnology

Dr. H.K. SRIVASTAVA
Department of Biotechnology
Ministry of Science & Technology
Government of India
Lodi Road, New Delhi - 110 003.

Introduction

Forestry in India is at the cross-roads. The effective forest cover is not more than 10.9 percent, which is too low to afford long-range ecological security and supply goods and services to people and industry. There is also a widening gap between supply of and demand for fuelwood, timber and pulpwood. Biotechnology is viewed here in a comprehensive context: as a multidisciplinary field in which biological systems are developed and/or used for the provision of commercial goods or services. Biotechnology increasingly offers opportunities for product and yield improvement in tree crop plants.

Application of Plant Tissue Culture for Forest Tree Improvement

Some major programmes need to be mounted to restore the forest cover by protecting forests in conservation areas and raise large-scale plantations and man-made forests. The role of biotechnology and tissue culture lie in enhancing the capability for the production of planting material of selected high-yielding types so as to boost production and productivity. Forest biotechnology has to be part of an overall forest genetics and tree breeding programme. the underlying philosophy in agro-forestry and industrial forestry has to be tree-crop farming.

Clonal Propagation: Plant tissue culture is an essential component of plant biotechnology. Apart from mass multiplication of elite plants, its also provides the means to multiply and regenerate elite plants from

J. Prakash and R. L. M. Pierik (eds.), Horticulture – New Technologies and Applications, 169–176.

genetically engineered cells. These elite plants may be cloned under aseptic conditions. Micropropagation has been used for clonal multiplication of orchids and ornamentals on a commercial scale for the last 25 years, but very little work has been done to this effect in forest trees. The only practicable method for their propagation has been through seeds, which is not favourable for high productivity due to genetic heterogenecity. By the time the tree has been evaluated for desirable traits, it passes the stage of vegetative propagation and loses the ability to root. In such cases, tissue culture is the most valuable aid for cloning selected trees. It is generally assumed that any form of clonal propagation for a quantitative character will lead to a greater gain than seed based propagation. So far success in the micropropagaion of elite trees has been largely restricted to those species that root readily **in-vitro**, the **in-vitro** propagation of difficult to root species deal mainly with embryonal or young seedling material. The role of tissue culture in clonal propagation of tree species varies with the situation. In hard to root species, it would be useful for entire propagation whereas in others it may help to acclerate the rate of multiplication. Considerable progress has been made over the last two decades on the development of tissue culture methodologies for trees and their possible application to forestry. In India, a few systems like Bamboo, Eycalyptus, Sandal and Teak have been worked out for **in-vitro** clonal propagation.

Micropropagation involves three main stages:-
a) Establishment of culture, (b) regeneration of plants, (c) transfer of plants from test tube to soil. The **in-vitro** plant regeneration may be (1) direct from explants (2) indirect initiation and differentiation of callus. Direct regeneration is helpful for clonal propagation whereas indirect method results in genetic veriability.

Direct Regeneration

Explants from young and juvenile trees are easier to grow as compared to old trees. However examples of direct regeneration from 100 year old trees of **Tectona grandis** and **Sequoia semperinens** are existing. Embryo culture has been employed as a useful tool for direct regeneration of trees where seeds are dormant, recalcitrant or where they abort at early stages of development.

Direct regeneration or clonal propagation is generally via bud culture or apical meristem culture. The latter generally involves only the actual apical dome of the shoot, preferably without a few leaf primordia. These are used for obtaining disease free clones. Clonal propagation via bud culture involves the entire rudimentary vegetative shoot. Explant for propagation is the bud of short node with axillary bud.

Indirect Regeneration from Callus Culture

The advantage of callus cultures is that a large number of plants can be produced from a single culture. Thus cell suspensions and protoplasts or pollen in one petri dish can produce a large number of plantlets directly or via embryogenesis. In contrast to the direct method, the regeneration of plants from callus may result in genetic variability, since after repeated subculturing callus cultures undergo genetic erosion, mutation and changes in ploidy. However use of callus culture for plant propagation in reforestation programmes has also been reported. Callus culture are generally used where quantity and not quality of the biomass is of importance. Winton and Huhtinen (1976) stated that "Polyploidy may be an important area of tree breeding that would benefit from the use of haploid plants. Conifer seedlings having more than the normal two sets of chromosomes generally are stunted and do not survive. Tetraploids have four sets of chromosomes, two from each parent and are thus allotetraploids, If further doubling of chromosomes occured in homozygous diploid plants, the resulting autotetraploids might have surprisingly vigorous characteristics. The effect of autotetraploids might be even greater among hardwood species." Trees of various ploidy levels like haploid rubber, diploid almond triploid sandalwoods, triploid poplars etc. have been regenerated from callus cultures. Their long term progress is yet to be evaluated. In some exceptional cases even after years of subculturing, no evidence of genetic variability was present. In Eucalyptus. Sussex (1965) and Piton (1969) did not observe polyploidy of aneuploidy in callus culture for 3 and 10 years respectively.

Somatic Embryogenesis

Induction of somatic embryogenesis whether by direct induction on the explant or through callus is a very important method of micropropagation. There may be derived or normal sexually produced embryos in plants that would otherwise abort or produce recalcitrant seeds. Somatic embryos may also be derived from callus cultures. Bajaj (1986) reported, that such embryos could be encapsulated in a seed coat like coating and these can then be used directly for large scale multiplication of trees. Somatic embryos can now be obtained in a number of tree species like conifers and papaya. Like somatic embryos , the pollen embryos have also been induced in a number of tree species i.e. Aesculus, Poinciana, Jacaranda, Rubber, and popular. Somatic embryogenesis offers tremendous potential for large scale production of plants at low cost. Somatic embryos in large numbers can be produced in suspension culture by using bioreactor technology to cut down cost. Thus mass plant propagation by tissue culture technology has developed into an important industry with considerable potential for future. In India some progress has been made in achieving high freqency regeneration of somatic embryos in Eucalyptus and Bamboo. Efforts are now being made to solve the problems of embryo synchronisation, proliferation, maturation, singulation and germination at production level.

Production of Haploids/Pure lines via Anther Culture/ Protoplasts

In Vitro techniques can be used for production of haploid plants from excised anther isolated pollen or haploid protoplast. Tree species are heterozygous and outbreeding and therefore production of pure lines may take years. In vitro haploid production can reduce this period. Haploids may also be utilised to facilitate the detection of mutations and the recovery of unique recombinants. The isolated pollen and the haploid protoplast are very good systems for mutation studies as well as mass production of haploids. The production of haploids from excised anthers is affected by a number of factors.

The significance of haploids in genetics and plant breeding has been realised for long time but their exploitation remains restricted due to their low occurence in nature.

Genetic Engineering in Forest Trees

Many research groups are already investigating on genetic engineering using recombinant DNA technology to bring improvement in forest tree species in terms of biomass production and disease resistance. By applying this technology to forestry, Durzan's group has demonstrated that **Agrobacterium tumefaciens** can be used for gene transfer in Douglas-fir. Similar work has also been done on Loblolly Pine and Poplar by other groups. Another group at the University of Kentucky has recently succeeded in transferring **Bacillus thuringiensis** (BT) toxin gene into conifer and further used restriction fragment length polymorphism (RFLP) in **Populus deltoides.** Genetic engineering has thus been extended from crop plants to forest trees, offering new ways to increase forest productivity, high timber quality and other desirable characteristics such as, resistances to pests, pathogens, stress and herbicides. The Ti-plasmid of A. **tumefaciens** is an established vector for genetic engineering of a wide range forest tree species. The incorporation of osmoregulatory gene and insect resistance gene (BT toxin gene) into plant genome will help in the development of drought, stress and pest resistant varieties.

Plant biotechnology and genetic engineering is a set of techniques designed to assist our abilities to change the genetic make up of plants. They can be used to overcome disease, pest and environmental constraints on production and to improve the quality of wood of forest tree crops. In so doing these techniques used in conjunciton with conventional breeding programmes could make dramatic contribution to sustainable forest biomass by producing improved tree crops that are more compatible with their environment. It is therefore critical to identify problems that have been difficult to solve with conventional approaches in forest genetics and breeding. The focus has to be on tree crop problems and the products and processes needed to solve them. Cell and Molecular biologies have to be exploited under new partnerships with the classical forest geneticists, breeder, agronomist and pathologist. These partnerships will ensure integration of the new techniques into forestry research and development programme, and to demonstrate their principles and application in forestry sector. The major goal of an integrated management plus genetical and molecular approaches must focus on the following: a) Tree disease and pest control, b)Tree tissue culture, micropropagation and transformation, c) Genetic mapping of nationally important tree crops with RFLP, d) Identification of genes for stress tolerance, salinity, drought and temperature, e) Compressing breeding cycle and enhanced harvest cycle by evolving fast growing species and f) Exploitation of heterosis/ hybrid vigour by uncovering genes responsible for male sterility, Cytoplasmic male sterility and restoration of fertility.

Forest Biotechnology - Work Done in India

Forest Biotechnology, Production of Biomass (Fuel, Fodder, Timber and Commercial and Industrial Woods) using tissue culture priorities are (i) use of wastelands for plantation which are the only lands available for this purpose (ii) plantations being based mostly on unselected wild-stocks which are highly heterogeneous as far as yield is concerned (iii) minimal involvement of forest genetics and tree breeding and (iv) outdated forest management practices, forests trees never being treated as tree-crops nor forestry taken as tree-crop agriculture. A task force set up by the Department of Biotechnology identified some species for undertaking work on tissue culture micropropagation, regeneration on a priority basis. They are: a) Multipurpose species - **Acacia nilotica, Alnus napalensis, Hardwickia binata, Madhuca latifolia, Prosopis cineraria, Tamarindus indica, Dendrocalamus,** b) High value timber species - **Tectona grandis, Shorea robusta, Dalbergia latifolia, Santalum album, Populus deltoides** and c) Desert Tree species - **Tecomella undulata, Prosopis cineraria, Anogeissus pendula, Ziziphus numalaria, Phoenix dactylifera, Commiphora wightii**

Among the tree species well studies in tissue culture for propagation are Sandalwood and Mulberry at the Bhabha Atomic Reserch Centre (BARC), Bombay, and Indian Institute of Science (IISC), Bangalore; **Populus, Prosopis** and **Bambusa** at the Tata Energy Research Institute (TERI), New Delhi; **Eucalyptus,** Teak and Bamboo at the National Chemical Laboratory, Pune; **Acacia,** Bamboo at the Botany Department of Delhi University and Jodhpur University. In the case of **Eucalyptus** (mature), Teak (Mature), Bamboos (Seedlings), **Santalum** and **Prosopis** (Mature) a reliable technology for propagatin on large-scale exists and can now be tested at a pilot plant scale. In Bamboo (mature) and **Acacia nilotica** (mature) further research is required to develop a viable technology. Among the remaining tree species of the priority list, successful propagation at mass scale level at which it can be utilised has not been recorded from **A. nilotica, H.binata** and **M. latifolia.**

One of the most critical aspects of afforestation and green cover is the availability of suitable planting material from elite stock of appropriate fast growing trees species with as wide a genetic background as possible. For this purpose, the Department of Biotechnology has set up two pilot scale production units with a capacity for producing million planting material per year. A tissue culture propagation system developed in the laboratory, resulting in few hundreds of plants, does not qualify for adoption in a large scale production programme unless an intermediate pilot plant unit is successfully demonstrated. These units have been given the mandate to: (1) demonstrate the feasibility of the tissue culture technique on a large scale with regard to adequacy of multiplication rates and health of plantlets produced (2) indicate whether cost reductions can be obtained at various stages of plant production for cost effectiveness (3) produce enough plants to conduct field trials of a suitable manner in different locations (4) create a work ethos suitable for large scale production.

Ideally the pilot plant facilities should be located near the area ear-marked for field trials. Since a primary aim of the pilot plant will be to test the feasibility of tissue culture, a pilot plant can be combined with a facility for plant production through conventional methods such as seeds or cuttings or rooted cuttings. This will facilitate comparisons of the economics and later, the field trial evaluation. Part of the infrastructure, like the greenhouse is common for both the methods.

The tissue culture method is attractive only if the benefits that are derived through this method will offset the difference in cost of production which is higher when compared to conventional vegetative propagation. In clonal propagation, homogeneity of the plantlets as well as the true-to-typeness with the selected elite mother trees is an essential prerequisite. An extensive field evaluation/ verification using multi-location trials with suitable statistical designs is necessary before proceeding with large-scale production and field planting. The controls used are the natural propagules like seeds or rooted cuttings. The evaluation of such a trial at all stages will determine the economic viability in the **in vitro** tissue culture methods and the feasibility for large scale application for localised production and plantations in field in the forestry sector.

Conclusion

Forestry is playing an ever important role in the world's society and economy. Forests today must be managed for all types of products, including recreation and conservation. There is an urgent need to make forests fully productive. The increasing need for greater forest production can only be met if more intensive forest management is practiced and this requires the application of tree improvement. Without tree improvement forest cannot come close to reaching the goal of optimum productivity. There seems to be a shortage of 211 Mm of fuelwood and if this has to be met on rotational cycle of 10 years (as in the case of Acacia nilotica and Prosopis cineraria) and with a productivity of 0.7 m per year per hectare from the waste lands, the country would need about 300 Mha to grow fuelwood alone. Timber, pulp and paper needs are far more realistic. Timber would require 7 Mha at a rotational cycle of 50 years and a productivity of 5m per hectare per year. While pulp and paper would need 1 Mha in a rotational cycle of 8 years (as in major pulp woods such as eucalyptus) and a productivity of 7m per hectare per year. Wood based industries would also need 1 Mha at 5 m per year productivity with a 50 year rotation. While fuelwood may be grown on wastelands, timber, pulp, paper, and other wood-based industries would need relatively fertile lands.

Large Scale Secondary Metabolite Production: The Sanguinarine Story

K.L. GILES
Department of Horticulture Science
University of Saskatchewan
Saskatoon, SK S7N 0W0 CANADA

Introduction

Sanguinarine is a benzo-phenanthridine alkaloid found in the rhizomes of Sanguinaria canadensis mixed with several other alkaloids. The preponderant color of the rhizome of this plant is red and pure sanguinarine is a bright orangy red color. Exposure to air will change this color to a brownish red over a period of time. Sanguinarine has been found to have very significant anti-microbial activity against the oral micro-organisms responsible for plaque build-up and periodontal disease. The irritation caused by periodontus, gum swelling and discomfort is currently only treatable by surgery. The gum pocket resulting at the side of the tooth is removed. This surgery is both cosmetically unattractive, expensive and can cause considerable discomfort. Treatment of periodontal pockets with sanguinarine has been shown to have significant benefits in terms of disease control.

Approximately 45 million people in North America are thought to suffer from the disease and strenuous efforts have been made to control it using a number of approaches. Sanguinarine has been incorporated into a slow release, biologically degradable compound that is placed in the periodontal pocket by a professional periodontist. The product stays in the pocket for a period of between 1 and 2 weeks, during which time sanguinarine is released into the pocket to directly contact the disease microorganisms. Very rapid alleviation of swelling and irritation is noticed and over the period of treatment, the diseased condition can be alleviated. This product is currently in phase 3 of F.D.A. trials, prior to its commercial release.

J. Prakash and R. L. M. Pierik (eds.), Horticulture – New Technologies and Applications, 177–180.

Supplies of pure sanguinarine necessary for the F.D.A. trials and approval can be produced by purifying the compound from the mixture of alkaloids, either from the rhizomes of sanguinaria or from the leafy tissue of some other species especially those of the Papaveraceae. However, the extraction and purification of the product is relatively expensive. Eilert et. al, 1985, demonstrated that tissue cultures of Papaver somniferum were capable of producing pure sanguinarine after elicitation by fungal homogenates. The productivity of this system was low, approximately 1% dry weight of cultures was extractable as sanguinarine after elicitation was completed.

Culture Scale-up

Significant work has now been completed on the scale-up of cultures of the Papaver somniferum cell line, and of sanguinarine production. Suspension cultures of the callus were initiated in 125 ml Erlenmeyer flasks in Gamborg's B5 medium. These cultures could be elicited with homogenates of an isolate of Botrytis cinerea and would routinely produce sanguinarine when elicited at the right stage of culture. It was soon noted that elicitation had to take place at a stage when the sugar levels were approaching zero, approximately 5 days after the initiation of the cultures. At this time sanguinarine could be detected 18 hours later and was maximum 24 hours after elicitation.

These suspension cultures were scaled-up through various stages at 2 litres, 5 litres, 10 litres and 20 litre cultures. Monitoring the oxygen demand during culture revealed that after elicitation there was a very profound increase in demand for oxygen and the aeration and agitation of the cultures had to be increased at that stage to ensure satisfactory productivity. A 300 litre, air lift bioreactor loop was also used for trial runs. This was used at the Alberta Research Council facilities in Edmonton. The cultures grew well and behaved in a production fashion similar to that which had been found in small bioreactors. A simpler 100 litre, air lift bioreactor was also constructed and used successfully for cell growth and production.

Cell-Line Improvement

Single cell isolates of the culture were taken and grown using a nurse callus technique to produce small calli. These calli were bulked up and subsequently used to produce suspension cultures. Considerable variation in the productivity of these isolated single cell derived suspensions was noted, some lines producing very little sanguinarine, others giving yields as high as 13% dry weight, a very significant improvement over the 1% dry weight from the mother culture. This yield increase clearly has very significant economic advantage in the production system. (Songstad, et. al., 1990)

Cryopreservation of Cell Lines

Because of the strategic importance of maintaining genetic stability in the cell lines, considerable effort has been put into the cryopreservation of this material. (Friesen, et. al., in press) Careful timing of cryopreservation relative to the subculture period was necessary here, as had been the case with elicitation sensitivity. A technique was developed for the successful cryopreservation of the cell line material providing for its recovery and subsequent production activity. For details see the above paper.

Discussion

The production of secondary metabolites in cell culture has been a much discussed topic over the last 15 or 20 years. There are still very few examples of where such production has resulting in an economically viable process. The production of sanguinarine, in the manner briefly described above, does provide economic production advantages. For the successful production of viable secondary metabolite programs there are 3 key factors that must be attended to. Firstly, the cell line must be vigorous genetically stable and grow easily and reliably in culture. This is important for scaling up the operation to a 200 or 300 litre scale. Secondly, the physiology of product production must be understood, the pH, oxygen demands, sucrose levels, etc. must be optimal. These factors can significantly influence design choices of the large-scale bioreactor and if incompletely understood optimization of the growth and production processes is impossible. Thirdly, product extraction and purification methods must be streamlined and, where possible, simplified. These steps are critical if maximum productivity is to be achieved. Poorly developed systems can decrease yield and down grade the purity and value of the product.

Molecular biology holds the key to future development in the secondary metabolite production field. Modification of individual enzyme steps in the biosynthetic pathway leading to product development will almost certainly have profound effects on secondary metabolite production processes in the not so distant future.

References

Friesen, L., Kartha, K.K., Leung, N., Englund, P., Giles, K., Park, J. and Songstad, D., 1990. Cryopreservation of Papaver somniferum (L.) cell suspension cultures. Planta Medica (in press).

Songstad, D.D., Giles, K.L., Park, J., Novakovski, D., Epp, D., and Friesen, L., 1990. Use of nurser cultures to select for Papaver somniferum cell lines capable of enhanced sanguinarine production. J. Plant Physiol. 136: 236-239.

CURRENT RESEARCH ON MICROPROPAGATION OF SULCOREBUTIA ALBA RAUSCH, SYRINGA VULGARIS L., HIPPEASTRUM HYBRIDS AND TULIPA HYBRIDS

R.L.M.Pierik
Department of Horticulture
Wageningen Agricultural University
P.O.Box 30
6700 AA Wageningen
The Netherlands
Publication 582

Factors affecting areole activation in vitro in the cactus Sulcorebutia alba Rausch

Abstract

A cultivation system was developed to study areole activation in explants of the cactus Sulcorebutia alba. Explants, containing three areoles, were cut from the cladodes, which were grown in vitro. The optimal culture medium for areole activation consisted of Murashige and Skoog's macrosalts at a strength of 1.5, Murashige and Skoog's microsalts (except iron), NaFeEDTA 25 mg/l, sucrose 2.5%, BA 0.25-1.00 mg/l, Daishin agar 0.5%, and a pH of 5.5. Cytokinin was an essential prerequisite for areole activation. The cytokinins PBA, 2iP and IAP were not suitable as they give rise to heavy callus formation. Optimal physical growth conditions for areole activation were a temperature of 27°C and an irradiance of 2.5-5.0 Wm^{-2} (photosynthetic active radiation).

Introduction

Relatively little is known about the micropropagation of cacti. Therefore a cultivation system was developed to study the micropropagation of cacti by areole activation. Sulcorebutia alba Rausch, originating from the highlands of Bolivia was chosen for study. Nutritional, hormonal, physical and plant factors were examined.

Material and methods

Adult cacti were grown in a greenhouse and explants isolated under sterile conditions in vitro, by means of areole activation, using the cytokinin 6-benzylaminopurine (BA). For the standard experiments on areole activation, explants with 3 areoles, from the central piece of the cladode (cactus body), were excised from in vitro grown cacti. On the basis of preliminary experiments explants were grown on a basic medium with Murashige and Skoog (1962) macro- and micro-elements at full strength (except iron), NaFeEDTA 25 mg/l, 3.5% sucrose and 0.5 mg/l BA. Cultures were incubated in a growth chamber at 27°C with a 16 h photoperiod provided by fluorescent light (Philips TL 38W/54, 5-8 Wm^{-2}). The effect of different concentrations of sucrose, agar and regulators, as well as different temperatures, pH, osmotic potential and irradiance level of the 16 h fluorescent light per day were studied varying always a single factor and keeping all the other factors constant. Areole activation was evaluated 8 weeks after

J. Prakash and R. L. M. Pierik (eds.), Horticulture – New Technologies and Applications, 181–188.

inoculation.

Results

The weight of the harvested cacti decreased strongly with an increase in agar concentration (optimal 0.5%). Optimal nutritional conditions for areole activation were: MS macrosalts at a strength of 1.5, 2.5% (w/v) sucrose and pH 5.5. Lowering the osmotic potential of the medium by adding mannitol decreased areole activation. Addition of a cytokinin was essential for areole activation. At 0.5 mg/l BA resulted in the highest number of normal and the lowest number of abnormal cacti per explant. There were no significant differences in number and weight of normal cacti between the cytokinins used. Although IPA and 2iP seemed to result in more abnormal tissue than BA, kinetin and PBA, these differences were not significant. Auxin increased the weight of normal cacti, but also the number and weight of abnormal cacti. Addition of 2,3,5-triiodobenzoic acid, abscisic acid, ethylene, and gibberellin all had a negative effect. Optimal temperature for areole activation was 27°C. Although the irradiance levels studied had no effect on the weight and number of cacti formed, at low irradiance elongation of the cladodes occurred and in darkness pronounced etiolation was observed. Cacti grown in darkness de-etiolated and formed normal cacti when transferred to light. Rooting of isolated cladodes was readily obtained on media without cytokinin and auxin.

Conclusions

Areole activation is strongly promoted by cytokinin and suppressed when the uptake of water and nutrients is hampered by increasing the agar, mannitol or sucrose concentration of the medium. The irradiance level used strongly influences the morphology of the cacti in vitro.

References

Dabekaussen, M.A.A., Pierik, R.L.M., Laken, van der, J.D. and Hoek Spaans, J. 1991. Factors affecting areole activation in vitro in the cactus Sulcorebutia alba Rausch. Scientia Hort. (in press).

Murashige, T. and Skoog, F. 1962. A revised medium for rapid growth and bioassays with tobacco tissue cultures. Physiol. Plant. 15: 473-497.

Pierik, R.L.M., Dessens, J.T. and Zeeuw, van der E.J. 1987. Activation of in vitro cultured areoles of the cactus Sulcorebutia alba Rausch. Acta Bot. Neerl. 36: 334.

Pierik, R.L.M., Steegmans, H.H.M., Molendijk, M., Dessens, J.T. and Zeeuw, van der E.J. 1987. Massale vermeerdering van Sulcorebutia species mogelijk via in vitro kultuur. Succulenta 66: 218-222.

Micropropagation of Syringa vulgaris L.

Abstract

The most efficient method of micropropagating root stocks and cultivars of lilac (Syringa vulgaris L.) is rejuvenation by repeated subculturing of shoots, followed by single-node culture. Optimal conditions for single-node culture are: Murashige and Skoog's (1962) macrosalts at a strength of 1.5, Murashige and Skoog's microsalts at full strength (except iron), NaFeEDTA

37.5 mg/l, sucrose 3.5%, Daishin agar 0.8%, IPA 1-2 mg/l (depending on root stock or cultivar considered), 21°C, and an irradiance of 4 Wm^{-2}. Rooting of rejuvenated shoots can easily be realized by subculture of shoots in Rockwool microplugs under semi-sterile conditions.

Introduction

Lilac (Syringa vulgaris L.) is a shrub normally planted for ornamental purposes in gardens, but another very important commercial application in the Netherlands, is the cultivation as 'forced shrub' for the production of flowering branches during winter. Lilacs are normally cloned by grafting cultivars onto the wild rootstock Syringa vulgaris L. Micropropagation through the single-node method is a very attractive alternative method of cloning lilac on their own roots. Our work on micropropagation (Pierik et al., 1988, 1990) will be summarized in this article.

Material and methods

Actively growing shoot tips of lilac (Syringa vulgaris L.) rootstocks and cultivars were selected and isolated in vitro in the spring from very old (adult) shrubs. Repeated subculturing was applied to obtain rejuvenation of this material before micropropagation through the single-node method started.
On the basis of preliminary experiments shoot tips and single-nodes were grown on the following culture medium (Pierik et al., 1988, 1990): Murashige and Skoog (MS, 1962) macrosalts at a strength of 1.5, MS microsalts full strength (except iron), NaFeEDTA 37.5 mg/l, sucrose 3.5%, IPA 1.0-2.0 mg/l, Daishin agar 0.8%, pH 6.0 before autoclaving. Cultures were incubated in a culture room at 21°C under a photoperiod of 16 h fluorescent light (Philips TL 54/38W, 4-5 Wm^{-2}). The effect of different factors (sucrose, MS, regulator and agar concentration), as well as different temperatures and irradiance levels of the 16 h fluorescent light (Philips TL 38W/54) per day were studied, varying a single factor and keeping all the other factors constant.
Shoot multiplication took place by applying the single-node method, whereas rooting experiments were caried out in Rockwool plugs from Grodan (Pierik, 1991) under semi-sterile conditions. Experiments on stem elongation and rooting were evaluated after 6 weeks.

Results

Optimal conditions for single-node culture and stem elongation for most root stocks and cultivars are: MS-macrosalts at a strenght of 1.5, MS-microsalts at full strenght (except iron), NaFeEDTA 37.5 mg/l, sucrose 3.5%, Daishin agar 0.8%, IPA 1-2 mg/l, pH 6.0 before autoclaving, a temperature of 21°C, and an irradiance of 4-5 Wm^{-2}. Irradiance had a very strong influence on stem elongation and number of leaves to be formed. A low irradiance is an essential prerequisite. Stem elongation could be promoted by increasing the temperature from 21° to 25°C, but at 25°C shoots became yellow and unhealthy.
In difficult-to-elongate cultivars such as 'Madame Florent Stepman' the type of cytokinin used was very important. Kinetin is very ineffective for lilac stem elongation, BA and PBA have a very weak stem-elongating action, 2iP and zeatin are quite effective, but the most effective cytokinins for stem elongation in the cv. 'Madame Florent Stepman' are the ribosides of 2iP and zeatin, IPA and zeatin riboside. Similar resuls were obtained for the cv. 'Hugo Koster'. In two other cultivars 'Herman Eilers' and Maréchal Foch', stem elongation could not be induced effectively by IPA and zeatin riboside.
Rejuvenated shoots of all rootstocks and cultivars could easily be rooted in so-called microplugs of Grodan Rockwool under semi-sterile conditions. In vitro the rooting of lilacs in agar media is not very effective and never reaches 100%, and root initiation and development

were very poor.

Conclusions

It can be concluded that micropropagation overcomes the many shortcomings of the conventional technology of lilac propagation by grafting. The classical rootstocks A2 and A3 and the important cv. 'Madame Florent Stepman' can now be cloned in vitro very rapidly and on a commercial scale.

References

Pierik, R.L.M., 1991. Micropropagation of ornamental crops. Acta Hortic., in press.
Pierik, R.L.M., Steegmans, H.H.M. and Sprenkels, P.A., 1990. Micropropagation of lilac (Syringa vulgaris L.). In: Bajaj, Y.P.S. (editor), Biotechnology in Agriculture and Forestry. 3. Medicinal and aromatic plants. Vol.20. Springer Verlag, Berlin. In press.
Pierik, R.L.M., Steegmans, H.H.M., Elias, A.A., Stiekema, O.T.J. and Velde, van der, A.J., 1988. Vegetative propagation of Syringa vulgaris L. in vitro. Acta Hortic. 226: 195-204.

Micropropagation of Hippeastrum hybrids

Abstract

An efficient micropropagation system for Hippeastrum hybrids was developed using floral stem explants as starting material for bulblet regeneration. Optimal factors for this regeneration are a temperature of 25°C, continuous darkness, a MS macrosalt strength of 1.5, a high sucrose concentration of 6-8%, and auxin (NAA 0.3 mg/l) and cytokinin (BA or 2iP 2.0 mg/l) supply. Bulblets, regenerated on these explants were subcultured in light on simple media without regulators to obtain rooting and leaf formation, essential prerequisites for an increase in bulb size. When a bulb size of at least 0.8-1.2 cm in diameter was reached, these minibulbs were quartered and subcultured again in the light on a relatively simple medium without regulators. Plants regenerated from the quartered minibulbs could easily be transferred to soil or used for further micropropagation. The in vitro produced plants appeared to be true-to-type.

Introduction

Regeneration of shoots from floral stem explants was first recorded by Pierik (1967). It was later shown, that isolated floral explants can be a valuable starting point for adventitious shoot formation, particularly in bulbous plants such as Nerine bowdenii, Nerine sarniensis, Eucharis amazonica, Freesia hybrids, tulip, and Hippeastrum hybrids. From 1985-1987 (Pierik et al., 1985; Koopman et al., 1987) research was initiated to develop a new and efficient micropropagation system for Hippeastrum.

Material and methods

Floral stems of two cultivars ('Hercules' and 'Cinderella') (4-5 cm etiolated stems forced in darkness at 25°C) were sterilized. Based on preliminary experiments discs subsequently cut into 1.5 mm thick discs; these discs were inoculated with their basal ends down on a bulb regeneration medium containing auxin (NAA 0.3 mg/l) and cytokinin (BA or 2iP 2.0 mg/l) and incubated at 25°C in darkness. The composition of the basic media was as follows: MS macro- and microsalts at full strength (except iron), NaFeEDTA 25 mg/l, sucrose 6-7%

(depending on the cultivar), Daishin agar 0.7%, pH 6.0 before autoclaving, vitamin B_1 0.4 mg/l and meso-inositol 100 mg/l. The basic culture medium was constant for all experiments, where only single nutritional, hormonal and other factors were tested. Further experiments were carried out changing the physical growth conditions, length of etiolated floral stem used and concentration of sucrose in the medium. Also in these experiments a single factor (light, temperature, etc.) was varied, while keeping all the other factors constant. Experiments were evaluated after 3 months and bulblets harvested. Bulblets, regenerated on floral stem explants, were subcultured on a MS macro- and microsalts with 4.5% sucrose without regulators. Using this medium bulblets regenerated roots, formed leaves, and increased their bulb size. They were grown under a regime of 16 h fluorescent light per day at 25°C. When the plants reached a bulb size of 0.8-1.2 cm, the minibulbs were split longitudinally into 4 segments and leaves and roots were removed. The bulb explants were inoculated on a fresh MS macro- and microsalts medium with 4.5% sucrose, but without regulators to produce new bulblets from the base of the 'miniscales'. Physical growth conditions were the same as described for plant formation. Plants regenerated from the quartered minibulbs were harvested after 3 months and then transferred to soil.

Results

Optimal bulblet regeneration in both cultivars took place on a medium with MS-macrosalts at a strength of 1.5 and a high sucrose concentration of 6-7% (w/v), and was strongly dependent on the concentrations of auxin (optimal NAA 0.3 mg/l) and cytokinin (optimal 2iP or BA 2.0 mg/l) used (optimal conditions differered depending on the cultivar). Optimal bulblet regeneration occurred at 25°C in continuous darkness, and when the length of the floral stem was 3-6 cm. Regeneration of bulblets was strongly inhibited if floral stem explants were placed inverted on the media, rather than inoculated with basal ends down. Optimal bulblet regeneration occurred with an explant thickness of 1.5 mm. Bulblet regeneration was strongly affected by the age (length) of the floral stems and also by the position of the explant in the stem. Regeneration potential is highest in stems with a length between 3-5 cm. Basal explants have a higher regeneration potential than explants from higher up in the stems.

After subculture of the bulblets to produce complete plants with a large enough bulb size, quartered minibulbs produced on average 1.3 bulblets (5.2 bulblets per minibulb). The newly formed bulbs were used to create a continuous bulb production system for Hippeastrum.

The in vitro produced plants appeared to be true-to-type and flowered normally after 2.5-3.0 years. Histological research (Smits et al., 1989) showed the regenerated bulblets were of multicellular origin and therefore the chance of mutants arising is small. In 1989 to 1990 the Dutch commercial tissue culture laboratories produced a total of 168,750 Hippeastrum plants using this new system.

Conclusions

The efficiency of regeneration of bulblets in Hippeastrum appear to be strongly dependent on quite a number of limiting factors: cultivar, stem length, position of the explant in the stem, thickness and orientation of the explants, light versus darkness, mineral nutrition, sucrose concentration, auxin and cytokinin.

References

Koopman, W.J.M., Steegmans, H.H.M. en Pierik, R.L.M. 1987. Snellere vermeerderingsmethode. Stengel Hippeastrum produceert vele bollen. Bloembollencultuur 98(31): 18-20.

Murashige, T. and Skoog, F. 1962. A revised medium for rapid growth and bioassays with tobacco tissue cultures. Physiol. Plant. 15: 473-497.
Pierik, R.L.M. 1967. Regeneration, vernalization and flowering in Lunaria annua L. in vivo and in vitro. Meded. Landbouw Hogeschool Wageningen, The Netherlands 67(7): 1-71.
Pierik, R.L.M., Steegmans, H.H.M., Sprenkels, P. and Bijlsma, J. 1985. Amaryllidaceae in kweekbuizen vermeerderd. Bloembollencultuur 96(45): 8-10.
Pierik, R.L.M., Blokker, J.S., Dekker, M.W.C., Does, de, H., Kuip, A.C., Made, van der, T.A., Menten, Y.M.C. and Vetten, de, N.C.M.H., 1991. Micropropagation of Hippeastrum hybrids. Proc. Eucarpia Meeting, Wageningen, Pudoc, Wageningen (in press).
Smits, P.C.H., Pierik, R.L.M. and Schel, J.H.M., 1989. An anatomical study of the in vitro bulblet regeneration from flowering stem explants of Hippeastrum cv. 'Liberty'.
Acta Bot. Neerl. 38: 348.

Regeneration of adventitious shoots from floral stem explants of tulip

Abstract

The regulation of adventitious shoot formation in tulip is restricted by the low regeneration potential of tulip scale explants. In this study the induction of adventitious shoot formation on excised floral stem explants of different cultivars were investigated. To enable the experiments to be carried out over a long time period, frozen tulips bulbs were used as a source for the in vitro culture of floral stem explants. Tulip bulbs of the cv. 'Apeldoorn' and 6 other cultivars were used. Shoot regeneration in cv. 'Apeldoorn' only occurred on media with a critical auxin and cytokinin concentration. Good regeneration was obtained: in continuous darkness at 18°C, when explants were inoculated with basal ends down on the medium and when 3 cm long floral stems were used. Small discs (1.5 mm thick) gave rise to the highest number of adventitious shoots. The other cultivars used did not show any or good shoot regeneration when the conditions optimal for cv. 'Apeldoorn' were used. Only the cv. 'Prominence' showed abundant regeneration of adventitious shoots. It is proposed that the optimal auxin/cytokinin ratio for shoot regeneration varies depending on the cultivar used. Preliminary experiments on bulblet formation on explants with adventitious shoots yielded poor results. Although regenerated explants were transferred to media without regulators and cold treatment (8-12 weeks at 5°C) was applied, only a low percentage of the cv. 'Apeldoorn' explants formed bulblets.

Introduction

Regeneration of shoots from floral stem explants was first recorded by Pierik (1967). It was later shown that floral explants can be a valuable starting point for adventitious shoot formation (Pierik et al., 1991). For that reason it was very attractive to use also floral stem explants of tulip for studying adventitious shoot formation in tulip. This was particularly attractive since the regeneration of adventitious shoots in tulip is restricted by the low regeneration potential of tulip scale explants. One of the first reports concerning succcess with micropropagation of tulip involved the isolation and regeneration of floral stem explants (Wright and Alderson, 1980). In this study the induction of adventitious shoot formation on excised floral stem explants of different cultivars, including the cv. 'Apeldoorn' was investigated. To enable the experiments to be carried out over a long time period, frozen tulips bulbs were used as a source for the in vitro culture of floral stem explants.

Material and methods

Frozen tulip bulbs (stored at -2°C) of cv. 'Apeldoorn' and 6 other cultivars ('Prominence', 'Inzell', 'Abra', 'Kees Nelis', 'Monte Carlo', 'L. van der Mark') were obtained at the end of November from a commercial grower. Bulbs were periodically forced by transferring them to 18°C in darkness and 3 cm length etiolated floral stems were obtained after 2-3 days. On the basis of preliminary experiments these explants were sterilized and subsequently cut into 1.5 mm thick discs inoculated with their basal ends down on the following culture medium: Murashige and Skoog (MS, 1962) macro- and microsalts at full strength (except iron), NaFeEDTA 37.5 mg/l, sucrose 4%, casein hydrolysate 1500 mg/l, meso-inositol 100 mg/l, 2iP 1.5 mg/l, NAA 1.0 mg/l, Daishin agar 0.7%, pH 6.0 before autoclaving. Again based on preliminary experiments cultures were incubated at 18°C in continuous darkness. Experiments were carried out by varying only one factor and keeping all the other factors constant. Experiments were evaluated after 2-4 months. Preliminary experiments were carried out with regenerated shoots of cv. 'Apeldoorn' to induce bulblet formation on the explants.

Results

Optimal shoot regeneration in the cv. 'Apeldoorn' took place a medium with MS macro- and microsalts at full strength (except iron), NaFeEDTA 37.5 mg/l, sucrose 4%, casein hydrolysate 1500 mg/l, meso-inositol 100 mg/l, 2iP 1.5 mg/l, NAA 1.0 mg/l, Daishin agar 0.7 %, pH 6.0 before autoclaving. Optimal regeneration occurred in continuous darkness (in comparison to light) at 18°C (in comparison to lower or higher temperatures) and when explants were inoculated with basal ends down on the medium (in comparison to inverted explants) and when 3 cm length floral stem explants were used. Small discs (1.5 mm thick) gave rise to the highest number of adventitious shoots. The other cultivars used did not show any or good shoot regeneration when the conditions optimal for cv. 'Apeldoorn' were used. Only cv. 'Prominence' showed abundant regeneration of adventitious shoots under these conditions, although cv. 'L. van der Mark' and cv. 'Kees Nelis' showed high adventitious root rather than shoot regeneration. No shoot regeneration was observed in the cvs 'L. van der Mark' and 'Kees Nelis', whereas the cvs 'Abra', 'Monte Carlo' and 'Inzell' showed weak shoot regeneration. It is proposed that the optimal auxin/cytokinin ratio for shoot regeneration varies depending on the cultivar used.
Preliminary experiments on bulblet formation on explants with adventitious shoots yielded poor results. Although regenerated explants were transferred to media without regulators and cold treatment (8-12 weeks at 5°C) was applied, only a low percentage of the cv. 'Apeldoorn' explants formed bulblets.

Conclusions

Optimal conditions for shoot regeneration of the cultivars 'Apeldoorn' and 'Prominence' produced no or poor regeneration for the other cultivars studied. Further investigation is required to determine the critical (limiting) factors for shoot regeneration in different tulip cultivars and bulblet formation on explants with shoots.

References

Wright, N.A. and Alderson, P.G. 1980. The growth of tulip tissues in vitro. Acta Hortic. 109: 263-270.
Pierik, R.L.M., 1967. Regeneration, vernalization and flowering in Lunaria annua L. in vivo and in vitro. Meded. Landbouwhogeschool Wageningen, the Netherlands 67(7): 1-71.

Acknowledgements

In the last 5 years numerous M.Sc. students have made a considerable contribution to the experimental work presented here without whose support this article would have been impossible. I am especially grateful to my assistants H.H.M. Steegmans and P.A. Sprenkels for carrying out many of the experiments. I am indebted Mrs. C.J. Kendrick for the English translation and correction.

Plant cell culture and food biotechnology: Current trends and future prospects

L.V. VENKATARAMAN and G.A. RAVISHANKAR
Autotrophic Cell Culture Discipline,
Central Food Technological Research Institute, Mysore - 570 013, India

INTRODUCTION

Plant cell culture methods have much to contribute to food biotechnology in terms of augmenting food productivity, improving the quality of food grains, producing food additives of high value and in postharvest applications of food materials. The advances in the area of plant cell cultures for food applications have so far been, primarily, on the preharvest front. Some postharvest modifications are now being attempted (Table 1). The processing industries are presently interested in tailoring the plants to produce improved raw materials in greater amounts, and in an economical manner. This paper gives an over-view of the achievements made in food biotechnological applications, using plant tissue and cell culture methods. It is divided into the following sections:-

Food value metabolites; Bioinsecticides ; Scale up of plant cell cultures and product cost; Improvement in processing characteristics; Modification of agronomic traits; Safety considerations; Emerging trends.

FOOD VALUE METABOLITES

Plant cells, cultured *in vitro*, produce a wide range of primary and secondary metabolites of economic value. The advantages of this technology over the conventional agricultural production are as follows: 1. they can be grown in controlled conditions free from agro-climatic constraints; 2. uniform production throughout the year is ensured, and 3. production is independent of seasonal variations and is rapid.

Production of phytochemicals from plant and cell cultures has, so far, focussed on pharmaceuticals. It is now realized that this technology can be used for obtaining the food additives (Table 2). In most cases, synthetic food additives are cheaper, but are not preferred due to their probable toxic effects. Hence cell culture system would be a viable alternative method of producing natural food additives.

J. Prakash and R. L. M. Pierik (eds.), Horticulture – New Technologies and Applications, 189–196.

FOOD COLOURS

The total world market for natural and synthetic pigments is estimated to be $150-200 million (Ilker, 1987). Most of the synthetic colours are obtained from the petroleum byproducts and many are, also, not safe for consumption. Hence there is need to find sources of natural food colours.

Selected cell lines of various plants have been used for the production of pigments e.g. Catharanthus roseus, Crocus sativus, Daucus carota, Euphorbia milli, Haplopappus gracilis, Lithospermum erythrorhizon, Prunus persica, Vitis vinifera. Among the food colours, anthocyanin production in cell culture has been widely worked out (Mizukami et al., 1988). The anthoyanins obtained from conventional sources are expensive, and cost $1,250/-kg. Canadian and Israeli biotechnology firms have now ventured into commercial exploitation of anthocyanin by cell culture technology. The possibility of obtaining novel anthocyanin, with 2-3 times greater stability at low pH, has been demonstrated (Vunsh et al., 1986). Crocin-digentiobioside ester of α-crocetin was obtained in the saffron callus (Visvanath et al., 1990). The in vitro production of stigma-like structures has been achieved (Himeno and Sano, 1987, Sarma, et al., 1990) by culturing immature ovaries and stigma. This process is being scaled up by Ajinomoto Co. of Japan. Shikonin, the red pigment from Lithospermum erythrorhizon cells, has been commercially produced by Mitsui Petrochemical Ltd., Japan. Currently Shikonin obtained by cell cultures is sold at $ 4000/kg, as against root derived shikonin costing $ 4500/kg.

FOOD FLAVOURS

There are about 4300 different flavours identified in foods. In tissue cultures, it is difficult to obtain the whole range of flavours responsible for the aroma, in plants such as coffee, tea, etc. Hence, it is important to concentrate on systems with flavours which are attributed to a single compound. Examples of this kind are 2-isobutylthiazole (tomato flavour), methyl-ethyl-cinnamates (strawberry), methylanthranilate (grape), benzaldehyde (cherry), menthol (mint) and safranal (saffron) (Whitaker and Evans, 1987). A range of flavours have already been obtained in plant cell cultures, e.g. capsicum, celery, chamomile, coriandrum, garlic, grape, mentha, onion, orange, peppermint, perilla, rose, saffron (Venkataraman and Ravishankar, 1989). Some very expensive flavours and fragrances, like jasmine, rose oxide, nootkatone and valencene need to be considered for cell culture production. The problem of obtaining the whole profile of flavour component in citrus has been solved to a great extent by the innovative approach of producing lemon vesicles in cultures (Tisserat, 1987). Similar approaches are needed for other flavour complexes.

SWEETENERS AND OTHER FOOD ADDITIVES

The demand for non-nutritive sweeteners is increasing rapidly, for use as food additives and for pharmaceutical application. USDA scientists have developed, from orange peel, some flavone glycoside derivatives which can be used as non-nutritive sweeteners. The other non-nutritive sweeteners of importance are stevioside from Stevia rebaudiana, thaumatin from Thaumatococus danielli, 8-9-epoxyperillartine from Perilla nankinesis and the hydroflurene diterpenoids from pine tree rosin. Transfer of the thaumatin gene to a microbial system has opened up the commercial production of thaumatin which is 3000 times sweeter that sucrose. Capsaicin, a pungent food additive is one of the widely investigated food compounds for the production by immobilized cell cultures. The enhancement of capsaicin production has been achieved by precursor biotransformation (Lindsey and Yeoman, 1985). Immobilization of placental tissues enhanced the yield of capsaicin (Sudhakar Johnson et al., 1990). Scaling up of these systems is currently in progress.

BIOINSECTICIDES

Bioinsecticides are gaining importance in recent years due to their target specificity and biodegradable nature. They are being increasingly applied in the storage of foodgrains. The well known biopesticides studied in tissue cultures are pyrethrins, nicotine, phytoecdysones (Kudakasseril and Staba, 1988). The development of high pyrethrin yielding cell lines has been achieved using genotypically superior plants of *Chrysanthemum cinerariaefolium* (Ravishankar et al., 1989). Thiophene production in root cultures of *Tagetes* has been reported. Hairy root cultures of this system have shown a potential for enhanced productivity. However, none of the systems have been scaled upto the stage of an industrial production process.

SCALE UP OF PLANT CELL CULTURES AND PRODUCT COST

Compared to microbial cells, plant cell cultures are difficult to scale up. The scale up of plant cells poses some problems due to the formation of clumps, slower growth, instability of cultures, shear sensitivity and low oxygen requirement. Different types of reactors have been used for scale up. Generally, the plant cells cultured *in vitro* produce the secondary metabolite, in post-exponential phase of growth. Hence, it is necessary to have two stage culture methods for scale up of cultures. The use of elicitors to enhance productivity of cell cultures is gaining importance. It is envisaged that an effective elicitor may reduce the operation from a two-stage level to a single-stage level thereby reducing the cost of production. It was estimated by Zenk (1978), that any compound costing more than $500/kg would be a viable candidate for production by cell culture method. However, Goldstein et al.,(1980) have calculated that market price of a commercially viable plant cell fermentation process will be as high as $1600/kg. Sahai and Knuth (1985) estimated that an increase of product concentration from 0.1% to 10.0% would reduce the cost of a hypothetical product from almost $6000/kg to $228/kg by using batch fermentation. However, Rosevear and Lambe (1985) believe that by using the immobilized cells in suitable reactors, the cost can be reduced to $20 to 25/kg. Probably, the prohibitive cost of production has resulted in attention being focussed on pharmaceutical products, rather than on metabolites of food value. However, several high value food additives have been identified, and are being examined for large scale production.

IMPROVEMENT IN PROCESSING CHARACTERISTICS

Tissue culture offers a valuable tool to enhance the processing characteristics of fruits, vegetables, and food grains. Tomatoes with a high solid content have been obtained through selection of somaclones, and by the protoplast fusion method (Brunt,1985). Such tomatoes enhance the output in processing for jam, jelly, ketchup etc. According to one estimate, in US alone, tomato processors would save $100 million per year as a result of one per cent increase in solid content (Kunimoto, 1986).

Moreover, by genetic engineering methods, low polygalacturonase containing lines of tomato have been obtained to enhance keeping quality and minimise transportation losses.

The processing quality of wheat depends mainly on the viscosity properties imparted by proper gluten concentration. Gluten consists of gliadin and glutenin. Gliadin fraction attributes viscosity to dough, while the glutenin provides elasticity which prevents over-extension of dough and its collapse during the rising and baking processes. The underlying interactions between high molecular weight subunits of glutenin cross-linked by disulphide bridges impart the above characteristics. It has been proposed that manipulating a portion of high molecular weight glutenin in wheat storage protein, by cross linking and by altering cysteine sites will be useful in producing high quality bread-making wheat varieties (Jones and Lindsey, 1988). Another exciting possibility is improvement of malting quality of barley. This is generally affected by the presence of the hordein fraction in barley protein, particularly β-hordeins and disulphide-linked compounds.

Rothstein et al., (1984) have cloned the wheat α-amylase gene into yeast cells which are able to synthesize and export active proteins. This strategy may find application for other enzymes involved in starch and protein hydrolysis (Jones and Lindsey, 1988). Other exciting possibilities would be alteration of trigylceride composition for the production of coconut type oil in soybean; to develop fruits and vegetables which withstand long distance transportation, with minimal losses. These examples, and the specific improvements already made, have opened up the possibilities of using plant tissue culture methods to improve food quality and to enhance industrial output.

IMPROVEMENT OF FOOD QUALITY

The improvement in food quality can be achieved by increasing the nutritional quality, and by decreasing the antinutritional factors. Increase in deficient amino acids such as lysine in cereals, threonine in wheat and rice, and tryptophan in corn, is possible by tissue culture methods. The general strategy followed for augmenting the deficient amino acid levels has been through screening of mutants by use of analogs of natural amino acids (Hibberd and Green, 1982). The selection of somaclonal variants of sugarcane resulted in increase of sucrose yield by 34% which was mainly due to increased sugarcane yield by 32%.

The elimination of antinutritional factors by screening somaclonal variants will be highly useful (Ravishankar and Venkataraman, 1990) The problem of solving lathyrism in India is a matter of great concern. The toxic amino acid from Lathyrus sativus causes crippling disease in humans. Some attempts have been made at the Indian Agricultural Research Institute, and a few universities, to eliminate toxic amino acid of L. sativus. Teutonico and Knorr (1984) have applied somaclonal variation method to eliminate oxalates from Amaranthus gangeticus. The use of antisense RNA method, to block the expression of undesirable pathway, needs to be adapted for elimination of antinutritional factors.

MODIFICATION OF AGRONOMIC TRAITS

Plant cell culture methods have played a great role in crop improvement, and continue to help breeders and agronomists as a supportive tool. Plants resistant to salt, drought, frost, diseases, and tolerant to herbicides and pesticides have been obtained (Ravishankar and Venkataraman, 1990). Meristem culture has been applied to eliminate viruses from sugarcane, banana, potato etc. Breeding of haploids in China has been responsible for obtaining improved varieties of rice which are presently grown in the eastern provinces of China, on 100,000 ha of land. These examples show that the potential for cell culture application is enormous, and will continue to play a major role in years to come for the improvement of agronomic qualities.

SAFETY CONSIDERATIONS

There is a general tendency to consider the natural food additives to be safer than synthetic ones. This may merely be a psychological phenomenon, but it may even be clinically true in some cases. The debate to grant GRAS (generally regarded as safe) status to biotechnologically derived food ingredients is continuing (Korwek, 1986). The acceptance of GRAS affirmation petition for α- amylase obtained by genetic engineering (FDA, 1986) suggests that atleast certain products can attain GRAS status. The products have to be examined and tested individually, and no blanket acceptance can be possible. It has been suggested that modification of food crops by genes, or vectors, require premarket clearance from Environmental Protection Agencies. Single gene resistance to broad spectrum herbicides is a major target of several genetic engineering programmes for plants. Such resistance might result in crop varieties becoming a weed problem in rotations. Some fear that altered biochemistry of an engineered plant variety might prove toxic to humans or other animals. For example: sorghum bred for bird-resistance (Hauptli, et al., 1985) proved to be rich in tannins which decrease protein utilization in humans and animals. Sorghum is a staple food in South India, and indiscriminate release of such crops can create serious health problems (Hauptli, et al., 1985). It is suggested that new herbicide-resistant or pesticide-tolerant crop plants need special toxicity tests, before release for agricultural purpose. A newly-resistant crop must be tested to ensure that the herbicide, pesticide, or their metabolized derivatives are not present, or are present only at acceptable low levels in plant parts used for food or feed. By far the greatest environmental threat of an engineered variety is, gene transfer from crop plants to weeds via natural sexual transfer e.g. transfer of genes for herbicide tolerance, increased photosynthetic efficiency, etc. Though these problems are real it is not attributable to rDNA methods. It is the responsibility of various organisations such as USDA, NIH (in USA), DBT (in India) to study the advantages and disadvantages of using such improved plant varieties. Suitable amendments in regulations are necessary to allow the cultivation of genetically engineered or tissue culture derived plants and plant products, in order to benefit the consumers, and also for economic gains of manufacturing companies.

EMERGING TRENDS

It is envisaged that the interdisciplinary approach of plant biotechnology will be applied in enhancing the food quality and quantity. However, the techniques of tissue culture and genetic engineering will be supplementary and sometimes complementary to the traditional technologies. The dependence on high quality seeds is increasing, hence, there will be a constant endeavour to improve seed quality, so as to enhance the economic gains for the farming sector. Cell culture research will pave the way for industrial exploitation of high-value phytochemicals of economic importance. The biotechnological application will lay emphasis on improving the processing characteristics of food materials.

REFERENCES

Brunt, J.V. (1985) Non-recombinant approaches to plant breeding. Bio/Technology 3: 975--980.

FDA. 1986. CPC International, Inc., Filing of petition for affirmation of GRAS status. Fed. Reg. 51: 10571.

Goldstein, W.E., Lasure, L.L. and Ingle, M.B. (1980) Product cost analysis. In: Staba,E.J. ed., Plant tissue culture as a source of biochemicals. pp. 191--234. CRC Press, Boca Raton.

Hauptli, H., Newell, N. and Goodman, R.M. (1985) Geneticlly engineered plants: environmental issues. Bio/Technology 3: 437--442.

Hibberd, K.K. and Green, C.E. (1982) Inheritance and expression of lysine plus threonine resistance selected in maize tissue culture. Proc. Natl. Acad. Sci (USA) 79:559--563.

Himeno,H. and Sano. K. (1987) Synthesis of crocin, picrocrocin and safranal by saffron stigma like structures proliferated in vitro. Agric. Biol. Chem. 51: 2395--2400.

Ilker, R. (1987) In vitro pigment production: An alternative to color synthesis. Food Technology 41: 70--72.

Jones M.G.K. and Lindsey, K. (1988) Plant Biotechnology. In: Walker, J.M. and Gingold,E.B. eds. Molecular biology and biotechnology, pp 117--147. Royal Soc. Chemistry, London.

Korwek, E.L. (1986) FDA regulation of food ingredients produced by biotechnology. Food Technology 40: 70--74.

Kudakasseril, G.J. and Staba E.J. (1988) Insecticidal phytochemicals. In: Constable,F. and Vasil, I.K. eds. Cell culture and somatic cell genetics of plants. pp. 537--552.Academic Press Inc. San Diego.

Kunimoto, L. (1986) Commercial opportunities in plant biotechnology for the food industry. Bio/Technology, 4: 58--60.

Lindsey, K. and Yeoman, M.M.(1985) Immobilized plant cell culture system. In:Neumann,K.H. ed., Primary and secondary metabolism of plant cell cultures, pp.304--315. Springer-Verlag, Berlin.

Mizukami, H., Tomita, K., Ohashi, H. and Hiraoka, N. (1988) Anthocyanin production in callus cultures of roselle (Hibiscus sabdariffa L). Plant Cell Reports. 7: 553--556.

Ravishankar, G.A., Rajasekaran, T., Sarma, K.S. and Venkataraman, L.V. (1989) Production of pyrethrins in cultured tissues of pyrethrum (Chrysanthemum cinerariaefolium Vis). Pyrethrum Post 17: 66--79.

Ravishankar, G.A. and Venkataraman, L.V. (1990) Food applications of plant cell cultures. Current Science, 89: In press.

Rosevear, A, and Lambe, C.A. (1985) Immobilized plant cells. In: Fiechter, A. ed. Advances in Biochemical Engineering, V 31, pp.37--58. Springer-Verlag, Berlin.
Rothstein, S.J., Lazarus, C.M., Smith, W.E. Baulcombe, D.C. and Gatenby, A.A. (1984) Secretion of a wheat α-amylase expressed in yeast. Nature 308: 662--665.

Sahai, O. and Knuth, M. (1985) Commercializing plant tissue culture processes:economics, problem and prospect, Biotechnol. Progress 1: 1--9.

Sarma, K.S., Maesato, K., Hara, T. and Sonoda, Y. (1990) In vitro production of stigma like structures from stigma explants of Crocus sativus L. J. Exp.Bot. 41: 745--748.

Sudhakar Johnson, T., Ravishankar, G.A. and Venkataraman, L.V. (1990) In vitro capsaicin production by immobilized cells and placental tissues of Capsicum annuum L. grown in liquid medium. Plant Science 70: 223--229.

Teutonico, R.A. and Knorr, D. (1984) Plant tissue culture: Food applications and the potential reduction of nutritional stress factors. Food Technology 38: 120--127.

Tisserat, P. (1987) Three UK universities commit to biotechnology. Bio/Technology 5:10.

Venkataraman, L.V. and Ravishankar, G.A. (1989) Plant cell cultures for food application. In: Raghavendra Rao, M.R., Chandrashekara, N., Ranganath, K.A. eds.Trends in food science and technology. pp. 126--134. AFST(I) CFTRI, Mysore.

Visvanath, S., Ravishankar, G.A. and Venkataraman, L.V. (1990) Induction of crocin, crocetin, picrocrocin and safranal synthesis in callus cultures of saffron - Crocus sativus L. Biotechnol. Appl. Biochem. 12: 336--340.

Vunsh, R., Matilsky, M.B., Keren, Z.M. and Robenfeld, B. (1986) Production of a natural red colour by carrot cell suspension cultures. In: Somers, D.A., Gengenbach, B.G., Biesboer, D.D., Hackett, W.P. and Green, C.E. eds. VI International Congress of Plant tissue and cell culture, p. 119, Univ. Minnesota, Minneapolis. USA.

Whitaker, J.R. and Evans, D.A. (1987) Plant biotechnology and production of flavour compounds. Food Technology 41: 86--101.

Zenk, M.H. (1978) The impact of plant cell culture on industry. In: Thorpe, T.A. ed. Frontiers of plant tissue culture. pp. 1--13. Univ. of Calgary Press, Canada.

Zenk, M.H. (1990) Plant cell cultures: a potential in food and bio-technology. Food Biotechnology 4: 461--470.

Table 1. Plant tissue and cell cultures for food applications

Pre-harvest

- Improvement in nutritive value
- Reduction and elimination of antinutritional factors
- Disease elimination/resistance
- Tolerance to herbicide, pesticide, drought, salt, frost, metals
- Resistance to insect, pests
- Increased yields
- Mass propagation

Post-harvest

- Production of flavours, colours, sweeteners
- Novel compounds/processes/products
- Easy harvesting
- Fruits and vegetables which withstand long transportation
- Enhanced keeping quality
- Improved and efficient processing

Table 2. Important food additives obtained from plant cell cultures

Food additive	Plant source
Colourant	
Anthocyanin, carotenoids	Daucus carota
Betalains	Beta vulgaris
Crocin	Crocus sativus
Flavour	
Angelica	Angelica sylvestris
Capsicum	Capsicum annuum
Celery	Apium graveolens
Garlic	Allium sativum
Menthol	Mentha piperita
Vanilline	Vanilla planifolia.
Pungent food additive	
Capsaicin	Capsicum annum
Sweeteners	
Stevioside	Stevia rebaudiana
Thaumatin	Thaumatococcus danielli

Tissue culture propagation : Problems and potentials

S.S. GOSAL AND H.S. GREWAL
Plant Biotechnology Centre
Punjab Agricultural University
Ludhiana-141 004, India

Key words : micropropagation, somatic embryogenesis, test tube corms

Introduction

Mass propagation of some ornamentals *in vitro* is one of the best and most successful examples of the commercial application of tissue culture technology. Multimillion dollar industries have been set up around the world to meet the growing demand for quality plants. The technique has great potential for rapid, large scale and true to type multiplication. Moreover, under several situations the plants thus obtained are devoid of pathogens. The rapid propagation *in vitro* is usually achieved through shoot bud/shoot proliferation (Pierik, 1990), induction of bulbs/ corms (Ziv, 1990) and through somatic embryogenesis (Ammirato, 1989). It is likely that improvement of somatic embryogenesis coupled with encapsulation technology and automation of multiplication systems will certainly boost the micropropagation industry.

Materials and methods

The present communication deals with the micropropagation of celery, chrysanthemum and gladiolus. In celery, the *in vitro* cultures were initiated from mature stem segments excised from field-grown plants and explants from aseptically grown seedlings. In chrysanthemum and gladiolus the cultures were initiated from nodal segments excised from field-grown plants. All the culture media were based on Murashige and Skoog (1962) salts. The cultures were incubated at 25±2°C under 16 hours fluorescent light. Micropropagated plantlets/corms have been successfully grown in the soil.

J. Prakash and R. L. M. Pierik (eds.), Horticulture – New Technologies and Applications, 197–200.

Results

Micropropagation through somatic embryogenesis - Celery. Among the 21 media combinations the best callus induction (61.3%) from seedling explants was obtained on MS containing 2,4-D (1.0 mg/l) and kin. (1.0 mg/l). Moreover, the calli subcultured, itself on the induction medium, exhibited very high frequency of somatic embryogenesis (Fig.1A). The maximum shoot regeneration frequency (50% with 100-150 shoots/flask) was achieved on MS medium containing kin. (0.5-1.0 mg/l) (Fig.1B). Profuse in vitro rooting on separated shoots was obtained on MS containing NAA (1.0 mg/l). The complete plants thus obtained have been successfully transferred to pots (Fig. 1C) and grown to maturity in the field.

Micropropagation through shoot bud/shoot proliferation - chrysanthemum. Nodal buds from cultivars Snowdon and Fredyule cultured on MS containing BAP (3.0-5.0 mg/l) exhibited the development of multiple shoot buds after 18 days of culturing. The shoot buds (Fig. 1D) were maintained and multiplied through periodic subculturing on MS medium supplemented with BAP (3.0 mg/l). The shoot bud clump sections regenerated into multiple shoots (Fig.1E) upon subculturing on MS basal medium within 15-20 days. Separated shoots were rooted in vitro on half strength MS containing IBA (2.0 mg/l) as well as in vivo by direct planting in sand after dip treatment with NAA or IBA. Micropropagated plants have been grown to flowering in the field (Fig. 1F).

Micropropagation through induction of in vitro corms/ cormlets - gladiolus. A three step procedure has been developed for mass propagation of gladiolus cvs. Hunting Song, Mayur, Sylvia and, Spic and Span through production of test tube corms/cormlets. The first step involves a rapid shoot bud proliferation (Fig. 1G) from nodal buds cultured on MS containing BAP (5.0 mg/l) and thus increasing the multiplication rate 5 to 6 folds with each sub culture after 4 to 5 weeks. Second step ensures normal shoot elongation from proliferated buds on MS basal medium. Whereas, the third step involves the induction of roots and corms/cormlets (Fig. 1 H,I) at the basal region of separated shoots cultured in liquid half strength MS containing IBA (4.0 mg/l) and sucrose (6%). The size of the mature corms/cormlets with brown tunica ranged from 4 mm to 22 mm in diameter (Fig. 1K). The number of large (16 mm to 22 mm, Fig. 1J) and medium (11 mm-15 mm) size corms/cormlets was more than the small (4 mm-10 mm) ones.

Discussion

The technique of micropropagation has been commercially exploited for mass propagation of some selected plant species *in vitro*. However, high costs of production limit its application in some other species. Therefore, reducing the cost of micropropagation by improving the limiting factors will certainly boost the micropropagation industry. Several factors including age and genotype of the donor plant, time of excision of explant, release of phenolics, degree of disease infection, vitrification and the frequency of rooting have been found to affect the success of micropropagation (Von Arnold, 1988). Micropropagation through somatic embryogenesis has certain specific advantages over the organogenesis (Ammirato, 1989). The single cell origin, non chimeral nature and genetic fidelity of plants derived from somatic embryos are some of the attractive features. Occurence of high frequency somatic embryogenesis have also been observed in some other celery genotypes (Orton, 1984). Thus the phenomenon of high frequency somatic embryogenesis coupled with encapsulation technology can be hopefully exploited for the mass propagation of improved celery genotypes. Several geophytes (corm and bulb) plants are known to produce protocorms *in vitro* (Ziv, 1990). In gladiolus, the development of corms/cormlets *in vitro* ensures better delivery system for mass propagation of this cut flower.

References

Ammirato P.V. (1989) Recent progress in somatic embryogenesis, IAPTC Newsletter 57:2-16.

Murashige T. and Skoog F. (1962) A revised medium for rapid growth and bioassays with tobacco tissue cultures. Physiol. Plantarum 32 : 1-9.

Orton T.J. (1984) Celery. In : Sharp *et al*. Eds., Handbook of plant cell culture, pp. 243-267. MacMillan Publishing Company, New York.

Pierik R.L.M. (1990) Rejuvenation and micropropagation. In: Nijkamp *et al*. Eds., Progress in plant cellular and molecular biology, pp. 91-101. Kluwer Academic Publishers, Dordrecht, The Netherlands.

Von Arnold S (1988) Tissue culture methods for clonal propagation of forest trees. IAPTC Newsletter 56 : 2-12.

Ziv M. (1990) Morphogenesis of gladiolus buds in bioreactors - implication for scaled up propagation of geophytes. In : Nijkamp *et al*. Eds., Progress in plant cellular and molecular biology, pp. 119-124. Kluwer Academic Publishers, Dordrecht, The Netherlands.

Fig.1 A-K. Micropropagation of celery, chrysanthemum and gladiolus. **A.** Germinating somatic embryos of celery, **B.** High frequency plant regeneration from embryogenic celery callus, **C.** Tissue culture-derived celery plants, established in soil, **D.** Shoot bud proliferation in chrysanthemum, **E.** Shoot proliferation from chrysanthemum shoot buds, **F.** Micropropagated chrysanthemum at the stage of flowering, **G.** Shoot bud proliferation in gladiolus, **H-I.** Showing the development of corms/cormlets *in vitro* in gladiolus, **J.** *In vitro* produced gladiolus corms/cormlets showing size upto 22 mm in diameter, **K.** Gladiolus corms/cormlets harvested from test tubes.

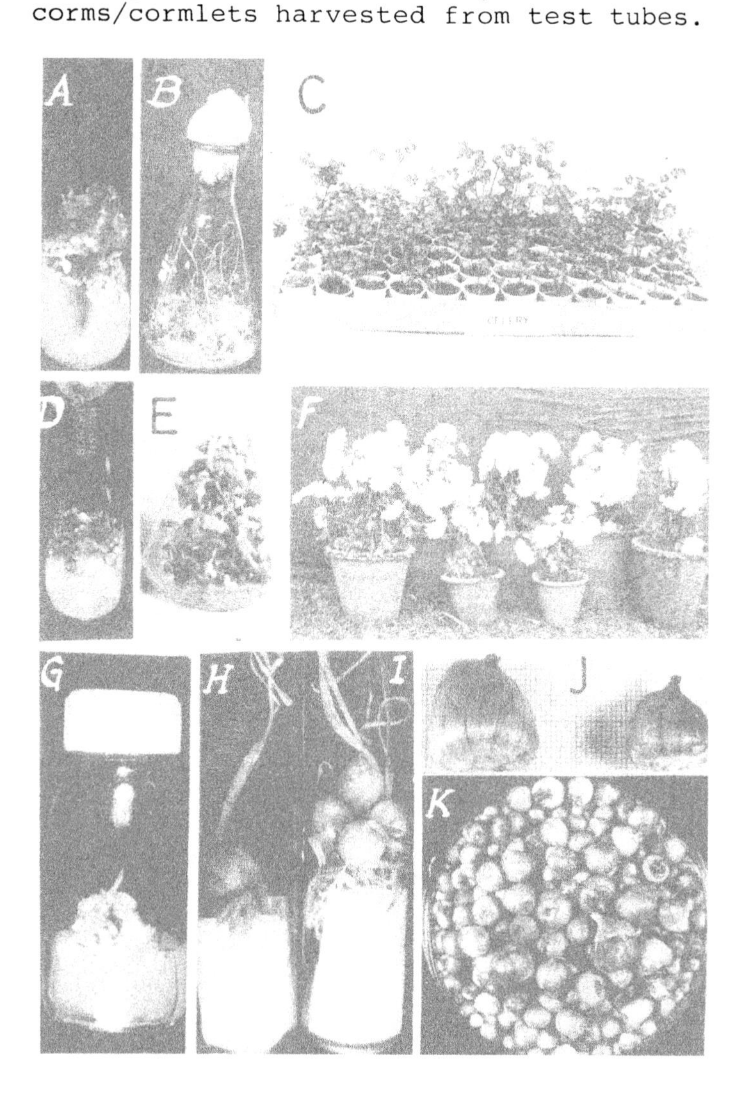

Micropropagation and plant conformity in Anthurium andreanum

FOJA SINGH AND SANGAMA
Indian Institute of Horticultural Research,
Hessaraghatta Bangalore 560 089

Key words Anthurium andreanum, micropropagation, leaf and spadix explants, regeneration, genotype, varieties.

Introduction Anthurium andreanum Lind is gaining importance as one of the most important commercial crop for cut flower production. There is an ever increasing demand for in vitro raised plants. Procedures for A. andreanum leaf explants tissue culture have been developed earlier (Pierik, 1976). Plant regeneration from spadix explants in A. scherzerianum has also been worked out (Geier, 1986). In the present study it was found that Spadix explants have a higher capacity for regeneration as compared to leaf segments, further plants derived from spadix segments were more stable than these derived from leaf explants. It was however not clear whether the variation observed in tissue raised plants were epigenetic or genetic in nature.

Methods explants were taken from healthy cultivars maintained in glass house. Leaf sections and spadix segments were disinfected with 70% alcohol for 10 seconds and sterilized with solution of Sodium hypochlorite (NaOcl) for 10-15 minutes. Sodium hypochlorite was removed from the explants by 4 consecutive rinses in sterile water for 30-45 minutes duration. Culture media were designed by modifying Nitsch 1969 media as follows. Standard callus media (SCM) NH_4NO_3 lowered to 200 mg l^{-1} and supplemented with benzyladenine (BA) 1 mg l^{-1}, and 2-4dichloro-phenoxyacetic acid (2,4D) 0.1 mgl^{-1} and coconut water (CW) 100 ml^{-1}. Standard shoot multiplication medium (SSM) consists of Nitsch medium supplemented with Coconut Water 100 ml^{-1} and BA 1mgl^{-1}. For rooting SSM

J. Prakash and R. L. M. Pierik (eds.), Horticulture – New Technologies and Applications, 201–204.

media without growth regulator was used. The cultures were incubated at 25+ 2° C. Callus induction and multiplication was achieved in continuous darkness while shoot multiplication and rooting was don
e under a 16 h light and 8 h dark period.

Results

Leaf explant Callus was producead on the margins of leaf explants after 50-75 days, direcst sprout formation was never observed. The intensity of callus and the frequency was highest in leaf segments cultured with mid veins. Out of 14 genotypes from which leaf explant were taken only 4 showed positive response indicating that genotype determines callusing and regeneration.

Spadix segments as explant Spadix segments were found to have a much higher capacity for callussing and regeneration as compared to leaf sections, however it was noticed that callus was rarely found on primary cultures but was induced and promoted by repeated dissection and transfer to fresh medium. The time taken for spadix segments to form callus was also much longaer (3-4 months) as compared to leaf sections.

Callus multiplication The callus multiplication was best seen in dark on SCM (solid) medium for a period of six months. Following the first transfer which was carried out after 4-5 months, the callus masses were divided and subcultured after every 3 months. At the first transfer the callus mass formed from one explant was used to establish one subculture however, during later transfers the multiplication rate was increased to 3 subcultures per culture.

Shoot multiplication & rooting The shoots regenerated from callus grown as SCM media were transferred to SSM in light where they turned green and produce new shoots. Further callus formation was totally suppressed, subculturing also resulted in more shoot formation which were both adventitious and axillary in nature. Rooting was observed when shoots were transferred to SRM medium, the first roots appeared within 20-25 days depending upon the extent of shoot and leaf development. After 50-60 days the roots were sufficiently developed to withstand transplantation to leaf manure/soilrite mixture. Plantlets bearing at leeast 3-4 roots gave 90-92% survival and were grown to flowering.

Plant conformity Cytological evaluation of 67 plants derived from leaf explant revealed 57 diploids, 3 tetraploids and 7 aneuploids, morphological variants were characterized by altered leaf shape, enhanced branching and low fertility. Plants derived from spadix segments were more stable, out of 47 plants evaluated 45 were diploid and 2 were aneuploid, morphological variation in these plants was very low and restricted to only leaf shape.

Conclusion It is evident from the results obtained presently that callus and shoot formation from primary explants depend on various factors like media composition especially the concentration of Ammonium nitrate, a low level of NH_4NO_3 (200 mg l^{-1}) proved beneficial for the induction of callus and regeneration, while higher level (720 mg) accelerated root formation. The effect of auxins and cytokinins was also evident in callus induction and shoot formation and in accordance with the earlier reports. B*a 1 mg 2,4-D 0.1 mgl^{-1} were most efficient in promising callus and shoot formation. Photoperiod played an effective role in callus formation and shoot development, darkness enhanced callus formation while light enhanced shoot devaelopment. Morphological variation which were restricted to single organ were more common in plants derived from leaf explant than from spadix segment. This is in accordance with earlier results obtained by Geier in Anthuruim scherzerianum (Geier . 1987). However in general it can be concluded that the extent of variability is clearly less than that usually occurs in seed propagated cultivars.

Acknowledgement The authors are greatful to Director IIHR for facilities, Sri V.R. Srinivasan for helping with preparing th manuscript on Wordperfect software and Mrs Dakshayini G. Prasad for typing the manuscript.

References

Geier T. (1986) Factors affecting plant regeneration from leaf segments of Anthurium scherzerianum Schott (Araceae) cultured in vitro. Plant Cell, Tissue and organ culture 6: 115-125.

Geier T. (1987) Micropropagation of Anthurium scherzerianum: propagation schemes and plant confirmity. Acta Horticulturae 212: 439-443.

Nitsch J.P. (1969) Experimental androgenesis in Nicotiana. Phytomorphology 19: 389-404.

Pierik R.L.M. (1976) Anthurium andreanum plant lets produced from callus tissues cultivated in vitro. Physiologia planterum 37 (1) 80-82.

Rapid in vitro propagation of virus-indexed Freesia

M.J. Foxe,
National Agricultural and Veterinary Biotechnology Centre,
University College, Belfield, Dublin 4, Ireland.

A tissue culture method has been developed to successfully establish a rapid *in vitro* propagation system for virus-indexed freesia (*Freesia refracta*). Using 6 freesia cvs., apical shoot-tips (0.5-0.7 mm) were established (Stage I) on an agar medium conaining Murashige-Skoog salts, 2.0 mg/l kinetin, and 0.1 mg/l naphthylacetic acid. Plantlets were transferred to a similar medium containing 4.0 mg/l kinetin and no auxin 6-7 weeks later for shoot proliferation (Stage II). Plantlets were subcultured on Stage II medium every 4-5 weeks as needed and yielded approximately 12 axillary shoots per culture. Roots developed readily in this medium suggesting that it may not be necessary to transfer the shoots to a specific root-inducing medium. Freesia cvs. were successfully transferred to the field and have produced true-to-type plants during a three year trial. Virus indexing was accomplished using several methods, including electron microscopy and enzyme-linked immunosorbent assay.

Introduction

Commercial freesia (*Freesia refracta*) can be infected by a number of virus diseases leading to significant reduction in yield and quality. Currently freesia is produced from corms or seed grown in greenhouses to prevent virus infection. These methods have disadvantages including a very slow rate of multiplication when corms are used and variation in plant form and flower colour when seed is used.

Tissue culture has the potential for virus-elimination and rapid increase of pathogen-free freesia stock. Freesia has been tissue cultured previously (Davies and Griffiths, 1971; Davies and Helsop, 1972; Hussey and Wyvill 1972; Bajaj and Pierik, 1974; Mori *et al.*, 1978; Pierik and Steegman, 1975; Stimark and Ascher, 1982), but each of the procedures described has certain disadvantages for large scale production including varietal stability.

This paper describes an efficient *in vitro* method suitable for producing virus-free true-to-type freesia plants.

J. Prakash and R. L. M. Pierik (eds.), Horticulture – New Technologies and Applications, 205–208.

Methods

The freesia cvs. Aurora, Ballerina, Blue Heaven, Fantasy, Rose Marie and Royal Blue were used in this study.

Flowering size corms were washed under running tap water, dipped into a solution containing zineb WP for 5 min, and allowed to dry for at least 1 hr before placing them in dark moist chambers for sprouting. Apical shoot-tips (0.5-0.7 mm) were excised from the sprouted corms, and immersed in a 20% chlorox solution containing 0.05 ml Tween-20/l for 20 min. Several culture media were tested for the establishment and rapid shoot proliferation of meristematic explants *in vitro*. These were based on Murashige-Skoog salts (1962) containing .100 mg/l myo-inositol, 0.4 mg/l thiamine, .30 g/l sucrose and varying kinetin and auxin levels. The kinetin levels tested were 1, 2, 3 or 4 mg/l and the auxin levels were 0, 0.1, 0.2, 0.5 and 1.0 mg/l naphthylacetic acid.

Cultures were maintained at 24-26°C at a light intensity of ca. 27 microeinsteins M^{-2} sec^{-1} for 16 hr.

Cultures were maintained for 6-7 weeks on the media before being transferred. Explants were removed, subdivided and transferred to new media in clumps, each containing approximately 3 shoots.

Rooted explants were rinsed under running tap water, dipped into solutions containing zineb WP and benomyl WP and transferred directly to a moistened soil mix consisting of peat, perlite and vermiculite (1:1:1) in 4 cm-diameter plastic pots. Plantlets were maintained under high humidity levels in moist chambers for approximately 7 days under artificial light (63 microeinsteins M^{-2} sec^{-1} for 16 hr) before transfer to the greenhouse.

Virus indexing of shoot-tips for bean yellow mosaic virus (BYMV) and freesia mosaic virus (FMV) was accomplished using both the enzyme-linked immunosorbent assay (ELISA) (Clarke and Adams, 1977) and immunospecific electron microscopy (ISEM) (Derrick, 1973).

Results

All 6 freesia cultivars were successfully established in culture and transferred to soil.

The best establishment of shoot tips was observed on culture medium containing Murashige-Skoog salts, 2.0 mg/l kinetin and 0.1 mg/l NAA. Rapid shoot proliferation was observed when plantlets were transferred 6-7 weeks later to a similar medium containing 4.0 mg/l kinetin and no auxin. Plantlets were subcultured on the shoot proliferation medium every 4-5 weeks and yielded 12 axillary shoots per culture. Roots developed rapidly in this medium indicating that it was not necessary to transfer the shoots to a specific root-inducing medium.

Viable dormant propagating units were produced 3 months after transfer to soil by all freesia cultivars tested in this study. The corms produced were transferred to the field and observed during a 3 year period for varietal stability. In all cases, plants produced from corms exhibited uniform growth and flower development. No abnormalities in flower colour or shape were observed.

Virus indexing
Either BYMV or FMV could be detected in leaves of tissue-culture derived plants growing *in vitro* or in soil using ELISA or ISEM. None of the plants derived from apical shoot-tips (0.5-0.7 mm) of the 6 cultivars was infected by either virus.

Discussion

It has been estimated that a Freesia corm will generate 3-6 corms per year when propagated by conventional means, indicating that the release of a new cultivar could take 8-10 years. The results of the *in vitro* propagation study demonstrate that disease-free freesia plants can be produced rapidly from apical shoot-tip explants. Following initial establishment of the shoot-tips shoot proliferation and rooting can be achieved by transfer to a simple medium containing only kinetin. The capacity for large scale multiplication is evident considering that subculturing can be carried out every 4-5 weeks and that transfer to soil can take place after 6-7 weeks on this medium. This is considerably more rapid than previously described systems (Stimart and Ascher, 1982; Bajaj and Pierik, 1974; Pierick and Steagmans, 1975).

Performance of virus-free tissue culture derived freesia should be superior to that of conventionally grown diseased material. However, the horticultural acceptability of such material must be determined by growing the crop in the field through to flowering. Field studies of virus infected and healthy lines of the 6 cultivars were carried out during a 3 year period, using corms derived from tissue culture. In all cases the tissue culture derived plants were superior in growth and quality to virus infected plants. The tissue culture derived plants were grown to maturity and flowered true-to-type.

The results clearly demonstrate the availability of a rapid *in vitro* method for production of disease-free freesia cultivars.

References

Bajaj Y.P.S. and Pierik R.L.M. (1974) Vegetative propagation of Freesia through callus cultures. Neth. J. Agric. Sci 22: 153-159.

Clarke M.F. and Adams A.N. (1977) Characteristics of the microplate method of enzyme-linked immunosorbent assay for the detection of plant viruses. J. gen. Virol., 34, 475-483.

Davies D.R. and Griffiths P. (1971) *In vitro* propagation of Freesia. A. Rep. John Innes Inst. 62: 45.

Davies D.R. and Helsop P. (1972) *In vitro* propagation of Freesia. A. Rep. John Innes Inst. 63: 64.

Derrick K.S. (1973) Quantitative assay for plant viruses using serologically specific electron microscopy. Virology, 56, 652-653.

Hussey G. and Wyvill C. (1972) Propagation of bulbous species by tissue culture. A Rep. John Innes Inst. 63: 64-66.

Mori Y., Hasegawa A. and Kano K. (1975) Studies on the clonal propagation by meristem culture in Freesia. J. Jpn. Soc. Hortic. Sci. 44: 294-302.

Pierik R.L.M. and Steegmans H.H.M. (1975) Freesia plantlets from flower-buds cultivated *in vitro*. Neth. J. Agric. Sci. 23: 334-337.

Stimart D.R.P. and Ascher P.D. (1982) Plantlet regeneration and stability from callus cultures of *Freesia* x *Hybrida* Bailey cultivar 'Royal'. Scientia Horticulturae 17: 153-157.

Effect of Culture Media and Growth Regulators on In vitro propagation of Rose

N. VIJAYA, G. SATYANARAYANA

There is no single medium to suit every crop. However, the salt mixture of Murashige and Skoog (1967) medium has proved satisfactory for many plants. For some plants, the level of salts in the MS medium is either toxic or unnecessarily high. The requirement of growth regulating substances varies with the system and mode of shoot multiplication. Skirvin and Chu (1979) on In vitro shoot tip proliferation of 'Forever yours' green house rose in high salt MS medium containing 2 mg/l BA and 0.1 mg/l NAA.

METHODS

Collection of explants

Experimental plant material obtained from field grown 'Andhra red' rose. To obtain lateral bud explants, the leaf scale covering the bud was first removed, a shallow incision of 1-2 mm was made into the stem and the bud was then excised with a small portion of adjacent stem tissue.

Unless stated otherwise MS (Murshige and Skoog) salts, 100 mg/l mio-inositol, 1 mg/l thiamine HCL, 1 mg/l pyridoxine Hcl, 1 mg/l nicotinic acid, 4 mg/l clycine, 30 gm/l sucrose and 8 gm/l Difco bacto agar was used for all the experiments. The p^H of media was adjusted to 5.7 ± 0.1 prior to autoclave sterilization at 120°C at 1.5 kg/cm^3 pressure for 20 minutes.

Shoot Multiplication Experiments

1. Shoot multiplication in relation to nutrient medium: Three media MS (Murashige and Skoog), LS (Linsmaier and Skoog) and B_5 (Gomborg et al.) were supplemented with 3% sucrose and 2 mg/l BA were tested for itheir effect on lateral bud growth and shoot proliferation. The shoot multiplication rates were also tested by altering the strength of MS salt concentration to one eighth, one fourth, half, full and double supplementedwith 3% sucrose and 2 mg/l BA.

J. Prakash and R. L. M. Pierik (eds.), Horticulture – New Technologies and Applications, 209–214.

2. In vitro shoot multiplication in relation to plant growth regulators: Three cyokinins, viz., N^6-benzyl adinine (BA), N^6- (2-isopentynyl)-adinine (2-lp), 6-furfuryl aminopurine (kinetin) were used for theireffect on lateral bud growth and shoot proliferation at concentration of 0.5, 1.0, 2.0, 4.0 mg/l. Different concentrations of Indole acetic acid (IAA), Indole butric acid(IBA), Naphthalen acitic acid (NAA), were tested in separate experiments for their effect on lateral bud growth and shoot proliferation. The effect of different concentrations of BA and Kinetin (1.0, 2.0, and 4.0 mg/l) in combination with IAA and NAA (0.0, 0.5 and 2.0 mg/l) were studied on lateral bud growth and shoot proliferation.

3. Multiple shoot formation: To obtain multiple shoots, individual shoots from the initial cultures were separated and transferred to the fresh medium after 5 weeks. This type of subcultures were repeated six times to evaluate the proliferation rate.

Root Initiation Experiments

1. MS salts: The In vitro derived shoots of 'Andhra red' rose were transferred to zero, ¼th, ½, 3/4th and full strength MS salts supplemented with 3 mg/l IBA.

2. Plant growth regulators: The effect of IAA, IBA, NAA and Seradix-B, on root initiation and transplant survival of in vitro shoots of 'Andhra red' rose were determined using ½th strength MS salts supplemented with 3% sucrose without cytokinins. IAA, IBA and NAA were used each at 0.0, 0.5, 1.0, 3.0 mg/l. The auxins were mixed with the medium before autocloving. As regards to seradix-B the in vitro shoots were dipped in the powder before they were transferred to the basal medium.

RESULTS AND DISCUSSION

Among the media(MS, LS and B) tested, MS medium was found suitable for lateral bud proliferation (Table 1). MS medium is generally found most suitable for shoot proliferation because of high salt concentration and the presence of NH_4NO_3. Alterations in MS salt concentration affect the growth of explants adversely (Table 2). Both at lower as well as higher levels of MS salt concentrations stunted growth and early senescence were observed.

It is clear from the resuls that BA was superior in stimulating shoot proliferation from lateral bud than 2-ip and kinetin (Table 3). With the increase in concentration of BA from 0.5 to 2.0 mg/l, there was an increase in percentage of cultures producing multiple shoots. The shoots produced with 2-ip treatment were small and pale green. But with kinetin, none of the cultures produced multiple shoots. The single shoots so produced were vigorous in growth and leaves were large in size with kinetin treatment compared to those produced in the media containing BA or 2-ip. Similarly, 2-ip and kinetin were reported to be ineffective in stimulating multiple shoot formation from freshly excised lateral buds of 'Improved Blaze' rose (Hasegawa, 1980).

Due to different auxin treatments, there were significant differences in the number of shoots produced per culture. Out of three auxins used, NAA was found to be more effective than IAA and IBA in the production of multiple shoots.

Having established that BA in isolation was superior to other cytokinins it was thought necessary to try BA in combination with IAA and NAA. The data revealed that BA alone increased percentage of cultures with multiple shoots and number of shoots per culture. An increase in shoot length was observed due to addition of IAA to the medium. At higher concentration of IAA in combination with BA friable callus was formed at the base of the shoot. It was observed that none of the cultures produced multiple shoots either with kinetin alone or in combination with IAA or NAA. With increased concentrations of auxins in combination with kinetin showed a tendency to produce callus in large number of cultures rather than inducing shoot formation. There was also direct rooting from lateral bud explants in some of the culture due to auxin treatments in combination with kinetin.

Root Initiation and Root Development

One-fourth strength MS medium was more effective for root initiation of in vitro shoots than the full strength of MS salts (Table 4). A reduction in nitrogen content in the modified medium may be predominent reason for the improved root initiation in in vitro derived shoots. According to Hyndman et al. (1982), alow nitrogen content in the medium enhance rooting. Although rooting could be induced at low concentration of MS salts, further shoot growth was not satisfactory. This may be due to less availability of salts in the rooting medium.When these rooted shoots were transferred to full strength MS medium just after the emergence of first roots the plantlets regained growth.

Table 1: Effect of nutrient media on bud break, shoot multiplication and shoot growth in Andhra red rose

Medium	Cultures with bud break after four weeks (%)	Cultures with multiple shoots (%)	Average number of shoots/ culture	Average length of shoot (cm)
MS	95.0	67.5	2.13	2.1
LS	57.5	35.0	1.00	1.8
B_5	45.0	20.0	0.63	0.5
S.E.	-	-	0.063	
CD at 0.05	-	-	0.175	

Table 2: Effect of MS salt concentration on shoot multiplication and average number of shoots in 'Andhra red' rose

MS salt conc.	Cultures with multiple shoots (%)	Average number of shoots per culture	Length of shoot(cm)
1/8 strength	30	1.15	0.3
1/4 strength	55	1.55	0.5
1/2 strength	65	1.90	1.0
Full sgrength	90	2.35	1.9
Double strength	45	1.55	2.1
S.E.		0.069	
C.D. at 0.05		0.194	

Table 3: Effect of cytokinins on shoot proliferation from lateral buds in Andhra red rose

Treatments (Conc.mg/l)	Cultures with multiple shots	Average number of shoots per culture	Average length of shoot (cm)
BA			
0.5	46.7	1.53	1.5
1.0	68.8	2.4	1.7
2.0	79.9	2.4	1.8
4.0	57.2	1.8	1.8
2iP			
0.5	33.3	1.4	0.5
1.0	46.6	1.8	0.5
2.0	33.3	1.5	0.3
4.0	13.3	1.0	0.3
Kinitin			
0.5	0.0	1.0	2.0
1.0	0.0	1.0	2.3
2.0	0.0	0.73	1.5
4.0	0.0	0.67	0.5
S.E.	-	0.065	
C.D at 0.05			
Cytokinins		0.0897	
Concentrations		0.104	
Interaction between Cytokinins and concs.		0.18	

Table 4: Effect of MS salt concentration on root initiation and number of shoots of cultured 'Andhra red' rose shoots

MS salt conc.	Cultures with roots (%)	Average number of roots per culture
'0' strength	0.0	0.0
1/4 strength	73.3	5.0
1/2 strength	46.7	4.7
3/4 strength	33.3	4.9
Full strength	20.0	5.0
S.E.		0.180
C.D. at 0.05		0.502

Table 5: Effects of IAA, IBA, NAA and Seradix-B on *in vitro* root initiation and transplant survival of 'Andhra red' rose shoots

Treatments (Conc. mg/l)		Cultures with roots (%)	Average number of roots per culture	Cultures successfully transferred to soil (%)
Control	0.0	13.3	3.00	0.0
IAA	0.5	13.3	3.50	50.0
	1.0	20.0	3.00	33.3
	3.0	0.0	0.00	0.0
IBA	0.5	26.7	5.25	50.0
	1.0	53.3	5.13	62.5
	3.0	66.7	4.80	70.0
NAA	0.5	26.5	3.75	25.0
	1.0	33.3	4.00	40.0
	3.0	53.3	5.00	50.0
Seradix-B		80.0	8.00	100.0
S.E.			0.130	
C.D. at 0.05				
Auxins and concentrations			0.205	
Interactions			0.354	

Successful rooting has been obtained on ¼th strength MS medium supplemented with 3.0 mg/l IBA or dipping the *in vitro* shoots in seradix-B and later culturing them on ¼th strength MS medium (Table 5). It was also found that the rooting could be induced only after the cultures had undergone 2 or 3 subcultures. Rooting could also be induced by addition of NAA (1.0-3.0 mg/l) or IAA (0.5-1.0 mg/l), but by this treatment the percentage of success was less and also the growth of plantlets was not satisfactory. *In vitro* flower formation was observed in some cultures in a medium containing NAA. This is due to the fact that NAA might have released the ethylene a natural growth regulator, from the shoots and have induced flowering.

The *in vitro* derived plantlets were transferred to pots containing sterilized vermiculite and subsequently transferred to the field for further observations. The growth and flowering of *in vitro* derived plantlets of 'Andhra red' rose was compared with the plants raised from cuttings. There was more branching in *in vitro* derived plantlets than in theplants grown from cuttings. However, the flower characters of *in vitro* plants were almost similar to plants raised from cuttings.

REFERENCES

Gamborg, Miller O.L.R.A. and Ojima K. (1968) Nutrient requirements of suspension cultures of Soybean root cells. Exp. Cell Res 50:151--158.

Hasegawa (1980) Factors affecting shoot and root initiation from cultured rose shoot tips. J.Am.Soc.Hort.Sci. 105: 216--220.

Hyndman S.E., Hasegawa P.M. and Bresson R.A. (1982) Stimulation of root initiation from cultured rose shoots through the use of reduced concentration of mineral salts. Hort. Science 17: 82--83.

Jacobs G., Allon P. and Bornman C.H. (1970) Tissue culture studies on rose: use of shoot tip explants II cytokinin: Gidberllin effects. Agroplantae 2: 25--28.

Leopold A.C. and Plumber T.H. (1961) Auxin - phenol complexes. Pl. Phy. 36: 589--592.

Linsmaier M.E. and Skoong F. (1965) A revised medium for rapid growth and bioassays with tobacco tissue cultures. Physiol. Plant. 15: 473--497.

Skirvin R.M. and Chu M.C. (1979) *In vitro* propagation of 'Forever Yours' rose. Hort Science 14: 608--610.

SHOOT TIP CULTURE METHOD FOR RAPID CLONAL PROPAGATION OF POMEGRANATE (*Punica granatum* L.)

D.M.MAHISHI, A.MURALIKRISHNA, G.SHIVASHANKAR and R.S. KULKARNI
Department of Plant Genetics and Breeding, University of Agricultural Sciences, GKVK, Bangalorre 560 065, INDIA.

Key words: Explant, Hormone, Polyphenol, Rooting Micropropagation

INTRODUCTION

Pomegranate is an important fruit crop and a native of Middle-East. Over years, it has spread to many geographical regions viz., Mediterranean, Southern Europe, Asia, Northern Africa and USA. Most of the present day elite genotypes of pomegranate are natural hybrids which would not breed true by seed propagation. Besides, conventional vegetative propagation methods are also relatively difficult and commercially less feasible. Hence, it would be desirable to develop a rapid and efficient method of micropropagation such as shoot tip culture.

METHODS

The shoot tips were excised from fresh sprouts of 3 years old pomegranate selection (HS-4) maintained under green house conditions. The explants were surface sterilized with 0.1% $HgCl_2$ for 15 minutes followed by 20% (v/v) NaOCl for 20 minutes. Surface sterilized explants were inoculated initially on Murashige and Skoog's (1962) basal salts medium (MSB) supplemented with 8 mg BAP and 1 mg NAA per litre. The media was also fortified with 600 mg/l of PVP-360 to reduce oxidation of polyphenols in cultures.

The cultures were incubated at 23±2°C with 16 hr light, 260 $umol.s^{-1}.m^{-2}$ under white fluorescent illuminiation. The axillary buds were activated and a growth of 20-30 mm was observed in 3 weeks. Subsequent subcultures of shootlets were carried out on Woody Plant Medium (Lloyd and McCrown, 1980) containing 6 mg/l IAA, 0.6 mg/l Kinetin and 440 mg/l $CaCl_2.2H_2O$ for rapid growth and elongation.

Shootlets measuring 30-40 mm in length were separated and transferred to different rooting media having half or full strength of MS basal salts at varying concentrations of BAP and NAA (Table-1). The rooting media was fortified with activated charcoal at the rate

J. Prakash and R. L. M. Pierik (eds.), Horticulture – New Technologies and Applications, 215–217.

of 3 g/l. Profuse rooting was observed by 4 weeks. The plantlets were further transferred to Peat, Perlite and Sand (1:1:1 v/v) potting mixture.

RESULTS

Growth and age of the explant and the method of surface sterilization were crucial in controlling polyphenol exudation and their initial establishment in culture. Explants from fresh sprouts produced less polyphenols compared to those from mature shoots. Addition of PVP-360 to the media and shifting of explants to fresh media 4 to 6 hours after initial inoculations were effective in reducing the adverse effects of oxidised polyphenols. The composition of initial culture media and subculture media were most effective in activation and elongation of axillary buds. Among the various rooting media tried, MSB supplemented with NAA and with or without BAP was very effective in inducing profuse rooting in 4 weeks. There was 80 per cent success in establishment of the plantlets in peat, perlite, sand (1:1:1 v/v) potting mixture under green house conditions.

Table-1: Successful combinations and concentrations of MSB and hormones that induce rooting of pomegranate shootlets grown *in vitro*

Media	BAP(mg/l)	NAA(mg/l)	Activated charcoal(g/l)	Rooting*
½MSB	5	0.5	3	++
MSB	5	0.5	3	++
MSB	-	0.1	3	+++
MSB	-	3.0	3	++++

* Each + indicates the extent of rooting induced

DISCUSSION

The use of tissue culture technique for the micropropagation of pomegranate encounters a major problem due to oxidative browning of the wounded explants (shoot tips). The brown exudate that diffuses into the media adversely affects the explant, which turn necrotic and fail to establish. Addition of 600 mg/l PVP-360 and shifting of explants onto fresh media within 4 to 6 hr after initial inoculation was found to be very effective in reducing the necrosis of the cultured tissue. Further, subculturing of the shootlets on modified WPM with 440 mg/l of $CaCl_2.2H_2O$ not only resulted in rapid elongation of the axillary buds, but also accounted for better root initiation in subsequent cultures. Amongst the rooting media tried, with full or half strength MSB supplemented with NAA (0.1 - 3 mg/l) was most effective in inducing profuse rooting with good vascular connections. The rate of multiplication of pomegranate through shoot tip culture method was several times more efficient when compared to the conventional method of airlayering.

ACKNOWLEDGEMENT

Financial assistance provided by The Karnataka State Council for Science and Technology (DR/KSCST-50/1987-88), Bangalore, India, is gratefully acknowledged.

REFERENCES

Lloyd, G. and McCrown, B., 1980 Commercially feasible micropropagation of mountain laurel (Kalmia latrifolia) by use of shoot tip culture. **Proc. Inter. Plant Prop.Soc. 30:421-427.**

Murashige,T. and Skoog, F., 1962 A revised medium for rapid growth and bioassays with tobacco tissue culture. **Phys. Plant. 15:473-497.**

Tissue Culture Strategies for Banana

R.DORE SWAMY AND LEELA SAHIJRAM
Tissue Culture Laboratory, Division of Plant Physiology and Biochemistry, Indian Institute of Horticultural Research, Hessaraghatta, Bangalore 560089, India

Keywords: banana, micropropagation, embryo rescue, shoot tip culture, floral apex culture

Introduction

The importance of banana as an international horticultural commodity needs no emphasis. Being a monocotyledonous crop it was thought to be intractable to in vitro techniques. Mohan Ram and Steward in 1964 demonstrated the possibility of raising callus cultures from the fruit tissue of various genomes. In recent years there has been a spurt of activity in developing tissue culture protocols for this crop (Cronauer and Krikorian, 1986). However, most of the protocols have been addressed to developing micropropagation methods which have now been exploited commercially. This paper describes results of our efforts in developing a multipronged tissue culture based biotechnology for amelioration of this important fruit crop.

Methods

Protocols were worked out to devise suitable strategies for: 1) micropropagation from shoot tips and floral tips, and 2) embryo capture and rescue. Material (suckers, male floral buds and hybrid embryos) were obtained from the gene pool collections and field plantings at IIHR. Customary methods of aseptic procedures were adopted for raising cultures (Dore Swamy, et al., 1983, Dore Swamy and Leela Sahijram, 1989, George and Sherrington 1984). Murashige and Skoog's medium (Murashige and Skoog, 1962) modified by omitting

J. Prakash and R. L. M. Pierik (eds.), Horticulture – New Technologies and Applications, 219–223.

$CuSO_4$ and substituting thiamine in place of the whole range of vitamins was used as the basal medium (BMS). The BMS was fortified with various concentrations and combinations of cytokinins and auxins as and when required. The medium was gelled with Gelrite (gellan gum) at a concentration of 0.25%. Potting of in vitro regenerated plants was carried out with sterilized potting mixture in plastic minipots or polyethylene bags.

Results

Micropropagation

Regeneration from shoot tips: Rhizome tips of cultivars Robusta and Dwarf cavendish (AAA) containing the apical and lateral meristems were found to be the best explant sources. BMS fortified with adenine sulphate (205 µM), benzyladenine (22.2 µM) and indole butyric acid (29.6 µM) was found to be suitable for proliferation of multiple shoots. On an average 6-10 shootlets issued from a single culture in about 8 weeks. A steady state in the production of new shootlets has been achieved by serial subculture. Shootlets excised from these cultures rooted on BMS containing indole butyric acid (4.9 uM). From one rhizome tip, 1500 to 2500 plantlets can be produced in one year time frame. A large number of plantlets thus obtained have been established in pots and subsequently transplanted to soil. Tissue culture derived plants flowered in 10 months upon transfer to field. No variants have been observed so far.

Regeneration of vegetative shoots from floral apices: Floral apices contained in the male bud of banana have been reported to be morphogenetically plastic. Hence, the possibility of deprogramming the seemingly determined floral apices and subtending flanks were explored in the cultivars Robusta (AAA) Rasthali (AAB) and Chandrabale (AAA). The male floral apices reverted to the vegetative state upon culture and produced a mass of green leafy shoots which have been kept in a state of active growth by serial subculture. Male flower primordia located on the peduncle distal to the apex also behaved similarly. Fully differentiated plantlets regenerated from both the explants have been established in soil.

Embryo rescue and culture

Attempts were made to culture excised hybrid embryos from a cross between Musa acuminata and M. balbisiana. A very high percentage of excised embryos cultured on BMS developed into plantlets which have been subsequently potted. Some of the excised embryos proliferated and produced multiple shoots when cultured on BMS containing adenine sulphate (205 µM), benzyl adenine(22.2 µM) and indole butyric acid (29.6 µM). Successive subculture of these has yielded several single embryo clones which can be further exploited for genetic manipulation of the crop.

Discussion

Morphogenetic potentialities of terminal and axillary meristems of banana are well documented (Cronauer and Krikorian,1986; Dore Swamy et al.,1983; Vuylsteke and De Langhe,1985). Micropropagation of banana is now being commercially exploited all over the world. One of the common criticisms in large scale rapid micropropagation is the production of phenotypic variants (Vuylsteke et al.,1988). Shoot proliferation in cultures of banana is a resultant of invigoration of the axillaries as well as de novo production of adventitious buds (Bannerjee et al.,1986). The production of variants may be attributed to the latter phenomenon.

Apical meristem culture is one of the powerful methods for recovering healthy plants from virus infected ones (Murashige,1974). However, in banana excision and culture of the apical meristem per se is fraught with difficulties. This is due to the peculiar anatomical structure of the meristem (Simmonds,1986). Floral apices located in the male bud of banana can be an alternative. Plantlets could be regenerated from excised floral meristem (Cronauer and Krikorian,1985) as well as the subtending peduncle (Dore Swamy and Leela Sahijram,1989). This method offers several advantages like ease of excision and relatively microbe-free explants.

Hybrid embryo culture is an important adjunct to banana breeding (Menendez and Shepherd,1975). However, no reports are available on microcloning protocols through multiple shoot development from hybrid embryos of banana. Results reported here clearly demonstrate that it is possible not only to rescue hybrid embryos but also to induce single embryo-derived clones through proliferating shoot clusters. This is of immense significance to the breeder of a triploid, sterile crop like the banana since large clonal populations of hybrid plantlets can be raised using this technique.

Thus, it has been shown that a multipronged strategy through tissue culture based biotechnology can help ameliorate an otherwise intractable crop like banana.

Acknowledgements

We thank Porfessor A.D. Krikorian of the State University of New York at Stonybrook,USA, for his valuable suggestions while he was here as a UNDP/FAO consultant in 1986. We thank Mr V.R. Srinivasan for having composed the manuscript on Wordperfect software. Technical assistance rendered by Mr B. Sudarshan is also acknowledged. Facilities provided by the I.I.H.R are also acknowledged.

References

Bannerjee N. Vuylsteke D. and De Langhe E.A. (1986). Meristem tip culture of Musa: histomorphological studies of shoot bud proliferation. In: Bajaj Y.P.S. ed. Biotechnology in forestry and agriculture. Vol.1:233-252 Springer-Verlag Berlin F.R.G.

Cronauer S.S. and Krikorian A.D. (1985). Reinitiation of vegetative growth from aseptically cultured terminal floral apex of banana. Amer.J.Bot. 72:1598-1601.

Dore Swamy and R. Leela Sahijram (1989). Micropropagation of banana from male floral apices cultured in vitro. Scientia Hortic.40:181-188.

Dore Swamy R. Srinivasa Rao N.K. and Chacko E.K. (1983). Tissue culture propagation of banana. Scientia Hortic.18:247-252.

George E.F. and Sherrington P.D. (1984). Plant propagation by tissue culture.Exegetics, Basingstoke, England.

Menendez T. and Shepherd K. (1984). Breeding new bananas. World Crops 27:104-112.

Mohan Ram H.Y. and Steward F.C. (1964). The induction of growth in explanted tissue of the banana fruit. Can.J.Bot.42:1569-1579.

Murashige T. (1974). Plant propagation through tissue cultures. Ann.Rev.Plant Physiol.25:135-166.

Murashige T. and Skoog F. (1962). A revised medium for rapid growth and bioassays with tobacco tissue cultures. Physiol.Plant.15:473-497.

Simmonds N.W. (1986). Bananas. Longman, London, England. 512pp.

Vuylsteke D. and De Langhe E.A. (1985). Feasibility of in vitro propagation of bananas and plantains. Trop.Agric.(Trinidad),62:323-328.

Vuylsteke D.Swennen R.Wilson G.F. and De Langhe E.A. (1988). Phenotypic variation among in vitro propagated plantain (Musa sp.cv.AAB). Scientia Hortic.36:70-88.

Gynogenic plants from ovary cultures of Mulberry (*Morus indica*)

G. LAKSHMI SITA AND S. RAVINDRAN
Plant Tissue Culture Laboratory
Department of Microbiology & cell Biology
Indian Institute of science
Bangalore, India

Key words : *Morus indica*, haploid, gynogenic plant, ovary culture, plant imporvement.

Introduction

In vitro induction of haploids now occupies an important place among the tecqniques available to the plant breeder. The importance of haploids for basic and applied research has been well documented by San and Gelebart (1986) and Chen (1987). The wide-spread use of reproductive tissues largely arose from the work of Guha and Maheshwari in 1964. Culture of ovules and ovaries remained unsuccessful during 60's. First successful gynogenic haploids were obtained from barley by San in 1976. Since then ovule and ovary cultures have been used for certain species as an alternate to anther culture for obtaining haploid plants. this approach had greater potential and normal plants were obtained in barley while 99% albinism was found in plants obtained by anther cultures. Induction of haploids from egg or other embryo sac cells in cultured unfertilized ovules has recently received increased experimental attention and haploid plants have been produced from unpollinated ovules or ovaries. Induction of hapliids and homozygous diploids in woody plants are much more valuable than in natural or artificial herbaceous plants. Reports of spontaneous occurence of haploidy is rare in angiosperm forest trees. It is impractical to obtain homozygus diploid lines by successive generations of inbreeding and because in most woody plant seeds cannot be obtained by self pollination. Also heterosis is much higher in good combination of pure lines than in crosses between two different varieties.

J. Prakash and R. L. M. Pierik (eds.), Horticulture – New Technologies and Applications, 225–229.

However, because of the long life cycle of trees, it is impossible to obtain pure lines by means of successive generation of inbreeding . It is not always easy to produce haploids by androgenesis due to the small size of the anthers. Although haploid have been produced in some economically important trees (by androgenis) like poplar, rubber and tea, so far no haploids have been produced by gynogenis.

Mulberry (Morus sps.) is important in silk industry as the chief feed of the silkworm (Bombax mori). The conventional method of propagation of mulberry is by stem cuttings. Perennial nature of a plant coupled with prolonged juvenile period limits the speed of improvement in this crop. Further dioecious nature of the important species in the genus is a barrier to genetic improvement by conventional hybridization techniques. Moreover in vegetatively propagated plants like mulberry it takes many years to evolve a desirable clone from economic and commerical point of view by conventional hybridization methods. In vitro techniques of obtaining haploids offer new openings for genetic manipulations. Haploid induction by anther culture was found to be difficult, hence with a view to induce gynogenic haploids ovary/gynoecium culture was taken up and experiments were conducted.

Materials and Method

Flowers of mulberry are unisexual and are found on separate plants. The female flower consists of four perianth lobes and pistil consisting of ovary, style and forked stigma. In the ovary there is a single ovule. Multiple or composite fruits are formed out of number of free ovaries of the whole inflorescence. The rachis, bracts and the fleshy perianth fuse and form the edible parts. Each ploygonal area represents a flower. This is called sorosis. Material was collected from female trees (cultivated variety) growing on the campus . Individual ovaries before and after fusion to form sorosis were cultured on Murashige and Skoog's (1962) medium supplemented with growth hormones. Standard aseptic techniques were used to establish the cultures.

Results and Discussion

Explants as described above were cultured on MS medium supplemented with auxins and cytokinin either singly or in combination. A total number of 65 plantlets were obtained. Plantlets regenerated directly without any callus phase. In some media like 1mg/L BAP and 1 mg/L Kinetin four plantlets were obtained from a single ovary. Usually within 3 weeks pantlets emerged. Fig.1 a,b,c,d show different stages of the emerging plantlets with well developed cotyledons and hypocotyl. Plantlets are complete with a well developed tap root system. Plantlet growth was found to be slow. Initially when transferred to media containing GA (1mg/L) and Kinetin (0.3mg/L) growth was found to be normal. After 15 days plantlets were subcultured to half strength MS salts supplemented with 0.5 -1.0 mg/L IBA. Profuse rooting was observed on transfer to half strength MS media. When plantlets are 5-7 cm having 3 - 5 nodes, nodal cuttings were made and these in turn were cultured on MS media containing 0.1 mg/L BAP and 0.2 mg/L kinetin. On these media well developed plants were formed. Ocassionally 3 - 4 multiple shoots were obtained on media containing cytokinins. These in turn can either be rooted or used as source material to get nodal explants. Cytological preparations were made from the root tips. Some of them showed the haploid number of chromosomes (n=14) while some were found to be diploid. It was difficult to get very good cytological preparations. However haploid number can be made out clearly. On media supplemented with 3 mg/L BAP + 1 mg/L kinetin + 0.5 mg/L NAA plantlets regenerated consistantly. On media supplemented with 1 mg/L BAP + mg/L kinetin alone four embryos were seen emerging. However these on subculture got contaminated. Plantlets obtained directly from ovaries, were subsequently multiplied by micropropogation using nodal cuttings. Plantlets obtained by the above described method are probably of haploid origin since the ovaries are unpollinated and also there are no male trees in the viscinity. Also no seeds are seen when the fruit is pressed with the fingers. These factors indicate the possibility of the plants arising from the female gametophyte. However, histological sections have to be taken to correctly identify the origin of these plantlets. The developmental stage of the embryo sac and definite physiological state are difficult to estimate, because the cytological techniques required for the good observations are ill defined. So far gynogenic plants have been reported in only a few cereals (San and Gelebart, 1986). However no gynogenic plants were reported from tree species. Gynogenic plantlets are known to regenerate directly from an embryo or embryoid and rarely from callus. The reason for obtaining more than one plantlet from a single ovary in the present study is due to their development from haploid cells of the embryo sac. Some of the plantlets obtained also are showing diploid number of chromosomes (2n=28). This could be due to diploidization during embryo development. Thus in lettuce and sunflower plantlets obtained wre diploid in nature (San and Gelebart 1986). During gynogenesis the rate of spontaneous diploids seem to be higher in many plants (San and Gelebart 1986) than that in androgenesis. Haploid mitosis were observed

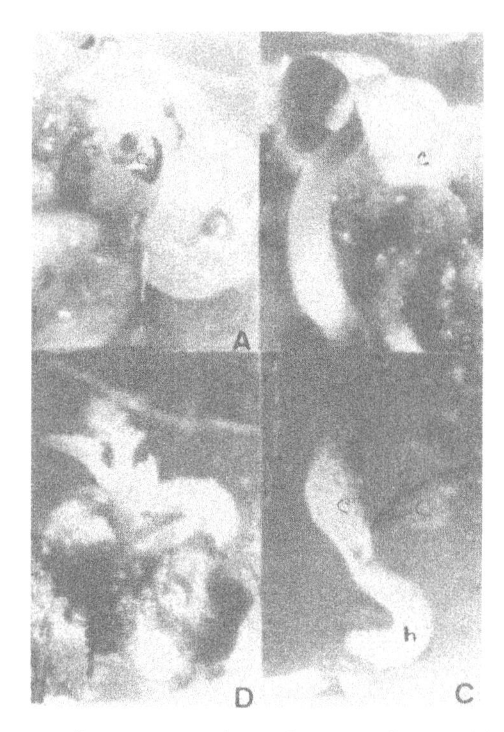

Fig.1: Induction of gynogenic plants in Mulberry. A B C D showing different stages of emerging gynogenic plants.

only in the first root tip of the plantlets. After transfer all root tips analyzed are diploid. In mulberry also similar observations were made by us. Gynogenesis in gymnosperms has been reported successfully in many species (Rohr 1987). Although the results have shown that regeneration by gynogenesis is potentially possible apogamous plants have remained small and did not survive after transfer from the in vitro environment to soil due to lack of proper vascular connections between shoot and root. On the other hand plantlets obtained in mulberry from ovary culture are complete plantlets with well developed shoot and root system which subsequently developed excellent lateral root system and established in pots.

So far experimental induction of haploid trees has been more successful by androgenesis rather than gynogenesis. The present report of gynogenesis plants is the firs report from an angiosperm tree. Results presented here suggest that plantlets directly regenerated are haploids or double diploids. Either way this is a novel way of getting gynogenic haploid plants. Advances in genetics have been much slower in tree species than in annual herbaceous species, mainly because of the lack of favaourable material for study. The present study opens up new alternatives for the productions of haploids.

References

1. Chen, Z. (1987) Induction of androgenesis in hardwood trees. In : Bonga, J.M. and Durzan, D.J. (ed) : Cell and tissue culture in forestry Vol. 2 pp 247-268. Martinus Nijhoff Publishers, Dordrecht.

2. Guha, S. and Maheshwari, S.C. (1964). In vitro production of embryos from anthers of Datura. Nature (London) 202, 497.

3. Murashige, T. and Skoog, F. (1962). A revised medium for rapid growth and bio-assay with tobacco tissue culture. Phyiol. Plant 15 : 493-497.

4. Rohr, R. (1987). Haploids (Gymnosperms) In : Bonga, J.M. and Durjan, D.J. (ed). Cell and tissue culture in forestry Vol. 2. pp : 230-246. Nijhoff Publishers, Dordrecht.

5. San, L.H. (1976). Haploids d'Hordeum vulgare L. per culture in vitro d' ovaries non fecondes. Ann. Antetor. Plant 26 : 751-754.

6. San, L.H. and Gelebart, P. (1986). Production of gynogenic haploids. In : Vasil, J.K. (ed). Cell culture and somatic cell genetics of plants. Vol. 3. pp : 305-322. Academic Press Inc. Orlando, Florida.

Induction of high frequency somatic embryogenesis and plant regeneration in Mandarins

M.I.S. GILL, B.S. DHILLON, ZORA SINGH AND S.S. GOSAL
Department of Horticulture and
Plant Biotechnology Centre,
Punjab Agricultural University,
Ludhiana-141 004, India

Mandarins constitute an important group among the citrus fruits. Kinnow (*C. nobilis* Lour x *C. deliciosa* Tenora) is now extensively grown commercially in North Indian plains, owing to its high yield and relative resistance to insect-pests and diseases. However, it has a tendency towards higher seediness, bitterness in juice, thick and tight rind and late maturity. Likewise, in *C. reticulata* Blanco cv. Local Sangtra, poor productivity, high incidence of sunburn and granulation warrant the incremental improvement in these two mandarins, via tissue culture. The paucity of the information, on the induction of high frequency somatic embryogenesis and plant regeneration in these cultivars necessitates this investigation.

Methods

In the present investigation, two citrus genotypes viz. Kinnow (*Citrus nobilis* Lour x *Citrus deliciosa* Tenora) and Local Sangtra (*Citrus reticulata* Blanco) were included. The seeds of these two genotypes were extracted from mature fruits after thorough washing with sterile water containing teepol. The seeds, thus obtained, were aseptically germinated on basal MS medium. Leaf, epicotyl, cotyledon and root segments excised from 3 weeks old seedlings served as an inocula. The different media used for induction of callus, somatic embryogenesis and plant regeneration were based on basal MS and modified Murashige and Tukcer (1969) salts, and various combinations and concentrations of NAA, 2,4-D, kin and BAP. The pH of the media was adjusted to 5.7 by using 1 N HCl or KoH. The cultures were incubated at 25±2°C, under 14 hr fluorescent light. The callus thus obtained were maintained and multiplied through subculturing after every 4 week intervals.

J. Prakash and R. L. M. Pierik (eds.), Horticulture – New Technologies and Applications, 231–235.

Results

The cultured seedling, explants, exhibited callus proliferation from the cut ends, within two weeks of incubation, in both the citrus genotypes. The callus induction frequency varied with the nature of the explant, genotype of the donor and the composition of media used. In Kinnow, the maximum callus induction (86.6%) was observed from the cotyledon segments, cultured on Modified Murashige and Tucker (1969) medium, whereas, in Local Sangtra, the maximum callus induction (90%) occurred from the epicotyl segments cultured on MS containing NAA (10 mg/l) and Kinetin (0.5 mg/l). The epicotyl derived callus cultures of Kinnow showed maximum somatic embryogenesis (40%) on modified Murashige and Tucker (1969) medium. However, in Local Sangtra, the maximum embryogenesis (86.6%) was obtained from epicotyl callus, subcultured on MS containing NAA (10 mg/l), Kinetin (1 mg/l) and vitamins (10 x). The half strength MS medium supplemented with BAP (5 mg/l) served as the best medium for plant regeneration from epicotyl derived callus in both the mandarins. However, the average number of shoots/culture was high (10-12) in Local Sangtra, as compared to Kinnow (4-6).

Discussion

Since the nucellar tissue, cultured *in vitro* produces true to type plants, it has, therefore, been preferred as an inocula, over the seedling explants. However, the use of nucellar tissue as an explant is tedious and has the disadvantages of seasonal availability, its association with juvenile characteristics (Chaturvedi and Mitra, 1974). Therefore, the present investigation was initiated using seedling explants viz. leaf, epicotyl, cotyledon and roots as an initial inocula. The inclusion of 2,4-D (2 mg/l) along with NAA (2 mg/l) showed a synergistic effect on callus induction, as compared to separate NAA or 2,4-D. The use of NAA and 2,4-D for callusing has also been infered (Raj Bhansali and Arya, 1978). The beneficial effects of high sucrose (5%) and vitamins have also been indicated (Murashige and Tucker, 1969). The callus induction in Local Sangtra was drastically reduced, when the kinetin concentration was increased from 0.5-5 mg/l). The increased concentration of NAA from 2-10 mg/l when used along with Kinetin (0.5 mg/l) stimulated somatic embryogenesis in Kinnow (Plate 1). Likewise, the somatic embryogenesis was provoked with the use of elevated levels of sucrose (5%) and vitamins, in both Kinnow as well as Local Sangtra. The half strength MS medium fortified with BAP (5 mg/l) proved better for differentiation of shoots as compared to full strength MS medium, in both the genotypes. The epicotyl derived calli of Local Sangtra and Kinnow exhibited fairly high degree of somatic embryogenesis upon transfer to MS medium fortified with NAA (10 mg/l), Kinetin (1 mg/l) and vitamins (10 x). Likewise, frequent somatic embryogenesis has also been observed in different species of citrus (Gmitter and Moore, 1986; Hidaka and Kajiura, 1988; Yu, 1989). Complete plants has been obtained from mature embryos of Kinnow (Plate 1) and Local Sangtra.

References

Chaturvedi H.C. and Mitra G.C. (1974) Clonal propagation of citrus from somatic callus cultures. Hort. Science 9 : 118-120.

Gmitter F.G. Jr and Moore G.A. (1986) Plant regeneration from undeveloped ovules and embryogenic calli of citrus : embryo production, germination and plant survival. Plant Cell Tissue and Organ Culture 6(2) : 139-147.

Hidaka T. and Kajiura I. (1988) Plantlet differentiation from callus protoplasts induced from citrus embryo. Scientia Hortic. 34 : 85-92.

Murashige T and Tucker D.P.H. (1969) Growth factor requirements of citrus tissue culture. Proc. First Int. Citrus Symp. 3 : 1155-1161.

Raj Bhansali R and Arya H.C. 1978. Differentiation in explants of Citrus paradisi Macf. (grapefruit) grown in culture. Indian J. Exp. Biol. 16 : 409-410.

Yu S.Q. (1989) Artificial embryogenesis and plant regeneration in citrus. pp. 191-193. In : Review of advances in Plant Biotechnology. (Eds. Mojeeb-Kazi and L.A. Sitch).

Table 1. Induction of callus, somatic embryogenesis and plant regeneration in Kinnow and Local Sangtra

Callus induction			Somatic embryogenesis		Plant regeneration	
Medium	Explant	%	Medium	%	Medium	%
			Kinnow			
MS_1	L	8.8		5.7	RM_1	0.0
	E	20.0	MS_1	10.0		5.7
	C	15.5		12.8		0.0
MS_6	L	73.3	MS_3	33.3	RM_2	11.1
	E	76.6		31.1		18.1
	C	83.3		38.8		10.0
MS_3	L	62.2	MS_6	20.0	RM_3	11.1
	E	71.1		29.1		23.5
	C	75.5		34.2		15.7
MT	L	66.6	MT	28.6	RM_4	6.8
	E	83.3		40.0		11.4
	C	86.6		32.2		8.3
			Local Sangtra			
MS_3	L	66.6	MS_3	53.5	RM_1	9.5
	E	90.0		79.1		16.6
	C	80.0		50.0		12.5
M_1	L	70.8	MS_{3+5}	61.1	RM_2	64.0
	E	87.5		83.3		66.6
	C	83.3		57.6		57.8
M_2	L	0.0	M_3	66.6	RM_3	80.9
	E	16.0		83.3		89.2
	C	12.5		58.3		83.3
M_4	L	20.8	M_1	63.6	RM_4	19.2
	E	33.3		86.6		17.3
	C	25.0		60.0		16.6

MS_1 = MS + NAA (2 mg/l) + Kin (0.5 mg/l)
MS_3 = MS + NAA (10 mg/l) + Kin (0.5 mg/l)
MS_{3+5} = MS + NAA (10 mg/l) + Kin (0.5 mg/l) + Sucrose 5%
MS_6 = MS + NAA (2 mg/l) + 2,4-D (2 mg/l) + Kin (0.5 mg/l)
M_1 = MS + NAA (10 mg/l) + Kin (1 mg/l) + Vitamins (10x)
M_2 = MS + NAA (10 mg/l) + Kin (5 mg/l)
M_3 = MS + NAA (15 mg/l) + Kin (5 mg/l)
M_4 = MS + NAA (20 mg/l) + Kin (1 mg/l)
MT = Murashige and Tucker (1969) medium
RM_1 = Half strength MS
RM_2 = Half strength MS + BAP (2 mg/l)
RM_3 = Half strength MS + BAP (5 mg/l)
RM_4 = MS + BAP (5 mg/l)
L = Leaf; E = Epicotyl; C = Cotyledon

Plate 1. A-E. Induction of somatic embryogenesis and plant regeneration in Kinnow mandarin

A. Showing callus proliferation from cultured epicotyl segments on MS+NAA (10 mg/l) + Kin (0.5 mg/l)

B. Showing the development of somatic embryos on callus subcultured on MS+NAA (10 mg/l) + Kin (1 mg/l) + Vitamins (10 x).

C. Shoot regeneration from embryogenic callus subcultured on half strength MS containing BAP (5 mg/l).

D. Different developmental stages of the somatic embryos.

E. Complete plantlets obtained from mature embryos.

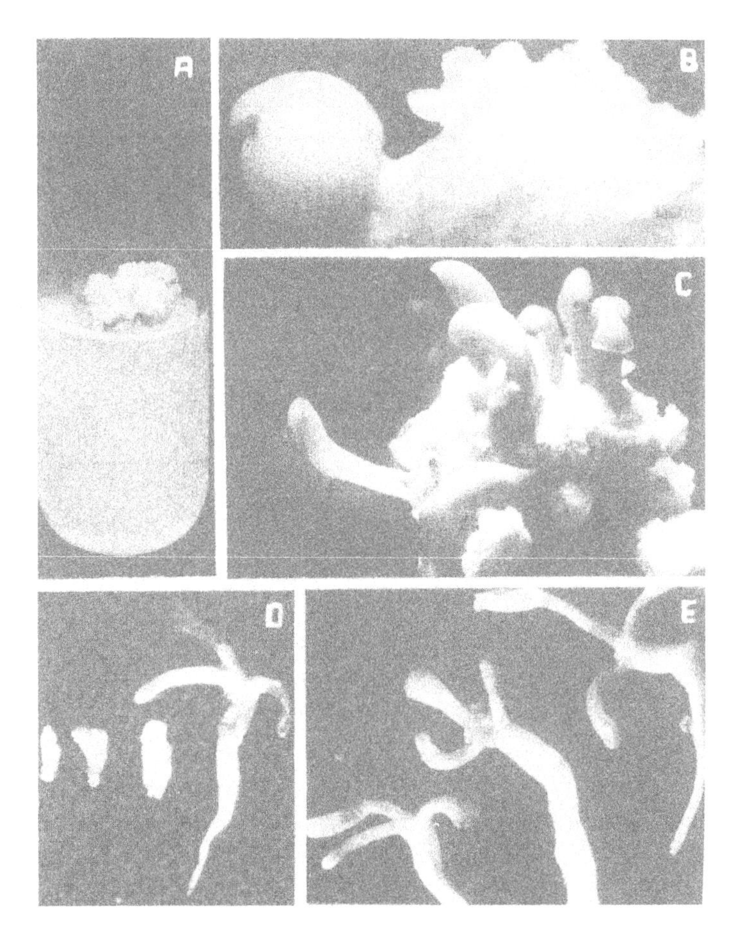

Tissue culture of medicinal plants: Morphogenesis, direct regeneration and somatic embryogenesis

V. SARASAN AND G.M.NAIR
Tissue Culture Laboratory, Department of Botany, University of Kerala, Kariavattom 695 581, India.

Key words *Piper longum*, *Hemidesmus indicus*, shoot tip culture, callus, regeneration, somatic embryogenesis

Introduction

Piper longum.L (Long pepper) and *Hemidesmus indicus* R.Br. (Indian Sarsaparilla) are two important medicinal plants found in the Indian forests (Kirtikar & Basu, 1935). *P.longum* (Piperaceae) is a unisexual perennial climber, fruits of which are used as a spice and contains alkaloids like Piperine and *Piplartine*, while *H.indicus* (Asclepiadaceae) has different steroids as active substances mainly stored in its woody root and other parts of the plant body and is used as a substitute for Sarsaparilla. An *in vitro* study has been reported in *Hemidesmus* on the levels of active substances in the whole plant as well as in the cultured tissues (Heble and Chadha, 1978). As there is an increasing demand for multiplying and conserving medicinal plants, *in vitro* studies are indispensable to meet the requirements. In this paper we are describing protocols for the multiplication of these two plants through shoot tip culture, direct and indirect organogenesis and somatic embryogenesis.

Methods

Shoot tips, stem pieces and leaf discs from green house grown mature plants of *Piper longum* L. and *Hermidesmus indicus* R.Br. were washed under running tap water for 3 hrs and sterilized using 0.1% mercuric chloride for 5 minutes. Culture medium consisted of MS nutrients (Murashige & Skoog, 1962) supplemented with 2 or 3% sucrose, different concentrations and combinations of hormones (Table 1). It was autoclaved at 121° for 15 min. at 1.06 kg/cm^2. All the cultures were incubated at $25 \pm 2^{o}C$ under 12 hr. photoperiod at light intensity of 30-45 $\mu E\ m^{-2}\ S^{-1}$

J. Prakash and R. L. M. Pierik (eds.), Horticulture – New Technologies and Applications, 237–240.

Table 1. Hormonal concentrations and combinations in mg/l used for various culture experiments in P.longum and H.indicus in MS medium.

Plant species	Media Used: Shoot tip culture	Direct regeneration	Callus induction	Morphogenesis	Somatic Embryogenesis
P.longum	1 BA 0.1 GA_3	3 BA 10%(v/v)CM	2-4 2,4-D 1 BA	3 BA & 1 BA, 2 PG	-
H.indicus	0.5 NAA 0.5 BA 0.5 GA_3	-	1 2,4-D 1 BA	-	basal MS or ½ basal MS

Results and discussion

Piper longum

Fifty two percent of the shoot tips explanted in 1 mg/l BA and 0.1 mg/l GA_3 produced multiple shoots (Table 2). Isolated shoots subcultured in the same medium showed vigorous growth though the multiplication rate was low (1:3). Internodal segments explanted in 3 mg/l BA and 10% Coconut milk produced adventitious buds directly from the entire surface of the explant when planted horizontally in the medium (Fig.1), while shoots developed from only the base of the explant when planted vertically (Fig.2). Leaf discs also responded similarly in this medium and the number of shoot buds developed was much more than any other explant (Table 2). Generally, multiple shoots are established only around the nodal portion. P.longum proved exceptional in induction of adventitious buds from the stem explant irrespective of a node, if placed horizontally in the medium. The position of the explant might have reverted the polarity of the cells and the exogenous phytohormones have evoked the cells to develop into adventitious buds.

Table 2. Response of various explants of P.longum in MS medium.

Explant	No.of explants cultured	Hormones (mg/l)	No.of explants responded	Response	Mean no.of shoots from an explant
Shoot tip	48	1BA,0.1GA_3	25	Multiple shoots	3
Internodal segment	88	3 BA,10% CM	75	Direct organogenesis	114
Leaf disc	88	"	67	"	289
Internodal segment	56	2 2,4-D 1 BA	48	Callus	84 (1 g fresh wt. callus subcultured in the regeneration medium)

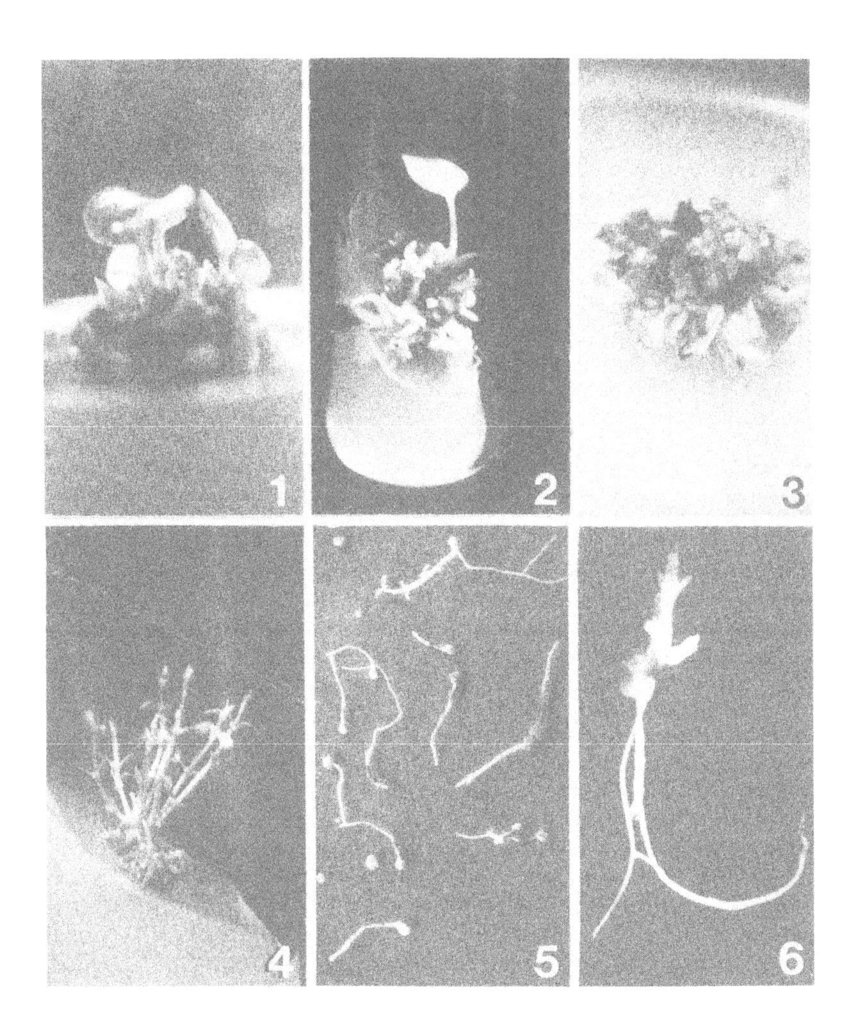

Figs. 1 - 3, Piper longum; 4-6, Hemidesmus indicus

1. Direct adventitious shoots from horizontally planted internodal segment in 3 mg/l BA and 10% CM; 2. Multiple shoots from internodal segment explanted vertically in 3 mg/l BA and 10% CM; 3. Shoot morphogenesis from black callus cultured in 2 mg/l PG and 1 mg/l BA; 4. Multiplication of shoots from node cultured in 0.5 mg/l NAA, 0.5 mg/l BA and 0.5 mg/l GA_3; 5. Somatic embryos in various stages of development in half MS basal medium; 6. A fully developed plantlet from somatic embryo.

Callus initiated from internodal segments explanted in 2 mg/l 2,4-D and 1 mg/l BA, subcultured to 3 mg/l BA containing medium developed shoot buds. Callus became black and lost morphogenetic potential when left in the medium for more than 3 months. However, shoot buds regenerated from these calli upon culturing in a medium containing 2 mg/l phloroglucinol and 1 mg/l BA (Fig.3). Phenolic compounds are known to participate in plant growth regulation (Kefeli and Kutacek,1977). Phloridzin and phloroglucinol are two compounds in this group which have significantly enhanced the growth of both shoots and roots of apple root stocks cultured in vitro (Jones and Hatfield, 1976 a & b). However, accumulation of toxic substances in callus cultures is usually found to inhibit growth and morphogenesis (Lam and Street, 1972). Our finding shows that phloroglucinol not only prevents the accumulation of toxic compounds in the cells and induce proliferation of cells, but also favours morphogenesis. Thus incorporation of phloroglucinol into the medium can be beneficial for favouring morphogenesis from long term callus cultures.

Hemidesmus indicus

Multiple shoots were induced from 83% of the nodal segments explanted in MS medium containig 6.5 mg/l NAA 0.5mg/l BA and 0.5 mg/l GA_3 (Fig.4). Isolated shoots were rooted in 1/2 MS supplemented with 0.2 mg/l IBA. Callus, initiated from stem pieces explanted in 1 mg/l 2,4-D and 1 mg/l BA, turned brown and lost regeneration potential. However, when subcultured to MS full strength or half strength basal media somatic embryos developed from this callus. Germination of the embryos was faster in liquid medium compared to solid medium (Fig.5). The germinated embryos established faster in vermiculite and, after 10 days, could be transplanted to the soil(Fig.6). Though several media were tried for germination of somatic embryos half strength MS favoured maximum germination. Subsequent growth became faster when transferred to full strength MS medium.

References

Heble M.R. and Chadha M.S. (1978) Steroids in cultured tissues and mature plant of Hemidesmus indicus R.Br. (Asclepiadaceae) Z. Pflanzenphysiol. 89, 401-406.

Jones O.P. and Hatfield S.G.S. (1976) Effect of phloridzin and phloroglucinol on apple shoots. Nature 262: 392-393.

Jones O.P. and Hatfield S.G.S. (1976) Root initiation in apple shoots cultured in vitro with auxins and phenolic compounds, J. Hort. Sci. 51: 495-499.

Kirtikar K.R. and Basu B.D. (1935) Indian Medicinal plants, Periodical Experts, Delhi.

Kefeli V.I. and Kutacek M. (1977) Phenolic substances and their possible role in plant growth regulation, In: Pilet P.E. ed., Plant growth regulation. Springer-Verlag, Berlin.

Lam T.H. and Street H.E. (1977) The effects of selected aryloxyalkane caraboxylic acids on the growth and levels of soluble phenols in cultured cells of Rosa damascena Z.Pflanzenphysiol. 84: 121-128.

Murashige T. and Skoog F.(1962) A revised medium for rapid growth and bioassays with tobacco tissue culture. Physiol. Plant. 15: 473-479.

Production of important phytochemicals through plant cell cultures and scale-up operations: Limitations and prospects

G.A. RAVISHANKAR and L.V. VENKATARAMAN
Autotrophic Cell Culture Discipline,
Central Food Technological Research Institute, Mysore - 570 013, India.

INTRODUCTION

The human population depends on plants, not only for its staple foods, but also for various chemicals derived from them. The phytochemicals of importance may be drugs, food additives, and a range of other chemicals needed for various end uses. The developments in modern biology have made it possible to exploit the productivity of plants in a desired manner, and also to modify them suitably. Despite the progress made in chemical synthesis, we still depend upon plant sources for a number of compounds, either because of complexities in chemical synthesis or because of their assured safety and in some cases for economic considerations. Moreover, due to the enormous population explosion the emphasis has been laid on increasing staple food crops, as a result land available for cultivation of plants for phytochemicals is diminishing. In the above context, it is relevant to develop alternative methods of production of phytochemicals by cell culture technology.

ADVANTAGES OF CELL CULTURE OVER CONVENTIONAL METHODS

Cell cultures can produce the whole range of metabolites that are produced by intact plants. The advantages of this method are as follows:- (i). uniform yield throughout the year ensured; (ii). no dependence on agroclimatic conditions, since growth occurs in a controlled environment; and (iii) possible to manufacture in a smaller area.

LIMITATIONS IN PRODUCTION OF PHYTOCHEMICALS

Though the plant cells are chemically totipotent, limitations exist in ensuring high productivity and scale up of the processes. Moreover, due to slow growth and often low yield potentials, plant cells cannot be used for processes yielding metabolites which can also be obtained from microorganisms. Nevertheless, there are a number of specific compounds belonging to alkaloids, steroids, flavours, insecticides, perfumes, pigments, etc., which are produced exclusively by certain plants. Several factors such as genotypic nature of the plants, nature of explant, stability of cultures, physical environment, media composition such as nutrients and hormones, influence the production of phytochemicals in cell culture.

J. Prakash and R. L. M. Pierik (eds.), Horticulture – New Technologies and Applications, 241–244.

IMPROVEMENT OF METABOLITE PRODUCTION

Enhancement of the yield of several metabolites has been possible by two-stage culture methods, in several plants, such as Catharanthus roseus (Sasse et al., 1982), Lithospermum erythrorhizon (Fujita et al., 1981), Digitalis lanata (Kreis and Reinhard 1988). Improvement of growth has enhanced the yield of biomass, for an overall increased productivity of phytochemicals. The enzyme marker has been developed to pick up cell lines with enhanced growth capabilities (Ravishankar & Grewal, 1990). Recent reports on the increase in secondary metabolite production, as a result of treating plant cells with microbe-derived molecules, have given a new impetus to industrial application of cell-culture technology (DiCosmo and Misawa, 1985). It is envisaged that elicitor application will reduce the two-stage culture system to a single-stage system, thereby reducing the cost of production of phytochemicals in large-scale cultivation processes.

The major drawback in biochemical regulation of secondary plant products is inadequate knowledge in respect of most of the pathways and the associated enzyme systems. It will be of great interest to isolate genes coding for specific enzymes, and express the same in a microbe, or enhance the copy number in the plant itself, to obtain over producers. An example of this type is the expression of thaumatin gene in a bacterium (Eden and Van der, 1985). Strictosidine synthase - a key enzyme in indole alkaloid pathway (Kutchan et al., 1988) is the first enzyme involved in alkaloid biosynthesis to be isolated. Similar studies are needed for many other important metabolites. Biotransformation of low-value compounds to high-value products is an area which has not been well explored. However, one important biotransformation which has reached commercial application is digitoxin to digoxin conversion using Digitalis lanata cells.

SCALE UP OF CELL CULTURES

In order to realize the biotechnological potential of cell cultures for high value phytochemicals, it is essential to have suitable bioreactors for large scale cultivation of cells. The major problems in scaling up of plant cell cultures are: slow growth, formation of cell clumps, shear sensitivity, and low oxygen requirement. An airlift bioreactor is useful, to avoid cell damage. When high cell densities are present, airlift methods are not adequate for mixing. Some have used a combination of airlift and stirred tank reactors, for growing cells in high densities. Roller drum type of reactor has been used for shikonin production in Lithospermum erythrorhizon. There are several designs of reactors used in experimental and production system (Kargi and Rosenberg, 1987). Scaling up of tobacco cell cultures in a 20,000 litre bioreactor has been reported (Noguchi, et al., 1977). Presently, only a few processes are well worked out for industrial production, viz: (i) Shikonin, a naphthoquinone pigment, from L. erythrorhizon by Mitsui Petrochemical industries, Japan; (ii). Berberine - an isoquinoline alkaloid from Coptis japonica; (iii). Production of ginsengosides by Panax ginseng cells in USSR (Zenk, 1990); (iv). The next product in pipeline in the USSR is ajamalicine, from Rauwolfia serpentina (Vollosovitch, 1988). The cost estimate of cell culture production varies from system to system. It is believed that only high-cost, low-volume products can be economical. Though it is difficult to generalize, one can attempt to scale up processes for phytochemicals costing over $ 750/kg. In an immobilized cell culture system which exudes desired metabolites, the cost of production can be reduced considerably.

WORK AT CFTRI FOR THE PRODUCTION OF FOOD ADDITIVES AND OF BIOINSECTICIDES FOR FOODGRAIN STORAGE

The plants under study are: saffron (*Crocus sativus*), pyrethrum (*Chrysanthemum cinerariaefolium*), and capsicum (*Capsicum annuum* and *C. frutescens*). Saffron has three important constituents - safranal (flavour), crocin (colour) and picrocrocin (bitter substance). Pyrethrum contains pyrethrins which are biopesticides, used in storage of food grains. Capsicum produces the pungent food additive, capsaicin.

The aim of tissue culture work in saffron is to produce the important metabolites in cultured cells, for scale up operation. Success in the production of pigment and flavour has been achieved (Visvanath *et al.*, 1990). The analysis of the metabolites indicated the enhanced production of picrocrocin in cultured tissues, as compared to natural stigma. The production of safranal from picrocrocin is possible by mild hydrolysis. Pyrethrins derived from cultured cells showed potent insecticidal activity (Ravishankar *et al.*, 1989). An yield enhancement of pyrethrins from 0.22% to 0.4% was possible by two-stage culture methods (Rajasekaran *et al.*, 1991).

Capsaicin production in immobilized cells and placental tissue of *C. annuum* and *C. frutescens* has been investigated. The immobilized placental tissues produced several fold higher amounts of capsaicin than the immobilized cells (Sudhakar Johnson *et al.*, 1990). In both the systems, capsaicin was actively produced, and released into the medium. The capsaicin production has been scaled up to the 2.0 litre level, in a column reactor process, using immobilized cells and placenta, entrapped in calcium alginate matrix. Cells treated with ferulic acid and valine produced 1600 µg capsaicin/g of fresh cells, over a 10 day period. Nitrogen stress resulted in two fold increase in capsaicin production in cultured cells (Ravishankar *et al.*, 1988).

CONCLUSION

Plant cell cultures can be used for compounds exclusively produced by plant systems. There is need to enhance productivity of cell cultures by understanding the biochemical regulatory mechanisms. Suitable elicitors for secondary metabolites need to be explored. Recombinant DNA studies may facilitate cloning of plant genes responsible for specific phytochemicals, into microbes for easy scale up of processes. There is need to explore specific biotransformations using plant cells as biocatalysts. Photoautotrophic cell cultures are to be developed for the production of phytochemicals. Future years will see many processes developed by cell culture technology, for specialized phytochemicals.

REFERENCES

DiCosmo, F. and Misawa, M. (1985) Eliciting secondary metabolism in plant cell cultures. Trends Biotechnol. 3: 318--322.

Eden, L. and Van der, W.H. (1985) Microbial synthesis of the sweet tasting plant protein thaumatin, Trends Biotechnol. 3: 61--64.

Fujita, Y., Hara, Y., Suga, C. and Morimoto, T. (1981) Production of shikonin derivative by cell suspension cultures of *Lithospermum erythrorhizon* II. A new medium for the production of shikonin derivatives. Plant Cell Rep. 1: 61--63.

Kreis, W. and Reinhard, E. (1988) 12 β-Hydroxylation of Digitoxin by suspension-cultured Digitalis lanata cells. Production of deacetyllanatoside C using a two stage culture method. Planta Medica 2: 95--110.

Kutchan, T.M. Hampp, N, Lottspeich, F. Beyreuther, K. and Zenk, M.H. (1988) The cDNA clone for strictosidine synthase from Rauwolfia serpentina. FEBS letters 2370: 40--44.

Martin, S.M. (1980) Mass systems for plant cell suspension. In: Staba, E.J. ed. Plant tissue culture as a source of biochemicals, pp. 149--166. CRC Press Inc., Boca Raton.

Noguchi, M., Matsumoto, T., Hirata, Y., Yamamoto, K., Katsuyama, A., Kato, A.S., Azechi, S. and Kato, K. (1977) Improvement of growth rate of plant cell cultures. In: Barz, W., Reinhard, E. and Zenk, M.H. eds. Plant tissue culture and its biotechnological application, pp. 85--94. Springer-Verlag, Berlin.

Rajasekaran, T., Rajendran, L., Ravishankar, G.A. and Venkataraman, L.V. (1991) Influence of nutrient stress on pyrethrin production in cultured cells of pyrethrum (Chyrsanthemum cinerariaefolium), Current Science (In press).

Ravishankar, G.A. and Grewal, S. (1990) ATPase activity as an index of growth capability of cultured cells of Dioscorea and Physochlaina. Biochem. Intl. 20: 897--901.

Ravishankar, G.A., Rajesekaran, T., Sarma, K.S. and Venkataraman, L.V. (1989) Production of pyrethrins in cultured tissues of pyrethrum (Chrysanthemum cinerariaefolium vis) Pyrethrum Post 17: 66--69.

Ravishankar, G.A., Sarma, K.S., Venkataraman, L.V. and Kadyan, A.K. (1988) Effect of nutritional stress on capasaicin production in immobilized cell cultures of Capsicum annuum. Current Science 57: 381--383.

Sasse, F., Knobloch, K.H. and Berlin, J. (1982) Induction of secondary metabolism in cell suspension cultures of Catharanthus roseus, Nicotiana tabacum and Peganum harmala. In: Fujiwara A. ed. Plant issue culture, 1982 pp. 343--344, Japanese Association for Plant Tissue culture, Japan.

Sudhakar Johnson, T. Ravishankar, G.A. and Venkataraman, L.V. (1990) In vitro capsaicin production by immobilized cells and placental tissues of Capsicum annuum L. grown in liquid medium, Plant Science 70: 223--229.

Visvanath, S., Ravishankar, G.A. and Venkataraman, L.V. (1990) Induction of crocin, crocetin, picrocrocin and safranal synthesis in callus cultures of saffron- Crocus sativus L. Biotechnol. Appl. Biochem. 12: 336--340.

Vollosovitch, A.G. (1988) Some peculiarities of alkaloid accumulation in tissue culture Rauwolfia serpentina Benth. 14. Int. Congress of Biochemistry, Prag, Abstract WE: C. 29--5.

Zenk, M.H. (1990) Plant cell cultures: a potential in food and bio-technology. Food Biotechnol. 4: 461--470.

Establishment of somatic cell culture and plant regeneration in grapes (*Vitis vinifera* L.)

ZORA SINGH, S.J.S. BRAR AND S.S. GOSAL
Department of Horticulture and
Plant Biotechnology and Tissue Culture Centre
Punjab Agricultural University
Ludhiana-141 004, India

Introduction

Perlette and Thompson Seedless cultivars are commercially grown for seedless table grapes in Northern plains of India. The synchronization of fruit maturity with the rainy season coupled with low total soluble solids and susceptibility to various fungal diseases, provides a greater scope of incremental improvement of these existing cultivars. Moreover, the improvement of seedless cultivars through conventional methods is tedious and laborious. Hence the somatic cell culture offers a greater scope for the incremental improvement of seedless cultivar of grapes. Since the first report on successful grape culture *in vitro* (Morel, 1944), subsequently, callus cultures have been initiated from various explants viz. stem, tendrils, petioles, flowers and berries of *Vitis* (Schenk and Hildebrendt, 1972; Standt *et al*.,1972; Hawker *et al*.,1973; Krul and Worley, 1977). Callus growth ceased after 40 days of incubation (Hawker *et al*., 1973). But shoot regeneration is problem in grapes. However, complete plants have been obtained from leaf callus (Favre, 1977; and Clog *et al*., 1990). The present investigation deals with the development of protocol for establishment of somatic cell culture and plant regeneration.

Methods

The young leaves, shoot apices, nodal and internodal segments were excised from mature vines of cultivars Thompson Seedless and Perlette. After washing and sterilization, the explants were aseptically cultured on Murashige and Skoog (1962) medium supplemented with different concentrations and combinations of NAA, 2,4-D,

J. Prakash and R. L. M. Pierik (eds.), Horticulture – New Technologies and Applications, 245–248.

and BAP. The cultures were incubated at 25±2°C under 14 hours fluorescent light.

Results

The growth response in Perlette and Thompson Seedless cultivars varied with medium combination. In Perlette maximum callus induction (91.6%) was from young leaves cultured on MS supplemented with NAA (10 mg/l) + BAP (0.5 mg/l). Whereas in Thompson Seedless the maximum callus induction was 86.3% from young leaves cultured on MS containing NAA (5 mg/l) and BAP (0.5 mg/l). The shoot apices of Perlette and Thompson Seedless cultured on MS containing NAA (10 mg/l) + BAP (0.5 mg/l) and MS supplemented with BAP (5 mg/l) induced highest callusing (86.6%) and (76.4%), respectively (Table 1). The nodal segments of Perlette and Thompson Seedless cultured on MS containing Kin (2 mg/l) + BAP (2 mg/l) and MS supplemented with NAA (10 mg/l) + BAP (0.5 mg/l) produced maximum callusing (89.4%) and (85.7%), respectively. Internodal segments resulted in 83.3% callusing in Perlette and 73.0% in Thompson Seedless when cultured on MS medium containing BAP (2 mg/l). The media containing 2,4-D failed to induce callusing in all the explants, except young leaves in Perlette. The green nodular calli of Thompson Seedless exhibited complete plant regeneration (11.0%) upon transferring on to MS medium containing BAP (2 mg/l) whereas the Perlette calli failed to regenerate into shoots.

Discussion

The genotype, nature of explant and chemical composition of culture media affected the callusing (Table 1). Likewise, the determining effect of various explants viz. leaf, stem or petiole segments have also been reported earlier (Krul and Worley, 1977). The media containing 2,4-D failed to induce callusing. In general, the calli were induced from the cut ends of the explants. The medium containing auxin produced friable, soft and creemish white to green calli. Upon subculturing, such calli turned brown and died within 4-6 weeks. Similarly, Krul and Worley (1977) reported that the callus culture of grapes began to turn brown after two subdivisions at 30 days interval. The long term maintenance of callus in presence of auxin is not always possible. Such observations have also been reported by Srinivasan and Mullin (1980). The medium containing cytokinins produced green, nodular and compact calli. When the calli of both the cultivars subcultured on MS medium containing BAP (2 mg/l) resulted in 11% complete plant regeneration in cultivar Thompson Seedless.

The calli of cultivar Perlette failed to regenerate. Like-wise successful organogenesis in internodal explants of grapevines has also been reported by Rajasekaran and Mullins (1981). Plant regeneration by organogenesis in *Vitis* has also been achieved by Clog *et al*. (1990). The methodology thus developed would become the basis of future work on the induction of somaclonal variation in grapes for obtaining incremental improvement, particularly in seedless cultivars of grapes.

References

Clog E., Bass P.and Walter B. (1990) Plant regeneration by organogenesis in *Vitis* rootstock species. Plant Cell Report 8 : 726-728.

Hawker J.S., Bownton S.J.S. and Mullins M.G. (1973) Callus and cell culture from grape berries. Hort Science 8 : 398-399.

Krul W.R. and Worley J.F. (1977). Formation of adventitious embryos in callus culture of 'Seyval' in French hybrid grapes. J. Amer. Soc. Hort. Sci. 102 : 260-363.

Morel G. (1944). Sur le development de tissue de Vigne cultives in vitro. C.R.Acad. Seances. Soc. Biol.Paris. 138 : 62.

Rajasekaran K. and Mullins M.G. (1981) Organogenesis in internode explants of grapevines. Vitis 20 : 218-227.

Schenk R.U. and Hildebrandt A.C. (1972) Medium and techniques for induction and growth of monocotyledons and dicotyledonous plant cell cultures. Can. J. Bot. 50 : 199-204.

Srinivasan C. and Mullins M.G. (1980) High frequency somatic embryo production from unfertilized ovules of grape. Sci. Hortic. 13 : 245-252.

Staudt G., Barner H.G. and Becker H. (1972) Studies on callus growth of di and tetraploid grapes in vitro. Vitis, 11 : 1-9.

Table 1. Percent callus and type of callus obtained from different explants of *Vitis vinifera* L. cvs. Perlette and Thompson Seedless cultured on MS (1962) medium containing different con--centrations and combinations of NAA, 2,4-D, Kinetin and BAP

Medium	Young leaves		Shoot apices		Nodal segments		Internodal segments	
	Percent callus	Type of callus	Percent callus	Type of callus	Percent callus	Type of callus	Percent callus	Type of callus
				Perlette				
M_1	40.0	-	-	-	-	-	43.7	LGNH
M_2	85.7	FCr	64.3	NCG	-	-	83.3	CrCN
M_3	66.6	FCr	65.0	NLG	-	-	58.8	CrCN
M_4	50.0	-	83.3	NLG	89.4	NLGH	52.6	CrCN
M_5	33.0	-	41.3	-	-	-	0.0	-0.0
M_6	0.0	-	0.0	-	-	-	0.0	-
M_7	80.0	FCr	55.5	NLGS	-	-	-	-
M_8	91.6	FCr	86.6	NLGS	-	-	-	-
				Thompson Seedless				
M_1	23.5	NWG	-	-	42.8	NCCrG	40.0	-
M_2	67.5	NCrGS	52.0	CrGSN	76.9	NCCrG	73.0	CrCN
M_3	78.1	NCrGS	76.0	CrGSN	83.3	NCCrG	68.0	CrCN
M_4	50.0	LGSN	71.4	WGSN	40.0	NCCrG	50.0	CrCN
M_5	0.0	-	0.0	-	0.0	-	0.0	-
M_6	0.0	-	0.0	-	0.0	-	0.0	-
M_7	86.3	CrGNS	70.0	LGNS	-	-	-	-
M_8	82.5	CrGNS	68.7	LGNS	85.7	WGHN	71.4	LGN

Abbreviations:
- \- : callus died
- 0.0 : No callus
- F : Friable
- Cr: Creemish
- C : Compact
- N : Nodular
- G : Green
- S : Soft
- H : Hard
- L : Light
- W : White
- M_1 : MS (1962) basal
- M_2 : MS+BAP (2 mg/l)
- M_3 : MS+BAP (5 mg/l)
- M_4 : MS+BAP (2 mg/l)+Kin (2 mg/l)
- M_5 : MS+2,4-D (2 mg/l)+BAP (0.5 mg/l)
- M_6 : MS+2,4-D (5 mg/l)+BAP (0.5 mg/l)
- M_7 : MS+NAA (5 mg/l)+BAP (0.5 mg/l)
- M_8 : MS+NAA (10 mg/l)+BAP (0.5 mg/l)

In-ovulo embryo culture in seedless grapes (*Vitis vinifera* L.)

S.J.S. BRAR, ZORA SINGH and S.S. GOSAL
Department of Horticulture and Plant Biotechnology and Tissue Culture Centre,
Punjab Agricultural University
Ludhiana-141 004, India

Introduction

Seedless table grapes are preferred by consumers, which demands breeding work to develop improved seedless cultivars. *Vitis vinifera* L. cvs. Perlette, Thompson Seedless and Beauty Seedless are stenospermic in which embryos abort after few days of anthesis (Stout, 1936). Embryos and endosperm breakdown occurs during third to fourth weeks after anthesis. Embryo development in aborting seeds ranged from few cell stage to globular stage (Stout, 1936). In-ovulo embryo culture prior to embryo abortion provide an attractive alternative to conventional breeding methods. In conventional method breeding of seedless grapes, the crossing of seeded cultivars with seedless cultivars resulted in 10÷15% seedless progeny (Winberger and Harmon, 1964). Theoratically by crossing seedless cultivars with seedless cultivars result in high percentage of seedless progenies in one breeding cycle. Two distinct methods, viz. culturing in liquid medium (Emershad and Ramming, 1984; Gray *et al*., 1987) and on solid medium (Cain *et al*., 1983; Spiegel-Roy *et al*., 1985) have also been reported. The present report deals with in-ovulo embryo culture of stenospermic Perlette, Thompson Seedless and Beauty Seedless grapes.

Methods

Berries were collected at 10, 15, 20, 30, 35, 40 and 45 days after anthesis from the vines of Perlette, Thompson Seedless and Beauty Seedless. The growth of berry and ovule along with number of shrivelled and viable ovules were recorded. Berries were dipped in 90% alcohol and flamed for short period. Ovules were cultured on MS (1962) medium supplemented with different concentrations and

J. Prakash and R. L. M. Pierik (eds.), Horticulture – New Technologies and Applications, 249–254.

combinations of NAA, IAA, kinetin, BAP, 2,4-D and gibberellin A_4/A_7 (GA4/7). The embryos of Perlette, Thompson Seedless and Beauty Seedless were also cultured on solid and liquid MS media containing IAA (2 mg/l) + BAP (0.5 mg/l) 30 days post anthesis. The ovules of Perlette, Thompson Seedless and Beauty Seedless 10, 15 and 20 days post anthesis were cultured on Nitsch's (1969) medium and White's (1954) medium fortified with 10 mM L-asparagine or L-cysteine. The cultures were incubated at 25±2°C and 14 hours fluorescent light.

Results

Growth and development of berries and ovules - The growth of berries and ovules after anthesis in all the cultivars is given in Table 1. The number of viable ovules decreased 20 days post anthesis in Perlette and 30 days post anthesis in Thompson Seedless and Beauty Seedless. In general, the shrivelling of ovules started 20 days after anthesis in all three cultivars.

Survival and growth of cultured ovules on MS solid and liquid medium - The ovules of cv. Perlette were cultured 30 days post anthesis on different media based on MS (1962) salts. The cultured ovules rescued growth within a week and increased in size. Per cent survival of cultured ovules (Table 2) was maximum (90.91%) on MS fortified with IAA (2 mg/l) and BAP (0.5 mg/l). The ovules of Perlette, Thompson Seedless and Beauty Seedless were cultured 30 days after anthesis on liquid and solid MS media containing IAA (2 mg/l) and BAP (0.5 mg/l). The per cent survival of the cultured ovules on liquid medium varied (77.7%, 76.9% and 76.2%) in cv. Perlette, Thompson Seedless and Beauty Seedless respectively (Table 3). The physical status viz. filter paper bridges put in liquid medium and agar solidified cultured media did not affect per cent survival much (Table 3).

Survival and growth of cultured ovules on Nitsch's and White's medium - The ovules of Perlette cultivar cultured 10 days post anthesis on Nitsch's medium fortified with 10 mM L-cysteine exhibited 70% survival and resumed growth. The survival of ovule was increased to 93.8% and 88.8% with better growth when cultured 20 days after anthesis on Nitsch's medium containing 10 mM L-asparagine and 10 mM L-cysteine respectively (Table 4). Among the ovules of Thompson Seedless cultured 10, 15, 20 days post anthesis, the per cent survival was higher when the ovules were cultured 20 days after anthesis. The survival of cultured ovules was 90.8% and 92.2% in Nitsch's medium fortified with 10 mM L-asparagine and 10 mM L-cysteine, respectively (Table 4). In both the cultivars, the addition of amino acid to basal Nitsch's medium has promoted the growth of in vitro cultured ovules. The ovules of Perlette and Thompson Seedless cultivars were cultured 10, 15 and 20 days after anthesis on White's basal medium containing 10 mM L-asparagine and 10-cysteine did not affect much the survival of the ovules. However, the growth of survived ovules was promoted with L-asparagine and L-cysteine (Table 4) and produced larger ovules. The promotion of growth of in vitro cultured ovules was more in cv. Perlette than in Thompson Seedless.

Discussion

The survival of the ovules cultured 30 days after anthesis on MS medium containing IAA (2 mg/l) + BAP (0.5 mg/l) was highest (90.91%) in cv. Perlette. The cultured ovules rescued growth within a week, increased in size and remained green throughout the experiment. The per cent survival of cultured ovule did not vary much in three different seedless cultivars. In general, the physical status with filter paper bridges put in liquid medium and on solid medium did not affect per cent survival much. The ovules of Perlette and Thompson Seedless were cultured 10, 15 and 20 days after anthesis on White's medium containing 10 mM L-asparagine and 10 mM L-cysteine did not affect survival much of ovules. However, the growth of survived ovules were promoted with L-asparagine and L-cysteine and produced larger ovules. The response of ovules to L-cysteine and L-asparagine suggest that these amino acids play important role in nutrition and development of immature ovules. Thus organic form of nitrogen may be readily incorporated during protein synthesis (Raghavan, 1966). Similar effects of L-cysteine and L-asparagine on growth of embryos cultured in vitro has been reported (Emershad and Ramming, 1984). In all the above mentioned experiments ovules resumed growth on all the media with a variable survival percentage. The ovules remained green throughout the experiment. Dissection of ovules revealed that all were hollow and contain no visible embryo and endosperm remnants even when the outer integument tissues remained green. The callus was produced by outer layer of the outer integument. Callus development from the outer integument of the cultured ovules may be due to higher levels of hormones during early stages of the ovules. However, a few workers have developed protocols for the growth of immature (45-70 days post anthesis) embryos to a stage where they can be successfully excised and grown into plants. But, under North Indian conditions the shrivelling of embryos occur within 20 days after anthesis in seedless cultivars.

In this experiment the percent survival increased with increased age of ovule at the time of excision. Similar results have also been reported by Tsolova (1990). The efforts to produce viable embryos and plants in vitro in stenospermic cultivars of grapes are in progress from a very immature embryos (10-20 days post anthesis).

References

Cain D.W., Emershad R.L. and Tarailo R.E. (1983) In-ovulo embryo culture and seedling development of seeded and seedless grapes (Vitis vinifera L.). Vitis 22 : 9-14.
Emershad R.L. and Ramming D.W. (1984) In-ovulo embryo culture of Vitis vinifera L. cv. Thompson Seedless. Amer. J. Bot. 71 : 873-877.
Gray D.J., Fischer L.C. and Matensen J.S. (1987) Comparison of methodologies for in-ovulo rescue of seedless grapes. Hort. Science 22 : 1334-1335.
Murashige T. and Skoog F.M. (1962) A revised medium for rapid growth and bioassays with tobacco tissue culture. Physiol. Plant. 15 : 473-497.
Nitsch J.P. and Nitsch C. (1969) Haploid plant from pollen grains. Science, 163 : 85-87.
Raghavan V. (1966). Nutrition, growth and morphogenesis of plant embryos. Biol. Rev. 41 : 1-58.
Spiegel-Roy P., Sahas N, Baron J. and Lavi U. (1985) In vitro culture and plant formation from grape cultivars with abortive ovules and seed. J. Amer. Soc. Hort. Sci. 110 : 109-112.
Stout A.B. (1936) Seedlessness in grapes. New York Agr. Expt. Sta. Tech. Bull. 238.
Tsolova V. (1990) Obtaining plants from crosses of seedless grapevine varieties by means of in vitro embryo culture. Vitis 29 : 1-4.
Weinberger J.H. and Harmon F.N. (1964) Seedlessness in vinifera grapes. Proc. Amer. Soc. Hort. Sci. 85 : 270-274.
White R.P. (1954) The cultivation of animal and plant cell. Ronald Press, N.Y.

Table 3. Percent survival of the ovules (30 days after anthesis) cultured on MS medium + IAA (2 mg/l)+BAP (0.5 mg/l)

Cultivars	Liquid medium			Solid medium		
	Cultured ovules	Survived ovules	Survival (%)	Cultured ovules	Survived ovules	Survival (%)
Perlette	162	126	77.7	118	77	65.3
Thompson Seedless	26	20	76.9	40	30	75.0
Beauty Seedless	42	32	76.2	-	-	-

Table 1. Average berry diameter, ovule length, number of ovules per diameter, number of shrivelled and viable ovule(s) in seedless *Vitis vinifera* L. cvs. Perlette, Thompson Seedless and Beauty Seedless cultured 10-45 days after anthesis

Cultivars	Days after anthesis	Av.berry diameter (mm)	Av.ovule length (mm)	Av.no.of ovules/ berry	Shrivelled ovules/ berry	Viable ovules berry
Perlette	10	5.88±0.32	2.91±0.47	3.1±0.30	0.0	3.1±0.30
	20	7.54±0.32	4.67±0.65	2.8±0.40	0.0	2.8±0.40
	30	10.09±0.54	5.35±0.39	2.8±0.60	0.3±0.64	2.5±1.34
	35	10.89±0.88	5.25±0.40	3.2±0.74	1.5±0.42	1.7±0.45
	40	10.93±0.54	5.05±1.09	2.6±0.49	1.2±0.60	1.4±0.66
	45	14.00±0.70	4.85±0.55	2.5±0.92	2.1±1.04	0.4±0.66
Thompson Seedless	10	4.04±0.21	2.70±0.50	2.1±0.94	0.0	2.1±0.94
	20	5.63±0.41	3.00±0.84	2.7±0.64	0.0	2.7±0.64
	30	8.43±0.35	3.25±0.64	3.2±0.87	2.9±1.4	0.3±0.90
	35	8.52±0.86	3.95±1.12	3.0±0.44	3.0±0.44	0.0
	40	8.81±0.45	2.95±0.93	2.8±0.87	2.6±1.02	0.2±0.40
Beauty Seedless	10	5.20±0.31	2.68±0.31	3.2±0.74	0.0	3.2±0.74
	20	7.54±0.32	2.90±1.11	3.8±0.74	0.0	3.8±0.74
	30	8.32±0.81	3.55±0.96	2.8±0.87	2.4±0.66	0.4±0.80
	35	9.17±0.48	3.76±0.60	2.9±0.70	2.9±0.70	0.0
	40	9.72±0.40	2.85±1.05	2.8±0.60	2.4±1.04	0.4±0.91

Table 2. Percent survival of the ovules cultured (30 days post anthesis) on MS medium containing different combinations and concentrations of IAA, NAA, 2,4-D, kin, BAP and GA_{4+7} in cv. Perlette

Medium	Cultured ovules	Survived ovules	Survival (%)
MS (1962) medium	112	40	35.71
MS+Kin (5 mg/l)	72	38	52.78
MS+2,4-D (2 mg/l)+Kin (0.5 mg/l)	100	0	0.00
MS+NAA (5 mg/l)+Kin (0.5 mg/l)	86	70	81.40
MS+NAA (10 mg/l)+Kin (0.5 mg/l)	136	18	13.24
MS+IAA (2 mg/l)+BAP (0.5 mg/l)	66	60	90.91
MS+IAA (2 mg/l)+BAP (0.5 mg/l) + GA_{4+7} (2 mg/l)	126	50	45.88

Table 4. Percent survival and growth of ovules of Perlette and Thompson Seedless cultured on White's (1954) and Nitsch's (1969) medium containing L-asparagine and L-cysteine

Cultivars	Days after anthesis	Medium	Cultured ovules	Survived ovules	Survival (%)	Growth response
Perlette	10	W_1	36	13	36.1	++
		W_2	72	22	30.5	+++
		W_3	72	17	23.6	+++
	15	W_1	36	26	80.5	++
		W_2	72	45	62.5	+++
		W_3	73	24	32.8	+++
	20	W_1	36	20	55.5	++
		W_2	72	43	59.7	+++
		W_3	73	34	47.2	+++
	10	N_1	33	17	51.5	+
		N_2	90	39	43.3	+
		N_3	90	63	70.0	+++
	15	N_1	65	41	63.07	+
		N_2	90	49	54.4	+
		N_3	80	42	52.5	+++
	20	N_1	70	42	60.0	+
		N_2	81	76	93.8	++
		N_3	90	80	88.8	+++
Thompson Seedless	10	W_1	54	0	0	-
		W_2	72	11	15.2	+
		W_3	72	7	9.7	+
	15	W_1	60	22	36.6	+
		W_2	72	31	43.0	++
		W_3	72	19	26.3	+
	20	W_1	36	17	47.2	+
		W_2	69	49	71.0	+++
		W_3	69	46	66.6	+++
	10	N_1	30	8	26.6	+
		N_2	90	20	22.2	+
		N_3	87	28	32.2	+++
	15	N_1	48	18	37.5	+
		N_2	91	44	48.4	++
		N_3	73	40	54.8	+++
	20	N_1	36	26	72.2	+
		N_2	87	79	90.8	++
		N_3	90	83	92.2	+++

W_1 - White's (1954) basal
W_2 - W+10mM L-asparagine
W_3 - W+10mM L-cysteine

N_1 - Nitsch's (1969) basal
N_2 - N+10mM L-asparagine
N_3 - N+10mM L-cysteine

Micropropagation of Davana (Artemisia pallens Wall.) by Tissue Culture

Md. UMER SHARIEF AND K.S. JADADISH CHANDRA*
Department of Post-graduate Studies and Research in Botany, University of Mysore, Manasagangotri, Mysore-570 006, India
Key Words: Micropropagation, Artemisia pallens, Multiple shoots, Regeneration

Introduction

Artemisia pallens, is an aromatic plant of the family Asteraceae commonly known as 'Davana' whose cultivation has been confined to South India only. It is a traditional herb prized for its fragrant leaves and flowers used in floral decorations and for worship in temples (Anonymous, 1985). The exquisite and delicate scent of Davana leaves is agreeable and welcome to everyone. The plant is an annual erect branched herb, 40 to 60 c.m tall with leaves that are small and much divided bluish green and under-surface being pale in colour. Significant genotypic differences have been reported in Davana with regard to plant height and percent oil content in the herbage (Farooqi et al., 1990). However, two distinct morphological types have been recorded, one with short stature having entire leaves at the base and somewhat dissected ones in the upper half portion and flowers early. The second type was tall with highly dissected leaves throughout and flowered late (Narayan et al., 1978). The leaves and flowers of A. pallens yields an essential oil known as 'Oil of Davana' used in high grade perfumes. The oil is a brown viscous liquid with deep mellow, persistent rich fruity odour. It is only in recent years Davana oil is getting a good export potential in India. During the year 1985-86 the Davana oil in India has earned a maximum of Rs. 2972.8 thousands out of the total of Rs. 147630.8 thousands earned by the export of nine different essential oils (Sudhir Jain 1987). Especially USA and Japan are showing an increasing interest for the oil, where it is used for flavouring cakes, pastries, tobacco and beverages. The main chemical component of the oil was found to be Cis-davanone, and recently, seven other sesquiterpene ketones have been reported in the extract of Davana (Cesar A.N. Catalan et al., 1990). Davana is grown as a short term crop from November to February, and as a ratoon crop extending upto April and May. The first flower buds appear by the end of January and seeds may be collected during the months of February and March. Tissue culture studies has been reported in A. absinthium (Shin et al., 1971), A. dracunculus (Garland and Stoltz 1980) and few other species. In the present paper we report the studies made on micropropagation of Artemisia pallens by tissue culture.

Materials and Methods

For experimental purpose it is necessary to use seeds of previous year crop, since rapid loss of viability has been observed in the old seeds (Range Gowda and Ramaswamy 1965).

* Principal Investigator of the UGC-Project No-F3-23/87-SR II. (New Delhi), 1987-90

J. Prakash and R. L. M. Pierik (eds.), Horticulture – New Technologies and Applications, 255–259.

The seed material for the present study was obtained from the horticulture department, Mysore. Seeds were soaked overnight in water, thoroughly washed and surface sterilized with 0.1% mercuric chloride for 10 minutes. They were rinsed thoroughly with sterile distilled water and inoculated aseptically on Murashige and Skoog's (1962) medium supplemented with growth regulators like KN individually, and in combination of 2,4-D and BAP at different concentrations. The MS media was further supplemented with adjuvants like biotin (1 mg/1) and casein hydrolysate (300 mg/1) which increase the vigour of shoot buds. The cultures were incubated at 25 $\pm$ 2°C under 12h daily illumination of 1000 lux with fluorescent light and 60-70% relative humidity. All the experiments were repeated thrice.

Results

The results obtained on the detailed studies of micropropagation of Davana are presented in the table 1.

Table 1. Effect of different growth regulators on seed culture of Artemisia pallens Wall

Sl. No.	MS Media and Growth Regulators	Intensity of callus devpt.	Multiple Shoots Nos.	Observation
1.	KN (1 mg/1)	-	4-5	Multiple shoots
2.	BAP (1 mg/1)	+	6-8	Multiple shoots
3.	2,4-D (1 mg/1)+BAP(1mg/1)	++	-	Callus formation
4.	2,4-D (2mg/1)+ BAP(2mg/1)	+++	-	Callus formation
5.	2,4-D (3mg/1)+ BAP(2mg/1)	++	-	Callus formation
6.	NAA (0.5mg/1)+ BAP(0.5mg/1)	+	-	Callus formation
7.	NAA (1 mg/1) + BAP(1 mg/1)	+++	20-25	Callus formation & Shoot differentiation
8.	NAA (1 mg/1) + BAP(0.5mg/1)	-	25-30	Multiple shoots
9.	NAA (1 mg/1) + BAP(0.3mg/1)	-	30-35	Multiple shoots
10.	IAA (1 mg/1) + BAP (1mg/1)	++	-	Rhizogenic callus
11.	IAA (2 mg/1) + BAP (2 mg/1)	++	-	Rhizogenic callus
12.	2,4-D (1 mg/1)+Ab.Acid(1mg/1)	+	-	Callus formation

+ Moderate, ++ High, +++ Intense, - Nil

The seeds inoculated on MS media with KN (1 mg/1) and also on 2,4-D +BAP (1 mg/1 each), germinated after 5 to 6 days. On the latter medium seedlings exhibited profuse callusing all over the surface after one week and the callus was maintained by regular subculture (Fig. 1). Subsequent subculture of such calli to the fresh medium with NAA + BAP (1 mg/1 each) resulted in the constant proliferation of callus and simultaneous regeneration of shoots (Fig. 2). The shoots attained suitable height only after a period of two to three weeks. The seedlings obtained on the media containing 2,4-D + BAP (1 mg/1 each), when subcultured on to MS + NAA (1mg/1) + BAP (0.5mg/1), direct multiple shoots were obtained (Fig. 3). Further lowering of BAP (0.3mg/1) facilitated vigorous and rapid development of shoots. The multiple shoots thus formed attained a height of 60 to 70 mms within 5 to 6 days (Fig. 4). The individual shoots of suitable size were aseptically excised at the base and transferred for rooting on full strength MS semisolid media with IAA, NAA, IBA and IPA at different concentrations and combinations. It was however noted that IBA (0.5mg/1) facilitated adequate rooting with profuse lateral roots after one week (Fig. 5). The rooted plantlets were maintained on the same media for 8 to 10 days. Later, they were removed, washed and transplanted to vermiculite : sand (1:1) mixture, and covered with plastic bags with constant misting. After acclimatization for one month the plants were transferred to the field conditions (Fig. 6). About 80 to 90% of survivability has been observed in *in vitro* regenerated plants.

Discussion

The present tissue culture studies has established a rapid method for propagation of *A. pallens*. This technique is advantageous especially in eliminating the ant-menace of the seeds, a serious problem commonly faced during the direct method of seed sowing in the nursery and in the open fields. Of the different concentrations and combinations of growth regulators used, NAA (1 mg/1) + BAP (0.3 mg/1) has shown to be intrinsically advantageous for the rapid induction of direct multiple shoots. Further biotin and casein hydrolysate added to the media increase the vigour but not the frequency of the multiple shoots. 2,4-D has proved to be essential for the continuous growth and maintenance of the calli without differentiation. It was also noted that these calli, in turn produced a creamy exudate into the medium, which however did not deter the callus culture. Similar type of response as to the induction of callus and multiple shoots has been observed by Garland and Stoltz (1980) in *A. dracunculus* var. *sativa* from leaf explants on LS medium with different concentrations of NAA and BA. Recently, Benjamin *et al.*, (1990) have also noted the formation of multiple shoots from seedling explants of *A. pallens* cultured on MS liquid media using altogether different growth regulators like BA + IAA and NAA. In the present material, adequate rooting with profuse lateral roots has been achieved in excised shoots, using, MS media with IBA (0.5 mg/1). But Benjamin *et al.*, have observed rooting in regenerated shoots of *A. pallens* using different growth hormones such as KN (1.0 ppm) + NAA (0.1 ppm) + IAA (0.1 ppm), NAA (0.1 ppm) or IAA (0.1 ppm). The present method of micropropagation by the induction of multiple shoots directly and also by callus differentiation are valuable tools for the rapid propagation of this important plant.

The authors gratefully acknowledge the UGC New Delhi, for financial assistance and University of Mysore, Mysore for providing facilities.

References

1. Anonymous (1985) Wealth of India - 1 (Revd.), CSIR publication New Delhi 434--442.
2. Benjamin B.D., Sipahimalani A.T., and Heble M.R. (1990) Tissue Cultures of *Artemisia pallens*: Organogenesis, terpenoid production. Plant Cell, Tissue and Organ culture 21: 159--164.
3. Cesar A.N.Catlan *et al.*, (1990), Sesquiterpene ketones related to davanone from *Artemisia pallens*, Phytochemistry 29 (8): 2702--2703.
4. Farooqi A.A., Dasharatha Rao N.D., Devaiah K.A. and Ravi Kumar R.K. (1990) Genetic variability in Davana (*Artemisia pallens* Wall.). Indian perfumer 34(1): 42--43.
5. Garland P. and Stoltz L.P. (1980) *In vitro* propagation of Tarragon HortScience 15(6): 739.
6. Murashige T. and Skoog F. (1962) A revised medium for rapid growth and bioassay with tobacco tissue cultures. Physiol. Plant. 15: 473--497.
7. Narayan M.R., Khan M.N.A. and Dimri B.P. (1978) Davana and its cultivation in India, Farm Bulletin, Central Institute of Medicinal and Aromatic plants, Lucknow.

8. Range Gowda D. and Ramaswamy M.N. (1965) Davana, Artemisia pallens Wall. in India. Perfumery and Essential Oil Record 56: 152--144.
9. Shin B.O, Mangold H.K. and Staba E.J. (1971) Lipid analysis of selected plant tissue cultures In: Les Cultures de Tissues de Plantes. Collaq. Int., C.N.R.S., Paris, No. 193: 51--69.
10. Sudhir Jain (1987) Perfumery in Soaps and Detergents, Pafai Journal, 30--33.

Micropropagation of *Artemisia pallens* (Fig. 1-6): 1. Callus induced from the seedlings on MS + 2,4-D (1mg/l each). 2. Differentiation of shoots from the seedling callus on NAA + BAP (1mg/l each). 3. Multiple shoots formed directly from the seedlings on NAA (1mg/l) + BAP (0.5mg/l). 4. Formation of vigorous and fleshy multiple shoots on NAA (1mg/l) + BAP (0.3mg/l). 5. Root induction in shoots on MS media with IBA (0.5mg/l). 6. Regenerated plants transplanted to the pots.

Clonal Propagation of Bamboo, Coffee and Mimosa

GOURI CHATTERJEE, GEETHA SINGH, POORNI THANGAM,
SATHYA PRAKASH AND JITENDRA PRAKASH
Indo American Hybrid Seeds,Biotech.Div.,P.B.No. 7099,B.S.K. II Stage, Bangalore - 560 070, INDIA

Keywords : Explant, Clonal propagation, Somatic embryogenesis

Introduction

The potential benefits of clonally propagated commercial planting stock has been well reviewed (Ammirato et al, 1984). In general, woody taxa are difficult to regenerate under in vitro conditions but recently some success has been achieved in a few leguminous and non - leguminous trees (Tomar and Gupta, 1988a; Vibha Dhawan and Bhojwani, 1985). Commercial cloning in vivo and in vitro of adult or mature woody plants is adversely affected by characteristics accompanying maturation such as, reduced growth rate and improper or total lack of rooting (Pierik, 1990). We report attempts undertaken at our laboratory to establish in vitro propagation from nodal cuttings in Dendrocalamus strictus, Oxytenanthera stoksii, Coffea arabica, and via somatic embryogenesis in Mimosa golois.

D. strictus, a drought resistant species finds its use in afforestation of arid lands and pulp industry. O.stoksii, a solid bamboo is used in construction. Coffea arabica is highly rated in the market when the planting stock supplied is of an elite clone with clonal purity. Mimosa golois, a product of grafting with Acacia root stock, is a rare ornamental species.

Materials and Methods

The explant sources of D. strictus, O. stoksii, C. arabica and M.golois were maintained in the

J. Prakash and R. L. M. Pierik (eds.), Horticulture – New Technologies and Applications, 261–264.

greenhouse. The standard aseptic techniques were followed when initiating the nodal explants from juvenile and mature plants of Bamboo, axillary buds and shoot tip explants of Coffee and nodal explants of Mimosa for axillary shoot multiplication and stem segments of Mimosa for callus induction.

The various parameters that were manipulated to establish suitable media for axillary shoot multiplication in the three crops were basal medium, phytohormones viz. BAP, Kinetin, GA3 concentration and combinations and adjuvants. The cultures were maintained at 25 ± 1° C and 16 hrs. light conditions throughout the experiment.

The observations that were carried out during the course of the experiment were used to standardise subculture cycles in the three crops. The main observational parameter was multiplication rate of the axillary shoots. The observations on phenolic oxidations, browning of explants, necrosis of plantlets and precocious leaf fall were also recorded. The parameters manipulated to establish suitable media for obtaining embryogenic calli were basal media, phytohormones viz. 2,4-D, NAA and BAP concentrations and combinations and pH gradients. For maturation of somatic embryoids, various concentrations of ABA were tried. For regeneration of plantlets from embryogenic calli, subculture cycles were optimised based on the observations recorded.

The varied parameters for rooting of plantlets were basal media, IBA concentration and adjuvants like activated charcoal. The **in vitro** rooted plantlets were acclimatised into established plants in the greenhouse.

Results and Discussion

Bamboo : Nodal explants showed pronounced bacterial contamination after sterilisation with 0.12% mercuric chloride. However, this was controlled by using 100ppm strepto-penicillin in combination with 0.12% mercuric chloride. Mature explants had to go through quick transfer cycles at 10-12 days interval as they exuded high concentration of phenolics.

Maximum multiplication of shoots from juvenile seedling explants of **D. strictus** was observed in the lowest node, the average being 25-30 after 18-20 subculture cycles in LS basal +3mg/1. BAP + 3% sucrose. This is in accordance with earlier reports (I.U. Rao et al., 1985 and I.V.R. Rao et al., 1989). Clusters having 3 to 5 shoots, were rooted in LS basal + 1mg/1 IBA + 0.25% activated charcoal. 60% of the rooted plantlets were acclimatised in the greenhouse.

Multiplication of matured explants of D. **strictus** and O. **stoksii** was a maximum of 8-10 shoots per node after 10-12 subculture cycles in LS basal + 3mg/l BAP + 0.2mg/l Kinetin + 4.5% sucrose. Sucrose seemed to play an important role in obtaining green healthy looking multiplying cultures. 4.5% sucrose extended culture life upto 20-24 weeks, which is otherwise 12-15 weeks. Further subculturing resulted in gradual death of culture as reported earlier in Bamboo callus cultures (Huang et al., 1989). Attempts to root these cultures at an early multiplying stage are in progress.

Coffee : To eliminate toxicity due to phenolic exudation in explants excised from mature plants, CA water (150mg/l citric acid + 100mg/l. ascorbic acid) was used. Among the various media used for initiation, MS basal + 1.5mg/l. BAP + 3% sucrose was found to be effective. After 2-3 subculture cycles in the same medium, the cultures transferred to MS basal + 3mg/l. BAP + 0.5mg/l GA3 responded best with the multiplying rate of 1:3.

Since the multiple shoots had condensed internodal regions, to facilitate rooting, shoots were transferred to media containing low BAP for elongation and later to 1/2 MS basal + 2mg/l. IBA + 0.25% activated charcoal. But the percentage of rooting was low. The rooted plantlets were acclimatised in greenhouse.

Mimosa: The axillary shoot multiplication was maximum in explants collected during April - May. The media that was found to show axillary shoot multiplication was MS + 0.75mg/l. BAP + 3% sucrose but the rate was not satisfactory. Hence, an alternative method namely somatic embryogenesis was attempted.

Embryogenic calli were induced in LSbasal + 2mg/l 2,4-D + 0.5mg/l. NAA + 0.1mg /l. BAP. The proliferation of embryogenic calli was found to be better when they were initiated in the same media but with a low pH and subsequently subcultured onto higher pH media. The pH gradient of 4 to 5.8 was found to be the best. Though the response of calli to pH gradient with respect to formation and development of embryoids is not in accordance with the earlier report (Smith and Krikorian, 1990) the drop in the pH seemed to cause a stress in the cells and subsequently promoted regeneration as reported earlier (Pechan et al., 1989).

To establish synchronous growth of the embryoids, ABA at 3mg/l. was found to be effective. ABA does not affect the embryoids at early stages of development but arrests growth of later stage embryos (Dane R. Roberts et al., 1990). Embryoids matured and reached the cotyledonary stage in ABA medium in 4-5 weeks. Plantlets devoid of roots were formed from embryoids in LS basal + 400 ppm glutamine. The percentage of plantlets obtained from embryoids was around 5 to 10%. Rooting occured only in a very few plantlets in LS basal or in low IBA containing media. The plantlets suffered from precocious leaf fall as reported earlier in another leguminous species (Vibha Dhawan & Bhojwani, 1985). This was overcome by supplementing the medium with 400ppm. glutamine.

References

1. Dane R. Roberts et al., (1990) Physiol. plant. 78:355-360.
2. Li-chun Huang et al., (1989) Envt. Expt. Bot. 29 (3): 307-315
3. P.M. Pechan et al.,(1989) Agricell report 13 (6):44
4. R.L.M. Pierik, (1990) Proceedings of VII Intl. Cong. Plant Tiss. Cult.
5. I.U.Rao et al.,(1985) Plant cell reports 4:191-194.
6. I.V.R. Rao et al., (1989) Propagation of Bamboo and Rattan through tissue culture. An IDRC Publication.
7. D.L. Smith and A.D. Krikorian, (1990) Abstracts VII Intl. cong. Plant Tiss. Cult.
8. M.R. Sondahl et al., (1984) In Hand book of Plant cell culture Vol. (3) Crop spp. ed. P.V. Ammirato et al.
9. Tomar U.K. and Gupta S.C., (1988a) Pt. Cell Rep 7: 70-73.
10. Vibha Dhawan and S.S. Bhojwani, (1985) Pt. Cell Rep. 4:315-318.

Greenhouse Environmental Control for Indian Conditions

David R. Mears
Biological and Agricultural Engineering Department
Cook College - Rutgers University
New Brunswick, New Jersey
U.S.A.

Introduction

It has been noted by several speakers at the seminar that there are very few functioning greenhouses in India with the exception of a few at various research centers. This is also the impression the author has developed during extensive travels to a number of research centers and agricultural production areas within India over the past several years. It has also been observed and noted that in many research centers the greenhouse facilities are not actively utilized. In some cases this is due to environmental control equipment being out of order and in others the problem is that the installed equipment is not adequate even if functioning properly. The most commonly observed problems are related to inadequacies of repair or design of cooling equipment.

As most of India has a normally warm climate, one's first expectation is that the opportunities for greenhouse technology would be quite limited. However, there are a number of circumstances within India in which we should regard effective greenhouse systems as essential. There are also a number of circumstances in which there are opportunities for greenhouse technology to make a significant contribution to the sustainability of agriculture and to the economic well being of farmers. It will be clear from a number of the papers being presented in this congress that the opportunities for profitable and sustainable agricultural production of high value plant products can be expected to increase throughout the coming decade and into the 21st century.

Greenhouse opportunities in India

There are a number of opportunities for greenhouse technology in India and the following discussion is by no means comprehensive. It must also be recognized that the degree of sophistication and level of investment per square meter of greenhouse must be appropriate to the application if the technology is to be economically justifiable. The following needs and opportunities are presented as a non-exclusive list and there is little doubt that the listing of opportunities will grow in the future.

There is a very great need throughout India for effective greenhouse facilities to support a range of basic agricultural research activities. A need exists in universities, government research stations and private research establishments. It should be emphasized that greenhouses are required in a variety of research projects on plant physiology, culture, plant development and production where controlled and replicatable environments are needed.

J. Prakash and R. L. M. Pierik (eds.), Horticulture – New Technologies and Applications, 265–272.

Many plant experiments require controlled and reproducible conditions even where protection against extreme weather conditions is not needed.

For research purposes, several greenhouse types are needed which can provide appropriate levels of control for various environmental parameters. In modern research greenhouses, technology can be utilized for the control of such environmental parameters as temperature, humidity, light level, light duration, carbon dioxide level, irrigation control, nutrient control, spacing etc. In some research activities, various levels of biological containment may be required, ranging from simple screening against ingress of local insects to containment to meet the regulatory requirements for research utilizing recombinant DNA techniques. For many research stations a variety of greenhouse facilities having adequate areas of space meeting selected technical requirements will be appropriate.

Modern tissue culture techniques are being rapidly adopted in many areas of the world, including India, for the multiplication of commercially important plants, as well as being utilized as a tool in plant research. In most cases plantlets produced in the tissue culture laboratory will require greenhouse facilities to provide an environment between the laboratory and the field to insure a high proportion of survival in the field. Knowledge of the specific environment required for this intermediate stage of production of plants from tissue culture will be acquired rapidly in the near future for a number of commercially important plants. As this technology develops it will be essential to evolve greenhouse environmental control systems suitable for providing the needed plant environments under existing external agro-climatic conditions in the area where the plants are to be produced.

Typically in the hardening off of tissue culture produced plantlets there are several intermediate environments between the laboratory growth room and planting out in the field. Immediately after coming out of the growth room the plantlets usually require light levels well below the maximum obtained in the greenhouse, relatively low temperature and a high level of humidity to avoid moisture stress. As the plantlets acclimate themselves the light levels can be increased and the humidity reduced. Therefore systems to provide several levels of controlled shading, misting and fogging are needed. While the costs of providing for precise environmental control on a per square meter basis are high, it is important to note that as plants are small and of high value each, the cost compared to the value of the product will not be unreasonable.

There are a number of opportunities in various agro-climatic zones in India for the commercial use of greenhouses in plant production. In many temperate areas of the world, vegetable farmers find that they can substantially increase income if they can start transplants in a greenhouse earlier than they can plant outdoors in the spring, enabling them to bring some produce to market early when prices are higher. Such an opportunity seems to exist in several areas in India, particularly in the Himalayan foothills regions where summer vegetables can be produced. Similarly, the use of greenhouses for starting of seedlings for plantation crops may be essential in some areas. In both of these situations a small greenhouse area can be utilized to produce a high number of quality plants that will be used to crop a much larger outdoor area.

The opportunities to utilize protected cultivation, especially low cost greenhouses, in the production of high quality and high value ornamentals and food items, should be carefully considered. The relatively affluent populations in many areas of the world, including India, provide an attractive market for high value plant products. There are strong trends in many societies indicating an increased interest in attractive house and garden plants, and as discretionary incomes increase, a strong agricultural business can be developed serving this market. Also, there is rapidly increasing awareness of the issues relating to good nutrition and a desire among many consumers for good quality fruits and vegetables that are free of pesticides. The opportunity for the greenhouse industry to supply a significant portion of this demand should not be overlooked.

Research priorities

A critical need for effective greenhouse utilization in most of the opportunities discussed above is knowledge and experience of the type of microclimate obtainable for various greenhouse technologies in specific locations and the responses of the plants to be grown to the microclimate. In this regard the greatest input of time and effort will be from the standpoint of determining plant responses to the environments that can be obtained under various local conditions. From the engineering side, the greatest need is for improved understanding of greenhouse cooling in general and adaptation of appropriate cooling technologies to various Indian conditions.

While a greenhouse is generally regarded as necessary to provide a warm environment in cold climates, it has also been shown that with properly designed cooling systems it is possible to improve plant growing conditions under excessively hot conditions. Adaptation of modern cooling technologies to Indian conditions will undoubtedly lead to increased opportunities for the production of high value plant materials in areas where the environment is extremely harsh. Protected cultivation also has the potential benefit of substantially increasing plant productivity per unit water consumption which is important in many areas where good quality water supplies are severely limited.

In order to determine and document optimum environmental control and cultural management practices required for various crops in different agro-climatic regions, research is needed by integrated interdisciplinary research teams. Engineers need to concentrate on the effects on the microclimate in the greenhouse of the various climate control technologies, especially with regard to cooling. Appropriate combinations of practices, including ventilation, evaporative cooling, controlled shading systems, fog systems and possibly nocturnal energy storage and selective transmission of glazing materials, need to be determined for various applications. Horticulturists and plant physiologists need to work out cultural recommendations for the crops of potential commercial or research interest under the microclimate conditions obtainable. Pest control practices must be developed for insects and diseases prevalent in each area. Markets need to be developed, and economic and social aspects of greenhouse enterprises must be studied under local conditions in different areas.

Environmental control, heating

For many, the notion of environmental control in a greenhouse tends to imply heating to enable crops to be grown when the outdoor temperature is too low for normal plant growth. The first reaction is often to the effect that as India is primarily a warm country there is little need to provide a warmer environment for plant growth and therefore little need for greenhouse technology. It is important to note that greenhouses are, in general, meant to provide a controlled environment for plant growing and therefore control of any environmental parameter related to plant growth may be of importance. Having made this observation, it is still very common to find temperature control of the first importance in greenhouse environmental control strategies.

While greenhouse heating is usually the most important and expensive environmental control activity it is also a relatively simple problem to solve from a technical standpoint. Factors that influence the heating requirements and cost include: structural design, the local weather, the crop temperature requirements, the type of heating system(s) employed, the energy source, energy conservation measures and management.

From the structural design standpoint the most important issues are the glazing system utilized and the design of the building. Regarding glazings, in general double glazing systems require only about 2/3 the heating capacity and fuel as single glazing systems. The properties of the glazing material itself are also important. The most common single layer glazing systems are glass and fiberglass for heated greenhouses and single layer plastic film for unheated rowcovers and low tunnels. Double layer glazing systems include air inflated polyethylene film and rigid structured sheets of acrylic or polycarbonate. Due to its low cost the air inflated double polyethylene glazing system is the most widely used. Small free standing structures require more heat per unit area of cultivation than large gutter connected multispan units as there is more exposed roof and wall area combined on the free standing units.

Heating systems commonly used in the greenhouse industry vary widely in cost and performance. Low cost unit air heaters are economical and popular for many commercial operations. For larger greenhouses perforated polyethylene ducts are used to distribute the warm air but, in general, the uniformity of temperature within the greenhouse cannot be as well maintained as is the case with other heating systems. Steam pipes for heat distribution and finned pipes with circulating hot water have received widescale application, particularly in classical glass greenhouses. Uniformity of heat distribution is generally better than with forced hot air systems and can be improved even further by the use of circulating hot water. Temperature control within the greenhouse is improved if the water temperature in the pipes is controlled at the level needed to maintain the greenhouse environment with constant water circulation. Placement of a portion of the heating pipes near the ground further improves uniformity of temperature control.

In recent years floor and bench heating systems have become popular in relatively warm climates as well as in the colder latitudes. By providing a basic heating system in the floor of the greenhouse or under the benches it is possible to maintain needed soil temperatures without resorting to an overheating of the air above the plant canopy. In colder latitudes a combination of low or root zone heating with a separately controlled air heating system has become popular as it gives the grower the maximum control of the plant environment at the minimum energy cost. For some crops and in some more moderate climates floor or bench heating has been found adequate without any provision for air heating. It is anticipated that floor or bench heating systems circulating warm water in low cost plastic pipe will provide adequate heating in many applications in India.

In the earlier energy crisis of the mid 1970's a great deal of effort was put into the development of energy conservation systems and one of the most effective was the use of movable night curtains to retain heat at night. Though the initial incentive for the use of curtain systems was for the purpose of energy conservation it soon became apparent to many growers that there was also a benefit to be derived from using the curtain systems to shade the crop at the peak of the daytime heat. As this benefit became more important selection of the material for the curtain system has been based more on the appropriateness of the properties of the curtain for daytime shading than for nighttime heat conservation.

Environmental control, cooling

In all areas where greenhouses are used, cooling is needed as the solar gain near the middle of the day exceeds the requirement of the greenhouse for heat, even in the coldest climates.

In many areas and for many crops adequate cooling can be obtained with ventilation alone. In many cases natural ventilation alone can be relied upon to provide sufficient cooling. Natural ventilation with side and ridge ventilators is most effective on smaller single span greenhouses. Ridge vents can be adequate for cooling multispan glass greenhouses in colder climates. In warmer climates fan ventilation will be required and the very popular polyethylene greenhouses are generally ventilated with fan systems.

Adequate airflow is the first requirement for any cooling system designed for the majority of Indian conditions. A basic rule of thumb is that at least one volume of air change per minute must be provided for large greenhouses where the fan to air inlet distance is about 30 meters. Higher relative airflow rates are required in shorter houses to provide adequate air velocity under the highest cooling requirement conditions. It is good practice to arrange for the fan system to operate in two to four stages so that airflow can be matched to the cooling requirement at any given time. With a properly designed ventilation system temperatures within a fully cropped greenhouse will not exceed more than three to five degrees hotter than outside ambient.

For many crops in most areas of India there will be times when it is desirable from the plant standpoint to keep the greenhouse cooler than ambient if possible. A popular method for achieving greater cooling than can be provided with ventilation alone is evaporative cooling. With a well designed evaporative system air temperatures approaching the ambient wet bulb temperature can be obtained. It should be noted that even in relatively humid locations significant benefits can be obtained with evaporative cooling. Temperatures in the middle of the day and early afternoon are significantly hotter than at dawn and as the specific humidity of the air does not usually increase very much during the day there is a substantial wet bulb depression during the hottest part of the day even in the humid tropics. Recent research and commercial experience in Florida, U.S.A. and in Taiwan confirm this.

The most common greenhouse evaporative cooling systems for a number of years have been pad and fan systems where the ventilation air is drawn into the greenhouse through a wet pad where evaporation and air cooling take place. Carefully manufactured labyrinth designs for wet pads are generally capable of more efficient cooling than Aspen type pad systems. In recent times several types of commercial fogging systems have become available in the market which have the advantage of distributing the effect of the evaporative cooling within the greenhouse as well as at the air inlet. One type of fogging system utilizes a special nozzle in which high pressure air and water are mixed to produce extremely fine fog. Other popular systems rely on high pressure water sprayed out through very fine holes to achieve small drop sizes. Good water quality and careful filtration are required to keep fog systems operating at full effectiveness and good system maintenance is essential.

It is the authors opinion that, in general, more emphasis is put on temperature reduction in the greenhouse than is warranted and that in fact the issue that should be addressed is the reduction of plant stress. While the ambient temperature is a major factor in plant stress there are other important parameters such as the direct radiant heat load on the plant, the ability of the plant to take up water and the water vapor deficit in the air. It should be possible to produce good quality plants under fairly high ambient temperatures if total plant stress can be reduced by addressing other environmental issues. In this regard controlled shading systems and fog cooling have been shown to be extremely effective management tools.

Environmental control, light

Light control is mentioned above with reference to cooling and reduction of plant stress. It has long been commercial greenhouse practice to control daylength in order to regulate plant development and provide for the critical timing of flowering of certain crops to meet specific marketing timetables. In fact the development of energy saving night curtains discussed above was based in large measure on technology used to black out the greenhouse to shorten photoperiod.

In addition to photoperiod control, the use of supplemental photosynthetic lighting to enhance plant growth is becoming an important commercial practice for some crops. For some high value crops where the extra yield can justify the investment and operating costs, or in cases where the timing of the crop is critical to marketing, supplemental light can be justified. High pressure sodium lamps are the most common in commercial practice. The effectiveness of the lighting on crop production is influenced greatly by the uniformity of the light intensity at the plant canopy. Therefore, a lamp reflector specially designed for horticultural lighting with a uniform distribution pattern within one to two meters of the light source is essential.

Environmental control, carbon dioxide

As long as light, water or plant nutrients are not limiting it is possible to enhance plant growth substantially by supplementing carbon dioxide levels significantly above ambient atmospheric levels. The addition of carbon dioxide is particularly important where supplemental lighting is being utilized and is also, in general, only justified in research activities or where very high value plant material is being produced. Carbon dioxide enrichment is not practical when the greenhouse is being vented for cooling as the added carbon dioxide will be almost immediately removed. The possibility of utilizing an energy storage system to provide cooling some of the time without venting has the potential of enabling the practical period of carbon dioxide enrichment to be extended.

Environmental control technology

In many areas the use of computers to control the greenhouse climate is becoming a popular and widespread practice. With the use of a computer based control system it is often possible to maintain more precise environmental conditions than is the case with simple thermostats, time clocks and other classical devices. It is also possible to develop control strategies where temperature, humidity, light level and other setpoints are adjusted as functions of the time of day and environmental parameters such as light levels. Another aspect of computer control of the greenhouse climate which is not as obvious as the above points, but none the less important is the data collection aspect of the computer based control system. Once the sensors for monitoring the environmental parameters are in place it is a simple task to record important data in the computer memory for analysis and display.

Many commercial producers find that the ability to record and analyze the environmental data related to the production of their crops enables them to refine and improve their management strategies resulting in an improvement in their productivity or the quality of their product. In research establishments the use of computer based data acquisition systems should be a priority even if the computer is not used for environmental control. In analyzing and interpreting plant data from greenhouse experiments it is essential to be able to correlate the data with the environment that was actually obtained in the greenhouse during the conduct of the experiment. Without having the actual data, there is the likelihood that it will simply be assumed that the temperatures obtained in the greenhouse were those indicated in the thermostat settings. In fact this is seldom the case in most research greenhouses. Also, for many experiments data on light levels, carbon dioxide levels, humidity and other parameters are as important as temperature alone.

Acknowledgement

New Jersey Agricultural Experiment Station Publication No. J-03232-37-90 Supported by State funds.

References

ANONYMOUS. (1987) Report of the 5th Meeting, Indo-US Subcommission on Agriculture, New Delhi, December 7-11.

ANONYMOUS. (1989) Active Foreign Agricultural Research Grants, (Foreign Currency Research Program {PL 480}), (U.S.-India Fund),(U.S.-Poland Joint Fund), (U.S.-Yugoslavia Joint Board). USDA/OICD. Washington DC, September 30.

Chadha, K.L., R. Gayen, D.R. Mears and B.P. Srivastava. (1988) Protected Cultivation and Greenhouses, Subproject paper for ICAR and USAID/New Delhi, June 7.

Chandra, P., J.K. Singh and G. Majumdar (1990) Prediction of the cooling pad temperature in a fan-pad cooling system used in greenhouse. Proceedings of the XI International Congress on the Use of Plastics in Agriculture, New Delhi, India.

Eshleman, W. D., C. D. Baird, and D. R. Mears (1977) A Numerical Simulation of Heat Transfer in Rock Beds, Proceedings of the 1977 Annual Meeting, American Section of the International Solar Energy Society, Orlando, Florida, June 6-10.

Giacomelli, G.A., M.S. Giniger (1984) Mist and fog evaporative cooling in greenhouses. Proceedings of the 18th National Agricultural Plastics Conference, Asheville NC.

Giacomelli, G.A., M.S. Giniger and A.E. Krass (1985) Utilization of the energy blanket for evaporative cooling of the greenhouse. ASAE Paper No. 85-4047.

Giacomelli, G.A., M.S. Giniger, A.E. Krass and D.R. Mears (1985) Improved Methods of Greenhouse Evaporative Cooling. Acta Horticulturae 174. pp 49-56. International Symposium on Greenhouse Climate and its Control. Waginengen, Netherlands.

Manning, T.O., D.R. Mears, R. McAvoy, and W.J. Roberts (1980) The flooded floor for waste heat utilization in greenhouses. Acta Horticulturae 115, International Symposium on More Profitable Use of Energy in Protected Cultivation, Dublin, Ireland. September 7-12.

Manning, T., D. Mears, R. McAvoy, and B. Godfriaux (1980) Waste heat utilization in the Mercer research greenhouse. Foliage Digest, Vol. III, No. 8, Apopka, Florida. September.

Mears, D.R., W.J. Roberts, J.C. Simpkins, P.W. Kendall, J. P. Cipolletti, and H. Janes (1980) The Rutgers system for solar heating of commercial greenhouses. Acta Horticulturae 115, pp 575-582 International Symposium on More Profitable Use of Energy in Protected Cultivation, Dublin, Ireland. September 7-12.

Mears, D.R. (1987) Alternative Energy Sources for the Future. Proceedings of the National Conference on Energy in Production Agriculture and Food Processing. Ludhiana, India. Oct. 30-31, 1987.

Mears, D.R. (1987) Greenhouses. Proceedings of the Conference on the Use of Plastics in Agriculture. Indian Council of Agricultural Research. New Delhi, India.

Roberts, W. J., D.R. Mears, and M. James (1980) Floor heating of greenhouses. Acta Horticulturae 115, pp.259-267, International Symposium on More Profitable Use of Energy in Protected Cultivation, Dublin, Ireland. September 7-12.

Roberts, W.J. and D.R. Mears (1980) Progress in movable blanket insulation systems for polyethylene-covered greenhouses. Acta Horticulturae 115, pp.685-692, International Symposium on More Profitable Use of Energy in Protected Cultivation, Dublin, Ireland. September 7-12.

Roberts, W.J. and D.R. Mears (1980) Warm water solar systems bring greenhouse savings. U.S.D.A. Yearbook, Section 4, Chapter 5-B.

Roberts, W.J. and D.R. Mears (1984) Heating and Ventilating Greenhouses. Extension Publication, Rutgers University, Cook College, New Brunswick, NJ.

Roberts, W.J. and D.R. Mears (1984) Floor Heating of Greenhouses. Extension Publication, Rutgers University, Cook College, New Brunswick, NJ.

Roberts, W.J., J.C. Simpkins and D.R. Mears (1984) Performance of IR Polyethylene Over Two Years' Trial. ISHS Engineering Conference, University of Hanover, Hanover, Germany.

Roberts, W.J. and D.R. Mears (1980) Energy use in greenhouses - how low can we go? 1980. Acta Horticulturae 115, pp.143-149, Invitational Keynote paper, International Symposium on More Profitable Use of Energy in Protected Cultivation, Dublin, Ireland, September 7-12.

Roberts, W.J. and D.R. Mears (1980) Research on conservation and solar energy utilization in greenhouses. ASHRAE Annual Meeting, Denver, Colorado. ASHRAE Transactions 86,2. June 22-25.

Roberts, W.J. and D.R. Mears (1980) Energy conservation and new energy sources in the greenhouse. Invitational paper. Proceedings of the Les Floralies International de Montreal. (English and French) August 22.

Sase, S. and L.L. Christianson (1990) Screening greenhouses - some engingeering considerations. ASAE Paper No. NABEC 90-201.

Walker, J.N. and D.J. Cotter (1968) Cooling of greenhouses with various water evaporating systems. Transactions ASAE Vol. 2.

Wu, C.H., H.S. Chang, J.T. Shaw and T.C. Kao (1990) Engineering protected cultivation systems in Taiwan. Annual report to the OICD/IRD, USDA of the Taiwan Agricultural Mechanization Research Center, Taipei.

NEW SEED-GROWN CUT FLOWERS FOR FLORICULTURE INDUSTRY

Cornelis Kieft, Director of Kieft Bloemzaden B.V., Blokker, Holland at the International Seminar on New Frontiers in Horticulture at Bangalore India on November 26, 1990.

The cut flowers we will discuss this morning belong to the group of Specialty Cut Flowers or Summer Cut Flowers and comprise an extensive range of genera, species and cultivars in many types, forms and colors. The other major group of cut flowers are called Standard Cut Flowers and in this group we list Roses, Carnations, Gerbera, Chrysanthemum, Lilies and many others. They are grown on a large scale, often under climate controlled conditions, in many countries, and they are well-known both to the trade and the public.

The Specialty Cut Flowers are not entirely new. They are being grown in Holland and several other European countries, in the USA and above all in Japan for many years. One of the reasons that Standard Cut Flowers have been dominating the international cut flower markets was the demand for mass marketing and consequently mass production after World War II, to which several of these flowers are well adapted as a result of the development of new varieties and improved growing, grading, packing and shipping techniques.

However, starting in the late seventies this picture has gradually changed in the advantage of the Specialty Cut Flowers which are becoming more and more popular every day.
One of the many reasons for this change in taste of the large public in the "spiritually underdeveloped countries" is the rising awareness of the existence of other flowers and plants which are also very suitable for use in the vase or in the garden.
Certainly many flower arrangers around the world, being somewhat tired of arranging always the same flowers and composing always the same type of bouquets have contributed to a large extent to the ever increasing demand of the Specialty Cut Flowers.

The Specialty Cut Flowers can be divided in flowers used for the fresh- and for the dried flower market.
The production of the Specialty Cut Flowers both for fresh- and dried use takes more and more place in Central- and South American countries for the supply of the North American market and in Africa, Kenya and Zimbabwe, in particular for the supply of the European market.
Very often flowers grown in Africa are flown into the Aalsmeer Auction and then exported again to the Middle East, the USA and even as far as Japan.

India, being halfway between the Far Eastern and the European markets could play an important role in the production and supply of Specialty Cut Flowers, and we certainly should not forget the possibilities of the domestic market.

J. Prakash and R. L. M. Pierik (eds.), Horticulture – New Technologies and Applications, 273–283.

Now let us have a closer look at some of the latest introductions of Specialty Cut Flowers:

ACHILLEA ageratum 'Moonwalker' "Sweet Yarrow"

Native to the south-western part of Europe, Achillea ageratum has been in cultivation since 1560. And that is quite a while by now!! Probably she was used as an aromatic herb and as a medicinal plant in those days.
'Moonwalker' may probably be used in that same way as well, but our selections were focused on the development of a cut flower.
By nature she is a long-living perennial and flowers easily the same summer when sown early in the year.
The flowerhead looks like a small version of the well-known A. filipendulina (Parker's Var.) which makes growing, grading, packing and handling in general a much easier job. Provided that the soil is well-drained, 'Moonwalker' will grow in any (poor) soil in both sunny and shady locations. Full-grown she reaches a height of 2-2½ ft.
'Moonwalker' adapts herself to the use as a fresh- and as a dried flower and carries a nice sweet aroma around her. Last but definitely not least 'Moonwalker' does an excellent job as a filler in mixed bouquets. Post harvest care: Flo 500.

AMARANTHUS cruentus Red Cathedral'

An Amaranth becoming increasingly popular with flower arrangers in Europe for use in large bouquets and arrangements, both fresh and dried. Also, like 'Green Thumb' and 'Pygmy Torch' an upright growing type, reaching a height of 3-5 ft with dark red flowers. A distinctive and eyecatching plant. Same growing instructions as 'Green Thumb' and 'Pygmy Torch'.

AMARANTHUS hypochondriacus 'Green Thumb' and 'Pygmy Torch'

Linnaeus (1707-1778), the famous Swedish botanist, must have had a very bad day when he named this Amaranthus species, because hypochondriacus means "of somber aspect" and hypochondria is not a pleasant way of living. In our opinion there is nothing somber or hypochondriac about these 2 varieties. On the contrary, they are fine and easy to grow cut flowers, used in fresh- and dry mixed bouquets.
In particular the 'Green Thumb' variety with her pretty greenish-yellow spikes is a very fashionable color and in high demand with modern flower arrangers. Height is 1½-2 ft, provided they are sown where to flower and sown thick so they grow close together, stretch and obtain the required stemlength. They will grow in practically any soil and prefer warm weather and a sunny location.

AMMI majus "Bishop's Weed" or "Queen Anne's Lace"

An annual of the Carrot family, native to the Mediterranean, which all of a sudden became an important cut flower. Mainly used as a filler in mixed bouquets. Ammi is grown on a large scale in East Africa and then shipped into Europe. It is often dyed in all kinds of (funny) colors. New selections are being worked on to obtain a better upright and compact growth and stronger stems. Easily grown, producing large heads of lacey, white flowers reaching a height of about 3 ft.

AQUILEGIA vulgaris 'Nora Barlow' "Columbine"

Coming more or less true from seed is this most unusual Columbine variety with blooms that start as highly decorative, tight double green buttons and open up into attractive rosy-red, pink and green flowers on 2-3 ft stems. The flowers do not have spurs which makes them easy to handle, to pack and to ship.
A fantastic, earlyflowering perennial, long-lasting cut flower much sought after by flower arrangers. Can also be used as dried flower.
Post harvest care: we obtained good results with Gard/Rogard.

ASCLEPIAS currasavica 'Red Butterfly' "Blood Flower"

Native to the beautiful island of Curaçao and other islands of the West-Indies, this is one of the most "multipurpose" plants we ever came across. A. currasavica has been in cultivation since 1655 and they must have forgotten to tell us about her. Anyway she is back now and make sure you see her soon!!
'Red Butterfly' is a fantastic cut flower for the greenhouse and outdoor in warmer climates. She prefers a semi-shade position and reaches a height of 2-3 ft. She needs to be grown on the dry side with very little N fertilizer. Also she can be grown as a beddingplant in cooler areas and then she will reach a height of 20-25 cm.
Last but not least 'Red Butterfly' does an excellent job as a patio plant, a couple of plants planted together in a large container. In frost areas you have to bring the pots inside in winter time.

ASCLEPIAS incarnata 'Soulmate' "Swamp Milkweed"

Although most people believe that every man and woman has a more or less developed soul, and we believe that as well, we have never seen one. We are pretty sure however, that the color of the soul is similar to this species, hence the name 'Soulmate'.
Besides that, 'Soulmate' turns out to be an outstanding cut flower and we believe that she definitely deserves much greater popularity and to be more grown, so everybody can become a 'Soulmate' with her or him (whatever you like). She decorates herself with highly fashionable rose-pink flowers which are borne on strong stems of about 70-90 cm tall. She was hiding in the swamps and other wet areas of the Eastern USA and got fairly well established there. Reason for which she prefers to be grown in wet soils but even in our trials in 1989 (dry summer) she did quite well. Our tests show a vase-life, without preservatives, of 10-14 days. A perennial, which flowers the same year when sown in Jan.-Febr.

BUPLEURUM rotundifolium 'Griffitii'

Originally a weed from the Mediterranean area which all of a sudden became an important cut flower and is grown now on a large scale in Europe and East Africa.
Bupleurum is an intriguing plant with upright, branched stems loaded with greenish-yellow flower umbels. Both flowers and the sickle-shaped leaves remind very much to certain Eucalyptus species used by flower arrangers. An excellent filler for mixed bouquets, provided it is cut at the right stage. An annual plant, easy to grow and reaching a height of 2-2½ ft.

CENTRANTHUS ruber 'Snowcloud' "Valerian"

A lot of confusion about the name of this plant. Back in 1557 when the plant became in cultivation, she was called 'Valeriana', then it became Kentranthus and now we believe her family name should be Centranthus. The well-known red-flowering Centranthus ('Pretty Betsy') is native to Southern Europe and now naturalized in many parts of the world.
'Snowcloud', her white-flowering relative, has not been travelling that much yet but there is no doubt that she will become a cosmopolitan in the very near future too. They both are delightful, easily grown, long-living perennials, well suited for beds and borders and, if grown in poor soil so that she remains dwarf, also an excellent rockgarden item. They flower the same year from an early sowing. 'Snowcloud' is also an excellent and highly fashionable cut flower with her clusters of pure white and fragrant!! flowers above bluish-green leaves. Fully grown in good soil, she reaches a height of 90-120 cm. 'Snowcloud', another Specialty Cut Flower For The Nineties!

DELPHINIUM 'Messenger Series' and D. 'QIS Series'   "Larkspur"

Larkspurs are important cut flowers both for the fresh- and the dried flower cut flower market. For outdoor production the 'QIS Series' (formerly called 'Sunburst') is far out the best selection presently on the market and is grown on a large scale in Europe, East Africa, Japan and other countries.
This 'Messenger Series' (D. ajacis) is brand new and about 2 weeks earlier than the 'QIS Series" (Sunburst) (D. consolida). Therefore 'Messenger Series' is the best choice for indoor (greenhouse) production in the northern part of Europe and the USA. They can of course also be grown outside.
Flowers of the 'Messenger Series' are fully double with short upright growing side branches that facilitate cutting, grading and packing to a large extent. Dark blue and white represent the first colors available in this earlyflowering series. More colors will be released in the near future. The earlyflowering 'Messenger Series' and the standard 'QIS Series' are the best choice for the Specialty Cut Flower Grower.

DELPHINIUM semi-barbatum (D. zalil, D. sulphureum)

A lot of names for this beautiful Delphinium which as far as we know is the only cultivated yellow Delphinium on the market and in high demand with specialty cut flower growers around the world. A short-living perennial with a delicate root system and native to semi-desert regions in Iran and surrounding countries with a height of 2-3 ft. A real beauty with bright yellow flowers in long dense racemes with a vase life of 10-14 days.
This Delphinium requires a dry and sunny location. Not an easy one to grow but definitely worthwile to give a good try. Seed always in short supply.

DIANTHUS caryophyllus 'QIS Series'   "Carnation"

The 'QIS Series' is an entirely new selection of the 'Chabaud Series', once widely grown at the French and Italian Riviera.
'QIS'Carnations produce large, fully double flowers on strong, 2-3 ft stems. They are best grown as annual and flower in 3-4 months from sowing. They can be grown both outdoor in warmer climates and indoor under glass or plastic using 1 or 2 layers of netting for support. Available in 6 separate colors.

FIBIGIA clypeata 'Roman Shields'

An interesting and useful perennial whose ancestors are native to Southern Europe from Tirol in Austria - where they used this plant to study yodelsinging, that famous mountain singing in the Alps - to Afghanistan.
Fibigia produces small, yellow flowers in spring, which is nice of course, but for cut flower use we are more interested in glaucous-green elleptic seedpods they develop afterwards and which we call 'Roman Shields', because they resemble a shield (and that is the meaning of clypeata). A small shield all right but still there is an increasing demand by flower arrangers for this material for use in dried flower arrangements but also for use in fresh mixed bouquets.
Easy to grow in practically any soil and as special feature a high resistance against hot dry weather. They require a cold period (winter) for flower induction. Vase life: for ever.

GAURA lindheimeri 'The Bride'

One way to describe this plant is to use the nice wording of Mr. Thomas H. Everett in the NYBG Encyclopedia of Horticulture:
"An unobtrusive plant of airy grace that will appeal more to those who appreciate the unusual and somewhat rare than seekers of brilliance."
We baptised this selection 'The Bride' and consider her a beauty, like all brides. She was rather unknown and we first heard from her from friends in South Africa who told us about the superb cut flower qualities of Gaura. And indeed, in our trials in Holland she showed promising qualities, which finally led to the selection and development of 'The Bride'.
Gaura is a perennial plant and prefers well-drained, moderately fertile soil and a sunny location. She will flower the first year from an early sowing. This is a good cut flower for dry growing areas because she withstands extreme drought and can live without irrigation for a long period of time.

GOMPHRENA 'Strawberry Fields' "Globe Amaranth"

The first true red Gomphrena, one of nature's finest everlastings. A beautiful, free and continous blooming plant, retaining its form and color indefinitely. Large blooms are borne in great profusion and are delightful in bouquets, either fresh or dried.
'Strawberry Fields' is also very useful for mass planting, borders and window boxes. For cut flower use we suggest to sow in rows 15 cm. apart, which forces the plant to stretch and to obtain sufficient stemlength.

Under greenhouse conditions we obtained a high production with strong 2 ft stems from the end of May onwards with sowing in early March. After cutting the plants back a second flush with the same stemlength appeared in August. Bottom heat is advisable to speed up germination for winter- and early spring crops in colder areas. Soil temperature should not drop below 15°C (59°F). Direct seeding gave very good results at a rate of 50 grams per 100 running meter (abt. 32 ft) with a distance of 15 cm (6 inches) between the rows. No netting required.

HELICHRYSUM cassianum 'Rose Beauty'

It is unbelievable that this beautiful species is not listed in Hortus 3 and other important reference books. This is a beauty from the desert of West-Australia and deserves to be more widely known and grown.
It is a perfect cut flower with trusses of small, starry, pink, everlasting flowers which can be used for both fresh- and dry mixed bouquets. A highly fashionable pastel color much in demand by flower arrangers.
'Rose Beauty' is easy to grow in practically any soil and in a sunny location. An annual plant with a height of 2 ft.

INCOGNITA calophyta 'New Gypsy'

All of a sudden there she was, amongst hundreds of other plants we collect from all over the world and plant in our trial fields each year. We don't know much about this beauty but we believe that she has a huge potential as a cut flower. We already gave her a name called 'New Gypsy' which gives an indication of our feelings that this strong grower may once cooperate with the well-known Gypsophila. She adapts herself to the use as fresh and a dried cut flower. In autumn her flowers turn into a kind of autumn color.

LIMONIUM gmelinii 'Perestrojka' "Statice"

This is a 'cool' one, native to Eastern Europe and Siberia (of all places), brrr ... We named her after and in honour of Mr. Gorbatsjov's crusade against oppression. Finally she got a chance to get out of there and now she is spreading around the world, becoming more popular every day as a very valuable, long-lasting cut flower and border plant. A perennial and a hardy one ...
Her color is beautiful soft-lilac blue. She is a strong grower, who produces some flowers the first year from an early sowing. She is suitable for use as a fresh- and as a dry flower and feels very comfortable with other flowers in mixed bouquets. We really clean her seed and with a litte bit of striptease, we manage to supply decorticated seeds of this beauty. A fine Specialty Cut Flower which deserves to be more widely grown and known. Height: 2 ft.

LIMONIUM perezii 'Atlantis' "Statice"

This native of the Canary Islands and naturalised in California is becoming more and more popular. 'Atlantis' is very useful as a dried and a fresh cut flower and is in high demand with flower arrangers. The color of 'Atlantis' ranges between light and dark lavender blue with large flower heads borne on strong stems and short upright growing side branches. They prefer well-drained, moderately fertile soil and a sunny location. They ship well. This variety 'Atlantis' is specially developed for cut flower purposes and reaches a height of 2-3 ft. As per the 1990 crop decorticated seed is available.

LIMONIUM sinuata 'QIS Series' "Statice"

The best Statice series presently on the market and available in 5 separate colors with large "combs". The white is a pure white color without "stoning", and therefore very much in demand for dyeing. The yellow color, botanically L. bonduelli, resembles the other colors by also having "wings" alongside the stem. Pale-blue is a highly fashionable pastel color and is becoming more and more popular.
Both dark-blue and deep rose are strong colors without shades.
At several occasions 'QIS' Statice obtained double the average day-price at the Aalsmeer VBA Auction. They are very vigorous in growth and suitable for both in- and outdoor production.

LOBELIA speciosa 'Compliment Scarlet' F-1 Hybrid

A new species that makes an excellent long-stemmed, longlasting cut flower. Brilliant scarlet spikes with green leaves. This is a 1989 Fleuroselect winner. Flowers the first year from an early sowing. An outstanding cut flower and a real eye-catcher in the flowershop with a vase life of about 10-14 days. 'Compliment Scarlet' is best grown as an annual in colder climates but is suitable for both in- and outdoor cut flower production.
In milder climates it can be grown as a biennial with sowing in autumn and flowering in early spring. Height 75 cm (2½ ft).

LYSIMACHIA 88-A2-051

The predominant colors in Lysimachia are white and yellow, but this is a dark wine-red species with a height of 3-3½ ft. It flowers already in May and we believe that with further selections this plant could become an interesting cut flower in the future.

NIGELLA orientalis 'Transformer'

The predominant colors in Nigella are blue, rose and white. This is a yellow one and a curiosity from a botanical point of view.
However, it is not only the color which makes this Nigella so special but above all the curiously shaped seedpods which are in high demand by artistic flower arrangers for use both in fresh- and dry mixed bouquets.

OXYPETALUM caeruleum 'Heavenborn'

This beautiful plant is really a gift from Heaven. She got herself established in Uruguay and Brazil and was waiting to be discovered. Fortunately we did. She has got spectacular sky-blue flowers becoming darker with age, a color seldom found in the plant kingdom.
She belongs with many other beautiful plants to the Asclepias family and we, and others found her an excellent cut flower. Dr. Allan Armitage from the University of Georgia asked us to make selections and to grow seed because this Queen will have a royal future as a cut flower. Basically she is perennial, but the best way is to treat her as an annual. Being native to a nice and warm part of South America, she does not like the cold of the North (who does?) and therefore in Northern Europe and USA she needs to be grown under cover. According to Allan, 70-75°F day temperatures are ideal. Since she has a slightly twining habit, we use 2 layers of netting to support the stems. She prefers high light conditions!! Shelf life appears to be 10-14 days but can probably be increased with preservatives. She needs to be cut when 2-3 flowers are open.

PLATYCODON grandiflorus 'Florist Series' "Balloon Flower"

A new series, especially developed for cut flower purposes and available in dark-blue, pink and white. Longliving perennials provided that the rootsystem is not disturbed. This series grows easily from seed with a good amount of flowers the first year after an early sowing. Other good cut flower strains are 'Komachi' from Japan with clear blue flowers and 'Shell Pink', rose pink. Hardiness range: zone 3-8. Height: to 3 ft.

SCABIOSA atropurpurea 'QIS Series' "Pincushion Flower"

A brand new series of annual Scabiosa, especially developed for cut flower use and available now in 3 separate colors. And more to follow ....
Contrary to the old strains this series has an upright growth with short and also upright growing side branches which facilitates the growing, grading and packing to a very large extent.
'QIS Series' are very uniform both in height and in particular in flowering time. Our tests showed a vase life of 10-14 days with preservatives and our local flower arranger, who always first tests our cut flower novelties, is delighted about the use of 'QIS' both in mixed bouquets and as a single or mixed bunch of Scabiosa.
They grow about 80-90 cm tall and are best direct sown in rows on the place where they have to flower. Sowing in plugs and then transplant into the field or greenhouse is also very well possible.
Another top quality cut flower, which deserves the title 'QIS'.
Post harvest care: we obtained best results with Florissant 400.

The brand name 'QIS' means

QUALITY......IN.......SEED

and is used exclusively for our very best cut flower selections. 'QIS' is a registered trademark of Kieft Bloemzaden B.V. and replaces the formerly used name 'Sunburst'.

OTHER GOOD CUT FLOWERS TO BE GROWN FROM SEED

- Ageratum houstonianum F-1 Blue Horizon - blue
- Anemone F-1 Mona Lisa separate colors + mixture
- Antirrhinum (Snapdragon) F-1 Rocket Series separate colors + mixture
- Aquilegia (Columbine) F-1 Music Series separate colors + mixture
- Armeria (Thrift) Kieft Pastel Shades, mixture of pastel colors
- Atriplex hortensis separate colors - for use as filler in mixed bouquets
- Callistephus (Annual Aster) Matsumoto Series separate colors + mixture
- Carthamus tinctorius Early Round Leaved White and Yellow Grenade
- Celosia argentea plumosa Rocket Series separate colors + mixture
- Chrysanthemum segetum Paradiso - golden yellow with dark centre
- Cirsium japonicum Rose and Pink Beauty
- Craspedia globosa Billy Buttons - yellow
- Dianthus barbatus (Sweet William) separate colors + mixture
- Didiscus caeruleus (Blue Lace Flower) - blue
- Eucalyptus Silverdollar - foliage for mixed bouquets
- Eucalyptus Silverdrop - foliage for mixed bouquets
- Euphorbia (Spurge) marginata - green veined silvery white foliage plant
- Eustoma (Lisianthus) russelianum Yodel Series separate colors + mixture available in normal- and pelleted seeds
- Freesia hybrida Royal Crown Series separate colors + mixture
- Godetia grandiflora F-1 Grace Series separate colors + mixture
- Helichrysum monstrosum separate colors + mixture for fresh and dry use
- Liatris spicata Floristan Violet and Floristan White
- Matthiola (Stocks) 100% double Flash Series separate colors + mixture
- Moluccella laevis (Bells of Ireland) - used as fresh- and dried cutflower
- Ranunculus asiaticus F-1 Victoria Series separate colors + mixture
- Saponaria vaccaria White Beauty and Pink Beauty
- Trachelium caeruleum Purple and White

Cut Flowers in the Western Hemisphere

ROBERT E. DANIELSON
BALL SEED COMPANY, WEST CHICAGO, ILLINOIS U.S.A.

The world floriculture business is getting larger rapidly. Southeast Asia is becoming a world player in the production and consumption of cut flowers.

For a country such as India to become a producer and exporter of cut flowers, they must realize the problems of the industry and develop the knowledge.

- You must produce and export quality cut flowers to compete on any market.
- The products must be free of disease and insect organisms.
- The products must be grown with proper nutrition and other culture requirements maintaining plant health.
- Micro climates must be selected for each species of plants that make it easy to grow quality flowers. Night and day temperatures are available at different altitudes and other micro-climatic conditions for culture of plants within the country.
- Knowledge of harvesting techniques, post harvest treatments and handling procedure are required.
- The transportation of cut flowers to market needs to be fast and cost effective. The transportation of cut flowers to a market could be one of the greatest obstacles to be overcome.

J. Prakash and R. L. M. Pierik (eds.), Horticulture – New Technologies and Applications, 285–291.

I do not know the potential of the domestic market for cut flower production in India. It is the best place to develop the technology to become exporters of cut flowers.

The three major crops of Roses, Carnations and Chrysanthemums make up the largest part of the market. These flowers have good shelf life, can be shipped and have good consumer value.

Each of these species are grown best at specific micro climates or would require environmental controls for heating or cooling, depending on location at which they were grown.

The United States imports 80% of the Carnations and Chrysanthemums consumed from South america, mostly from Columbia.

The United States also imports 50% of the roses it consumes and this is increasing.

The low cost of Columbia's production in micro-climates at various elevations within the country and good air transportation to the United States market makes it cost effective. The industry in South America has developed over the last 20 years. The quality of the flowers produced and shipped to the United States has been very good.

The success of growing any cut flower crop will require locating them in the best natural climate for the plant species. As a rule, cut flowers should be grown at cooler night temperatures with warm days and bright sunlight.

It would not be good to locate production where you have heavy rainfall and high humidity which would cause disease problems of the flowers or foliage.

Rose and Chrysanthemums should be located in a high light area with a deep loam or sandy loam soil with good drainage. Night temperature should average between 15° C - 18° C and day temperature should not exceed 29° C for the best production.

Carnations require cooler night temperature for quality production; night temperature should be 8° C and 11° C. Day time temperatures should range from 18° C to 24° C.

In the United States, along the Southern California coast, a micro-climate exists of cool nights and warm sunny days, where quality cut flowers are grown year around with very minimal structures required.

Growing structures are required for seasonal colder weather. Structures need not be elaborate or costly, but should be built to hold poly film and screen shading where necessary. This protects the crop from rain, wind and too bright of light conditions for the developing blooms.

Watering and fertilizing is usually applied through some type of watering tube directly to the soil. Overhead watering cannot be used on cut flowers being harvested without some damage.

Water soluble fertilizers are used where the watering system permits.

Most crops have specific disease problems and if they are not controlled, they can become limiting.

SPECIFIC PROBLEMS

ROSES - Problems are Black Spot and Mildew. Because of this, roses need to be produced where the foliage can be kept dry.

CARNATIONS - problems are fusarium and virus. These build up in the soil. Because of this, all soil should be sterilized and plants obtained from cultured index stock.

CHRYSANTHEMUMS - Chrysanthemums are a short term crop of only 16 weeks and need replanting for each maturing crop. A disease free indexed stock block needs to be established for production.

It will be very important that cut flower growers and shippers master the techniques for maintaining the quality and performance of products on a world market, if they are to compete.

Factors Affecting Post Harvest Quality

Flower Maturity - Minimum harvest maturity for a cut-flower crop is the stage at which harvested buds can be opened fully and have satisfactory display life after distribution. Many flowers are best cut in the bud stage and opened after storage, transport or distribution.

<u>Temperature</u> - Respiration of cut flowers, an integral part of growth and aging, generates heat as a by-product. Furthermore, as the ambient temperature rises, the respiration rate increases. Rapid cooling and proper refrigeration are thus essential for maintaining quality and satisfactory vase life of cut flowers.

<u>Food Supply</u> - Starch and sugar stored in the stem, leaves and petals provide much of the food needed for cut-flower opening and maintenance. These carbohydrate levels are highest when plants are grown in high light and with proper cultural management. Flowers are preferably harvested in the early morning, because temperatures are low, plant water content is high, and a whole day is available for processing the cut flowers.

<u>Water Supply</u> - Cut flowers, especially those with leafy stems, have a large surface area, so they lose water and wilt very rapidly. They should be stored at relative humidities above 95% to minimize water loss, particularly during long-term storage.

<u>Bacterial Plugging</u> - Substances produced by the bacteria, and the bacteria themselves, can play the water-conducting system. For this reason, it is important that buckets be cleaned and disinfected regularly, and that flower-holding solutions contain germicides to prevent the growth of micro-organisms. An acidic solution also inhibits bacterial growth.

<u>Physiological Plugging</u> - When a plant is cut or injured, the cells at the cut surface may respond to the damage by closing off the wound, drastically reduce water flow, and lead to early wilting.

<u>Water Quality</u> - Hard water frequently contains minerals that make the water alkaline. Alkaline water, which does not move readily through cut-flower stems, can substantially reduce vase life. This problem can be overcome either by removing minerals from the water (by using a deionizer) or by making the water acid.

<u>Ethylene</u> - Certain flowers, especially carnations and some rose cultivars, die rapidly if exposed to minute concentrations of ethylene gas. A number of cut flowers produce ethylene as they age.

Ethylene gas is produced in large quantities by some ripening fruits, and it is also produced in high concentrations during combustion of organic materials (e.g. gasoline, firewood, tobacco). Storage and handling areas should be designed not only to minimize contamination of the atmosphere with ethylene, but with adequate ventilation to remove any ethylene that does occur. Finally, refrigerated storage is beneficial in that ethylene production and ethylene sensitivity of the product are reduced greatly when temperatures are low.

Mechanical Damage - Bruising and breakage of cut flowers should be avoided. Flowers with torn petals, broken stems or other obvious injuries are undesirable for aesthetic reasons.

Disease - Flowers are very susceptible to disease, not only because their petals are fragile, but also because the secretions of their nectarines often provide an excellent nutrient supply for even mild pathogens. The most commonly encountered disease organism, gray mold (Botrytis cinera), can germinate wherever free moisture is present.

POST-HARVESTING MANAGEMENT TECHNIQUES

Harvesting - Harvesting is normally done by hand using shears or a sharp knife.

Grading - The designation of grade standards for cut flowers is one of the most controversial areas in their care and handling. Objective standards such as stem length, which is still the major quality standard for many flowers, may bear little relationship to flower quality, vase life or usefulness. Straightness of stems, stem strength, flower size, vase life, freedom from defects, maturity, uniformity, and foliage quality are among the factors which should also be sued in cut flower grading.

Bunching - The number of flowers in the bunch varies according to growing area, market and flower species. Spray-type flowers are bunched by the number of open flowers, by weight or by bunch size. Bunches are held together by string, paper-covered wire or elastic bands. Materials used for sleeving include paper (waxed or unwaxed) and polyethylene (perforated, un-perforated and blister).

CHEMICAL SOLUTIONS

Rehydration - Wilted flowers, placed in water to restore turgidity, should be rehydrated with deionized water containing a germicide. Wetting agents (0.01 to 0.1%) can be added, and the water should be acidified with citric acid, HQC, or aluminum sulfate to a PH near 3.5.

Pulsing - The Term "pulsing" means placing freshly harvested flowers for a relatively short time (a few seconds to several hours) in a solution specially formulated to extend their storage and vase life. Sucrose is the main ingredient of pulsing solutions and the proper concentration ranges from 2 to 20%, depending on the crop. Some cut flowers are also pulsed with silver thiosulfate (STS) to reduce the adverse effects of ethylene.

Packing - There are many shapes of packing containers for cut flowers, but most are long and flat. This design restricts the depth of the flowers in the box, which may in turn reduce physical damage of the flowers. In addition, flower heads can be placed at both ends of the container for better use of space. Most packers anchor the product by using enough flowers and foliage in the box so that it holds itself firmly. To avoid longitudinal slip, packers use one or more "cleats". These are normally foam or newspaper-covered wood pieces that are placed over the product, pushed down, and stapled into each side of the box.

Cooling - By far the most important part of maintaining the quality of harvested flowers is ensuring that they are cooled as soon as possible after harvest and that optimum temperatures are maintained during distribution. Most flowers should be held at 32°- 35° C (0-2° C). Chilling-sensitive flowers (anthurium, bird-of-paradise, ginger, tropical orchids) should be held at temperatures above 50° F (10°C).

Forced-air cooling of boxes with end holes or closeable flaps is the most common and effective method for pre-cooling cut flowers; cool air sucked or blown through the boxes. In general, packers use less paper when packing flowers for pre-cooling.

Retail Handling - In order to ensure maximum quality and satisfaction to the final customer, it is vital that retail florists handle flowers properly on receipt. On arrival, packed boxes should be unpacked, or placed immediately in a cool-room. After unpacking, flower bunches should be re-cut and placed in a cool-room for at least a few hours.

In this talk, I have tried to identify some of the problems and opportunities of the Cut Flower industry:

- Quality products produced in low cost structure in specific micro-climates.
- Labor intense industry, specialized in producing specific crops.
- The key requirements of post harvesting conditioning and packaging to maintain quality.
- The problems of transportation and competing on a world market.

Decision Support for Integrated Greenhouse Production Systems

K.C. TING, W. FANG and G.A. GIACOMELLI
Biological and Agricultural Engineering Dept.
New Jersey Agricultural Experiment Station
Rutgers University-Cook College
New Brunswick, New Jersey, U.S.A.

Key Words: Decision support, Systems analysis, Intelligent database, Greenhouse, Operations research

Introduction

A greenhouse is a system integrating a number of interrelated hardware and software components. Some of them are the structure, the environment within the structure, the crop growing facility, the materials handling equipment, the environmental control apparatus and strategies, the information processing devices, and the management and labor forces. The degree of complexity of a greenhouse depends on the operating objective of the system. To ensure the compatibility of all the components and the workability of the entire system, a systems level approach can never be overemphasized (Giacomelli et al., 1987; van Weel and Giacomelli, 1990).

From the engineering point of view, the process of system design and management is first to identify workable systems and then to select, among the workable systems, the optimum. A workable design is defined as the system which performs the assigned tasks and satisfies the imposed constraints. In selecting an optimum system, workable systems are evaluated based on predetermined criteria, and the system which has the highest rating becomes the optimum.

Greenhouse production requires a higher investment per unit area as compared to the traditional open-field agricultural production. However, the return on investment is expected to be higher for greenhouse production systems, if they are properly designed, constructed, and managed. The complexity of an integrated greenhouse system combined with the sophisticated design process requires that its planning and operation include a large number of decisions to be made. Appropriate information is an essential basis for every successful decision-making process. In this case, the information needs to come from various disciplines such as engineering, horticulture, operations research, computer science, and statistics. Furthermore, the greenhouse operators' practical knowledge as a result of many years of experience is particularly valuable.

In view of the importance of useful information during planning, design and operation of greenhouse systems, work has been done towards the development of a computerized

J. Prakash and R. L. M. Pierik (eds.), Horticulture – New Technologies and Applications, 293–298.

decision support system. The activities of decision support were envisioned as to gather, store, retrieve, analyze, present, and interpret information to aid in the process of decision-making. A series of projects have been completed at the Department of Biological and Agricultural Engineering, Rutgers University, and the results were used as the building blocks of this decision support system. This paper will present an overview of this development work.

Methods

Engineering Economics

User-friendly, menu-driven, screen-editing microcomputer software was developed by Ting et al. (1989) to evaluate the economic feasibility of greenhouse systems. The software has a generic greenhouse engineering economic analysis algorithm plus a number of supporting subprograms for calculations of costs and revenues associated with specific design/operation conditions. The potential uses of engineering economic analysis applied to greenhouse operations are to:

- conduct case studies of an existing or proposed projects.
- perform sensitivity analyses for alternative design/operations.
- conduct parametric studies to cover ranges of input variables.
- facilitate investment risk analyses.
- incorporate the economic analysis program with other technical models, such as crop production, controlled environment, greenhouse automation and mechanization, etc.
- provide a means of evaluating quantitative values of an objective function in a systems optimization process.

As an example of the software application, a greenhouse production system employing a single truss tomato cropping scheme assisted by supplemental lighting was investigated for its economic feasibility. This single truss tomato production system was developed by McAvoy and Giacomelli (1985) and Giniger et al. (1988). In this analysis, a 4047 m^2 air-inflated double-polyethylene greenhouse located in New Brunswick, New Jersey was used as a base module. The base line information on the initial and operating costs and the experimental design may be found in the article by Ting et al. (1989). The operating cost for supplemental lighting and the corresponding monthly revenues were calculated using the model by Giniger et al. (1988).

Simulation of greenhouse internal transport

An animated computer model has been developed by Fang et al. (1990a) to simulate greenhouse internal transport systems. The computer model was written in a combined continuous-discrete simulation language SIMAN and its associated animation package CINEMA developed by Systems Modeling Corp., Sewickley, Pennsylvania, USA. SIMAN is a FORTRAN-based simulation language for modeling general systems. CINEMA animations are based on simulation models written using the SIMAN simulation language. The model was used to study the performance of materials handling operations within a greenhouse. The potential bottleneck of a transport system could be visually detected on the computer monitor. Statistical analyses on the system parameters, such as the status and utilization of machines, workers and waiting lines, and throughput time of an operation were performed during the simulation. From these data, the interaction between machines and workers within a greenhouse system was studied.

Resource allocation for greenhouse production

A procedure for studying the profitability of greenhouse potted plant production systems subject to resource constraints was developed (Fang et al., 1990b). The constrained condition and resources were the crop production schedule, greenhouse space, labor, and budget. A database containing the information for determining the required resources and operating costs for growing various crops was established. The database also provides the estimated revenue from sales of the crops, on a per pot basis. An algorithm was developed to determine first the feasibility of a given production plan and then determine the quantities of crops to be grown in order to yield an optimum profit. The result of this algorithm may serve to optimize allocation of resources for year-round production. The algorithm along with the crop database was incorporated into a user-friendly micro-computer program.

The optimization scheme includes a crop database, a crop module program and, an optimization module program. The crop database provides information, for various crops, material and energy costs in production, space and labor requirements at different stages, and predicted revenue. The crop module program allows the user to access the crop-related databases, calculates the space, time, and labor requirement at each stage for each crop, performs the gross margin calculations, and prepares data files containing production schedules for future use within the optimization routines. The major functions of the optimization module program are (1) to perform a case study, i.e., to determine the feasibility of a proposed production plan; and (2) to solve optimization problems, i.e., to determine the quantity of crops to be produced in order to maximize profit. A production plan's feasibility is evaluated by testing whether the proposed conditions satisfied the imposed constraints. The optimization algorithm seeks out the workable plan, which gives the optimum objective function value. The objective of this optimization process is to maximize the profit for a year-round production cycle. The constraints are the crop culture practice, and the available space, labor and budget.

Strategic planning of greenhouse production systems

A microcomputer model consisting of many module programs (including all of the above mentioned programs and more) and databases was developed to support the planning process of a greenhouse system (Fang, 1989). Each module can be used separately or in combination to facilitate the design, the renovation, and/or the decision-making process of a greenhouse crop production system. The modules are: (1) the crop module, (2) the device module which can be used to analyze the economic competitiveness of any two proposed mechanized systems, (3) the layout module which provides a tool for optimal bench and layout design under given constraints. A method to represent any shape of a greenhouse layout is proposed, (4) the simulation module which is helpful in studying the dynamic behavior of a greenhouse internal transport system, (5) the optimization module, and (6) the engineering economic analysis module.

Results and Discussion

Economic analysis

A greenhouse economic analysis program, EEGA, was developed. A user can enter the costs and revenues data, led by the program, to obtain results in both tabular and graphic forms. The output of the program gives the return on investment and breakeven period of a greenhouse project. EEGA was used to study greenhouse operations which employed supplemental lighting assisted single truss tomato cropping systems. A total of 159 different cases were analyzed. The operating parameters considered in the study were (1) reference harvest date, (2) number of crops per year, (3) months with supplemental lighting, and (4) supplemental daily photon fluence. It was found that tomato market price was highly influential in the analysis. Even a moderate increase in market price may result in a significant increase in the overall rate of return on investment.

Internal transport

An interactive, animated, microcomputer simulation model was developed, which has the following features:

1. There are multiple choices of greenhouse layouts.
2. Pertinent system parameters such as type of operation, capacity of each bay, transport speed of worker, etc., can be altered.
3. The animated graphics are optional.
4. The time varying data of system parameters are provided during the simulation and a report summarizing the system's performance is recorded into computer storage files.
5. The animated graphics display the real time statistical information about the utilization of machines and workers, transport cycle time of the workers, and the operating time of potting machine and harvest station.
6. After a simulation run, all recorded data may be analyzed and the results may be displayed in graphic and/or tabular forms.

Using this simulation technique, the behavior of internal transport system for potted plant production was investigated. The internal transport of greenhouse crops was found to be affected by interactions of workers and machines, and based on the greenhouse layout, production procedures and management strategy (Giacomelli et al., 1990).

Resource allocation

The organization charts of the crop database/module program and the optimization module program may be found in the article by Fang et al. (1990b). This crop module program was written in dBASE III Plus (Ashton-Tate, Inc., Torrence, California, USA) programming language and compiled by using CLIPPER (Nantucket Corp., Los Angeles, California, USA). The optimization module program was written and compiled using QuickBASIC (Microsoft Corp., Redmond, Washington, USA). The modular organization of the crop database and the optimization program provides the system flexibility for future modification and expansion of crop information. It can become a valuable tool for greenhouse managers to aid their ever challenging task of production planning. It is especially useful for the managers when performing relative comparisons among alternatives. The accuracy of the program output may be improved by updating the crop database with more reliable information obtained in the future research.

Strategic planning

The result of the effort in incorporating all the computer program modules for decision support is an integrated software package called CASE. The software remains to be modular and contains dynamic information modules and static procedural modules. This user-friendly software is self-contained, i.e., it can be used without relying on any other software.

The information modules are dynamic because they can be modified with ease by the user. The modules include the databases related to crops, greenhouse layouts, plant transport devices, and engineering specifications. The procedural modules are the algorithms which manipulate the data in the information modules for specific analyses. They are static because normally there is no need for changes by the user. Specifically, the procedural modules provide the following functions:

1. user interface,
2. database manipulation,
3. crop transport simulation (man-machine-layout interactions),
4. year-round production planning (schedule feasibility and optimization),
5. engineering design, and
6. project life-time economic analysis (return on investment)

Based on the above description, it is obvious that this decision support software is useful in many ways. The well-documented and readily-accessible databases are valuable information sources. The software may be used in many phases of a greenhouse project, such as planning, design, and operation. The automated data-driven paradigm makes it a powerful tool for what-if type feasibility study which is an important process in providing support for decision-making. Furthermore, the software can be used as an effective research and training aid in the areas of controlled environment agriculture.

Conclusions

This is an information age; but, frequently, information needs to be processed into knowledge to become useful. Therefore, information alone will not necessary be helpful. As a matter of fact, too much "raw" information can do very little more than just being overwhelming. However, information together with the ability to effectively make use of the information will be a great advantage in a decision-making process. There are two major ways of manipulating information/knowledge to reach a certain conclusion. They are procedural analysis and heuristic approach. Both ways have been proven to be effective and continue to be under development.

Recent social, economical, and technological developments have increased the complexity of controlled environment agriculture. It has become very critical that a greenhouse be designed and managed as an integrated system. The advancement made to improve any part of this system will most likely affect the other parts. Systems analysis is a well developed technique which can be applied to process the information related to greenhouse production for drawing conclusions at the systems level.

The use of computers is growing at an ever-increasing rate. Computers offer a vast memory capacity and a rapid computation speed. Both of these features make computers ideal tools for system analysis involving a large amount of information processing. This is the underlying concept of an automated decision support system. The challenge has been how to make such a decision support system deliverable to the users. The results obtained from the development work described in this paper have moved us several steps towards this goal.

Acknowledgement

New Jersey Agricultural Experiment Station Publication No. J-03232-32-90. Supported by State funds.

References

Fang W. (1989) Strategic planning through modeling of greenhouse production systems. Ph.D. Dissertation, Biological and Agricultural Engineering Dept., Rutgers University, New Brunswick, NJ, U.S.A.

Fang W., Ting K.C. and Giacomelli G.A. (1990a) Animated simulation of greenhouse internal transport using SIMAN/CINEMA. Transactions of the ASAE 33(1):336-340.

Fang W., Ting K.C. and Giacomelli G.A. (1990b) Optimizing resource allocation for greenhouse potted plant production. Transactions of the ASAE 33(4):1377-1382.

Giacomelli G.A., van Weel P.A. and van der Shilden M. (1987) Expert systems development for greenhouse potted plant production. ASAE Paper No. 87-5012, St. Joseph, MI:ASAE.

Giacomelli G.A., Fang W., Ting K.C. and van Weel P.A. (1990) Behavior of internal transport system for potted plant production. The XXIII International Horticultural Congress, Firenze, Italy, August 27-September 1, Abstract No. 2343.

Giniger M.S., McAvoy R.J., Giacomelli G.A. and Janes H.W. (1988) Computer simulation of a single truss tomato cropping system. Transactions of the ASAE 31(4):1176-1179.

McAvoy R.J. and Giacomelli G.A. (1985) Greenhouse tomato production in a transportable, potted plant cropping system. Presented at the Symposium on Mechanization, ISHS, Bonn, Germany, May 1985.

Ting K.C., Dijkstra J., Fang W. and Giniger M. (1989) Engineering economy of controlled environment for greenhouse production. Transactions of the ASAE 32(3):1018-1022.

van Weel P.A. and Giacomelli G.A. (1990) Systematic model for the design of integrated greenhouse production systems. The XXIII International Horticultural Congress, Firenze, Italy, August 27-September 1, Abstract No. 2271.

Greenhouse Pot Plant Production in the United States

ROBERT E. DANIELSON
BALL SEED COMPANY, WEST CHICAGO, ILLINOIS, U.S.A.

The individual pot plant producers in the United states are becoming larger and fewer. They specialize in the production of Holiday Crops; Poinsettias for Thanksgiving and Christmas; Lilies for Easter; Azalea and Cyclamen for Valentine Day and Easter; Chrysanthemum and miscellaneous garden type plants for Mothers Day. Foliage plants are produced in the Summer for sales when the people finally move back inside for Winter.

The other part of the Greenhouse Plant Production is the bedding plant industry which produce plants for outdoor planting. In the Southern and Western area this tends to be year around. In the North, Spring and Summer sales.

Large pot plant growers develop markets that may be within 200 miles or as far away as 1500 miles, with temperature controlled trucks making the delivery in 2 days.

Florida ships tropical foliage all over the USA and Canada, taking advantage of temperature and sub-tropical climate in Florida. California produces Poinsettias and Chrysanthemum pot plants for the Midwest and Eastern market.

Greenhouse construction and environmental control vary with different areas of the country.

In the Midwest and Eastern areas of the country, Dutch type glass houses are built with many labor saving and automation devises. The houses are computer controlled for climatic conditions. This is in the colder area of the country with up to 5 months of Winter conditions.

In the South and Western areas of the country, they are more likely to build lower cost structures using poly film and screen type houses. Plants are often times grown on the ground without benches.

J. Prakash and R. L. M. Pierik (eds.), Horticulture – New Technologies and Applications, 299–305.

Since it is almost impossible because of plant quarantine restrictions to import pot plants, this business is more stable than the cut-flower industry where nearly everything is imported.

The growing media for pot plant production centers around the main component of peat moss from Canada. Various additives are used to improve drainage, aeration and nutrition.

The additives used are perlite, vermiculite, rock wool, pine bark, styrofoam, sand and volcanic scoria. The industry continues to look for a low cost synthetic growing material.

The most common watering method is with the Chapin water tubes with individual pot watering. The smaller size pots, 3" and 4" are often on a capillary matt and watered overhead until flowering; then individually watered to prevent flower damage.

Some growers have installed ebb and flow watering for smaller size pots. The water is recycled by pumping up the level of the water to about 2 inches, standing in water, then pulling the plug and letting the water flow back into the tank.

Most fertilizers used are water soluble in the watering solution and program on a constant fertilizer program of 200 ppm nitrogen, 100 ppm Phosphorous and 200 ppm Potassium, with ample Calcium and trace elements added at minor rates. Usually feeding starts immediately after planting because of low nutrition in the growing media and extends until flower shows color.

Growth regulators are used to control height and quality of the plants. Alar (B-9) is most widely used for height control on all plants except poinsettias. Cycocel is used to control height on Poinsettias and Geraniums.

Disease and pest control compounds are very closely controlled by the government because of the environmental and people hazards. The application methods has changed using mist type equipment. This equipment uses low volume and high concentration which penetrates to all areas of the plant and greenhouse. People do not need to enter the area to spray the plants.

The source of plant material for finish plant production usually comes from plant specialists. Most of the plant material is culture indexed for systemic pathogens, virus viroids, and clonal indexed for variety performance.

The Ball Seed Company specializes in selling culture index plants from producers to growers across the North American Continent. The Ball Seed Company salesmen are technical and product trained to help the growers, their customers, to improve their performance and profitability.

Foliage plant industry has become very specialized using tissue culture plant material, most of the finished plants are produced in California and Florida.

Some Foliage plant species are imported from Central America, Puerto Rico and Holland. These plants are imported without soil and are inspected at port of entry.

Most pot plants today are sold through large super market outlet stores and in large numbers. Prices are discounted over what you would pay for a plant from a retail flower shop.

Greenhouses Environmental Controls

Heat

- Heat Forced Air Heater -- Gas or Oil Fire
- Hot Water or Steam Radiation Pipes from Boilers

Cooling

- Wet Pad and Exhaust Fans for Cooling
- Horizontal Air Flow Fans for air movement
- Black out curtains for day length control
- High pressure sodium lamps supplemental lights during low light periods
- Computerized control systems for environmental control

Growing Media

PROVIDES FOR:

1. An anchor for the plant root system;
2. Water reservoir for plant absorption;
3. Air space for oxygen for root respiration. Roots and micro organisms produce CO_2 which needs to be liberated into the atmosphere and oxygen needs to be resupplied.
4. A reservoir of plant nutrients that can be absorbed by the plant roots.

At no time should the plant be placed under stress for water or fertilization levels.

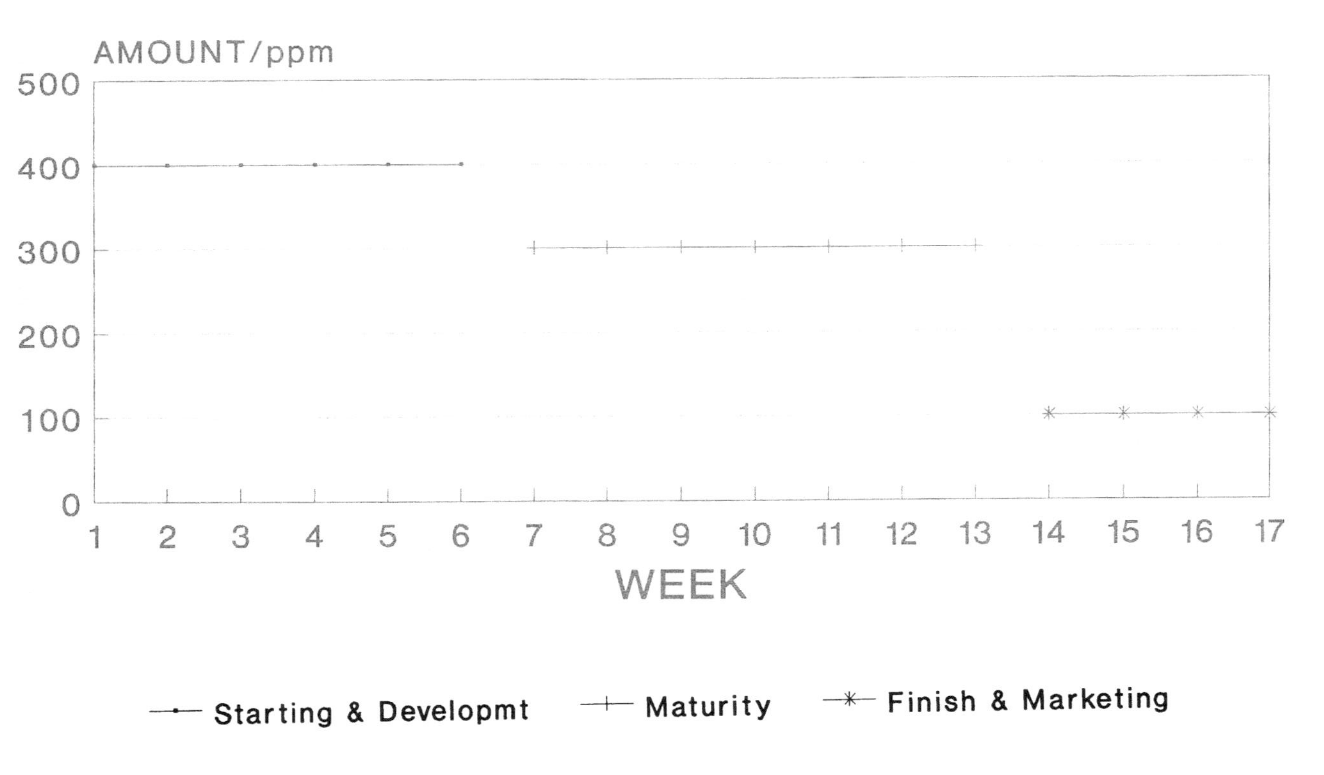
Fertilizer & Growth Cycle
Flowering Pot Plants
AMOUNT/ppm
500
400
300
200
100
0
1
2
3
4
5
6
7
8
9
10
11
12
13
14
15
16
17
WEEK
Starting & Developmt
Maturity
Finish & Marketing
Balanced Fertilizer
20 - 10 - 20
N P K

Poinsettia

Color:	White - Pink - Red
Flowers:	On short days
	-Natural short days - September 25th
Market time:	Thanksgiving - November 25th
	Christmas - December 25th
Propagation:	Vegetative Cuttings
	Stock Plants under long days
Photo period:	Vegetative - 14 hours light
	Flower - 14 hours dark
Option Temperature:	Days - 21 C - 27 C
	Nights - 16 C - 21 C
Flower year around:	Use long day/short control
Growth regulators:	Cycocel used to control height
Special culture:	Pinch for branched plants
Source of plants:	Ecke Poinsettia (Encinitas, CA)

Pot Plant Chrysanthemum

Many flower types & Colors

Flower on short days	-	Short days begin August 15th above the Tropic of Cancer
Natural season	-	September through December. Depends on response group 7 through 13 weeks to flower
Long day requirement	-	14 hours of light
Short day requirement	-	14 hours of dark
Optimum temperature	-	Night 16 C - 18 C
		Day 21 C - 29 C
Propagation	-	Vegetative cuttings grown from certified stock plants. Disease and virus free
Growth regulator	-	Alar applied to control height
Special culture	-	Pinch plants to promote branching
Source of plants	-	Yoder Brothers (Barberton, OH)

Kalanchoe - Hybrid

Color: Red, Orange, Salmon, Pink, and Lavender
Flowers on Short Days
Market Time: Year around
Propagation: Short tip cuttings 2 or 3 nodes
Stock plants grown under 14 hour days
Flower: Requires 14 hours of dark for 42 days
Optimum Temperature: Night 18 C - 21 C
Days 22 C - 27 C
Special requirements:
Heat delay at high night temperature
Plants may or may not require pinching
Alar may be used to finish more compact plants
Light: 3,500 - 4,500 Foot Candles (50% shade)

Robotics Applications to Transplanting of Plug Seedlings

K.C. TING and G.A. GIACOMELLI
Biological and Agricultural Engineering Dept.
New Jersey Agricultural Experiment Station
Rutgers University-Cook College
New Brunswick, New Jersey, U.S.A.

Key Words: Robotics, Transplanting, Plugs, Seedlings

Introduction

The production of seedlings for transplants, especially plug seedlings, has become a very beneficial procedure in commercial greenhouses. One necessary task in a plug production system is the transplanting of seedlings from one container to another to allow further growth and development. This transplanting operation is currently done by human labor. The seedlings are young plants of various shapes and sizes, depending on the species, age and the containers used for their growth. The large volume of plug production every year (estimated 2.75 billion in North America in 1989) emphasized the need for automating the plug transplanting operation. And, inherent uncertainty and variability associated with the seedling production conditions prompted our study of flexibly automated transplanting using robots.

At the inception of this research activity, a decision was made to adopt a commercially available robot system as the base manipulating mechanism. The topics of study to be undertaken were: (1) the grasping, transport and planting of seedlings, (2) the sensing capability for detecting the presence of a seedling, (3) the layout and materials flow pattern within a transplanting workcell, and (4) the effects of design parameters on the productivity of a workcell, taking into consideration the stochastic factors.

Methods

A typical seedling is composed of an aerial portion (stem and leaves) and a root zone portion (roots distributed in growth medium). The root zone portion is confined in a cellular space and these cells are located on a tray in a regular pattern. The numerous geometry and structural materials of trays, and the types of growth media used for seedling production provide almost unlimited possible combinations. During transplanting, the seedlings are extracted from those trays and planted into larger cells within a different tray. A robot was recognized as being capable of locating and transporting the seedlings. However, a key component to the success of robotic transplanting was identified to be an end-effector. The function of an ideal end-effector would be to grasp, hold, insert and release a seedling regardless of its species and/or container. To aid in the development of the end-effector, some design concepts were established (Ting et al., 1990b). The guidelines used for the gripper design were:

J. Prakash and R. L. M. Pierik (eds.), Horticulture – New Technologies and Applications, 307–309.

1. The gripper be structurally simple, and powered by low cost actuators.
2. The gripper handle a wide variety of seedlings in different sizes of containers.
3. The gripper avoid the aerial portion and operate on the root zone portion.
4. The "fingers" of the gripper be able to effectively grasp the plug with minimum damage.
5. There be a sensor to detect the presence of a seedling on the gripper.

In order to have a high assurance that a transplanted tray would be filled with seedlings, the gripper was equipped with a capacitive proximity sensor. This sensing capability was instrumental in solving the problem of inherent uncertainties associated with the transplanting operation. The uncertainties were mainly caused by the missing seedlings on the source tray and the occasional unsuccessful seedling extraction by the gripper. A series of experiments were conducted to identify the mechanical and horticultural factors which would affect the successful extraction percentage (Yang et al., 1989).

The methods of presenting source trays and removing transplanted trays within a robot workcell were investigated to study their feasibility and effectiveness. A bi-level, indexed-motion layout and materials flow pattern was established (Ting et al., 1990a). In this workcell, the source and the transplanted trays would pass each other at different elevations. The movement of each respective tray could be programmed to advance any specified distance when necessary. The purpose was to minimize the required robot travel time during transplanting.

A computer model "WORKCELL" was developed to simulate the productivity of SCARA robot-based seedling transplanting workcells (Ting et al., 1990c). WORKCELL was a stochastic model which took into consideration the probabilities of empty cells in the source tray and unsuccessful extractions by the gripper. The output of the model gave the throughput times of transplanted trays produced by the workcell. The model was utilized to study the effects of workcell design parameters on its productivity.

Results and Discussion

A series of end-effectors were designed based on the fundamental concept that a seedling was to be manipulated by its root zone portion by grasping with a pair of needles. The first design was a gripper consisting of two slanted needles. The needles were kept parallel when penetrating or releasing the root zone portion of a seedling. A mechanism was included in the gripper so that the tips of the needles could be rotated toward each other for firmly grasping the root zone portion during seedling extraction and transportation. In the second gripper design, the rotating actions used in the first design for opening and closing the needles were changed so the two needles were extended/retracted inside two collars. Therefore, for grasping a seedling, the two needles became fully extended to penetrate the root zone. To release a seedling, the two needles were simply retracted into their individual collars. After an extensive testing of this second gripper design, it was found that there were two characteristic angles important to the effectiveness of the gripper. They were (1) the included angle between the needles when they were fully extended (alpha) and (2) the angle between the plane defined by the two needles and the horizontal (beta). On the third design version of the gripper, both alpha and beta angles were made fully adjustable.

It was found the capacitive proximity sensor installed on the gripper was capable of detecting a seedling held by the gripper. The physical location and sensitivity of the sensor could be adjusted to accommodate different seedling shapes and sizes. The sensor output was compatible with the input signal to the robot controller. Therefore, the robot would complete a transplanting process only when a seedling was correctly held by the gripper.

Many mechanical and horticultural factors were found to affect the percentage of successful transplanting, during the laboratory testing of the gripper. The mechanical factors were: (1) the gripper needles' angles, (2) plug extraction speed, and (3) the sensor sensitivity. The horticultural factors included (1) empty cells on the plug trays, (2) plant species, (3) root connections, (4) adhesion between roots and cell walls, (5) root zone moisture, and (6) the number of plugs in one cell.

A menu-driven, animated, stochastic, micro-computer model was developed for simulating SCARA robot-based plug transplanting workcell. A total of sixty four input variables which relate to the design and operation of the workcell may be specified by the user. The model output provides quantitative descriptions of workability and productivity of the workcell under study. The model software can be delivered in the form of compiled Quick Basic; therefore, it can be executed on most DOS machines. This stochastic model has been validated by comparing its output with actual robot operations. In addition to being a useful tool for computer-aided-design of workcell, the software can be readily utilized to systematically investigate "what-if" type scenarios. Furthermore, this model can be used to facilitate the engineering economic analysis of this robotic application in plug transplanting.

The practical experiences gained by incorporating an industrial robot within a workcell for an agricultural application such as seedling transplanting can be utilized for other plant production tasks. The basic layout and materials handling of the living product within the system may not change. However, the specific task will require a compatible end-effector to allow sufficient capacity for the timely completion of the task, as well as to assure that no change occurs to the seedling inhibit its future growth. Our studies have demonstrated that the needles gripper can successfully transplant seedlings ranging from the small "plug" (1 cm diameter root zone) up to large seedlings (within 15 cm pots). However, the many variations of robotic mechanisms which may be available do not include the interface between the robot and the plant material, as most have found, nor is the experience to integrate them presently available in the horticultural industry. In addition, the need to consider the robotic or "hard automation" unit as a critical mechanical component of an entire integrated crop production system cannot be ignored, if this basic technology is to be successfully applied to intensive greenhouse systems.

Acknowledgement

New Jersey Agricultural Experiment Station Publication No. J-03232-33-90, supported by State funds.

References

Ting K.C., Giacomelli G.A. and Shen S.J. (1990a) Robot workcell for transplanting of seedlings part I - layout and materials flow. Transactions of the ASAE 33(3):1005--1010.

Ting K.C., Giacomelli G.A., Shen S.J. and Kabala W.P. (1990b) Robot workcell for transplanting of seedlings part II - end-effector development. Transactions of the ASAE 33(3):1013--1017.

Ting K.C., Yang Y. and Fang W. (1990c) Stochastic modeling of robotic workcell for seedling plug transplanting. ASAE Paper No. 901539. St. Joseph, MI:ASAE.

Yang Y., Ting K.C. and Giacomelli G.A. (1989) Factors affecting robot workcell for flexibly automated seedling transplanting. ASAE Paper No. 897055. St. Joseph, MI:ASAE.

Immunological and Molecular Approaches to the Diagnosis of Viruses Infecting Horticultural Crops

D.D. SHUKLA, M.J. FRENKEL, N.M. MCKERN and C.W. WARD
CSIRO, Division of Biomolecular Engineering, 343 Royal Parade, Parkville, Victoria, 3052, Australia

Plant viruses are estimated to cause economic losses world-wide of $15 billion per annum, and the development of effective control strategies is dependent on the availability of reliable methods of detection and identification (Ward and Shukla, 1991). A number of immunological and molecular techniques have been developed over the years for the detection of plant viruses. The sensitivity of some of the recent techniques, viz enzyme-linked immunosorbent assay and nucleic acid hybridization, has reached the level that any known plant virus can now be detected even in a single infected seed or a single virus-carrying insect (Table 1).

In spite of these developments in detection technology, some plant viruses still can not be accurately identified by the methods currently available. Members of the potyvirus group of plant viruses fall into this category of viruses. The difficulties associated with the identification of potyviruses are not due to any inherent problems with the techniques themselves, but are due to the complex nature of potyviruses. The potyvirus group is the largest of the 34 plant virus groups currently known. It contains at least 180 distinct members representing over one third of all known plant viruses (Ward and Shukla, 1991). In 1974 the members of the group were reported to infect 1,112 species of 369 genera in 53 plant families (Edwardson, 1974). Since then many more species, genera and families have been added as hosts of potyviruses. Consequently, members of the potyvirus group cause important diseases in agricultural, pasture and horticultural crops. Some of the well known potyviruses causing losses in horticultural crops include: bean common mosaic, bean yellow mosaic, bearded iris mosaic, carnation vein mottle, clover yellow vein, iris fulva mosaic, iris mild mosaic, leek yellow stripe, lettuce mosaic, narcissus yellow stripe, papaya ringspot, passionfruit woodiness, pea mosaic, pepper mottle, potato virus A, potato virus V, potato virus Y, tulip blotch, tulip flower breaking, watermelon mosaic and zucchini yellow mosaic viruses.

It has been pointed out repeatedly by taxonomists and reviewers that the taxonomy of the potyvirus group is very unsatisfactory and that successful resolution of potyvirus detection and identification is a major challenge for plant virologists (Hollings and Brunt, 1981; Francki, 1983; Francki et al., 1985; Harrison, 1985; Milne, 1988). This unsatisfactory state of potyvirus taxonomy has been due to the large size of the group, the apparent vast variation among the members and the lack of suitable taxonomic parameters that will distinguish distinct viruses from strains (Francki, 1983; Francki et al., 1985). The use of conventional approaches such as host range, symptomatology, cross-protection, cytoplasmic inclusion morphology and serology has suggested the presence of a 'continuum' between strains of two or more distinct potyviruses, implying that 'species' and 'strain' concepts can not be applied to potyviruses (Bos, 1970;

J. Prakash and R. L. M. Pierik (eds.), Horticulture – New Technologies and Applications, 311–319.

Hollings and Brunt, 1981; Lana et al., 1988). In contrast, the recent use of nucleic acid and coat protein amino acid sequence data from our lab has clearly demonstrated that potyviruses can be divided into distinct members and strains. The sequence data, in combination with information on the structure of the potyvirus particle and immunochemical analyses of overlapping synthetic peptides covering the entire coat protein, have established the molecular basis of potyvirus serology, explained many of the problems currently associated with the application of conventional approaches and provided several novel approaches for the accurate identification and classification of potyviruses (Shukla and Ward, 1989a, b; Ward and Shukla, 1991). The following account of approaches to identification and classification of potyviruses is based on these studies.

Potyvirus Genome Organization

The potyvirus genome consists of single stranded, positive sense RNA. The aphid-, mite-, and whitefly-transmitted potyviruses have a single molecule of RNA of approximately 10,000 bases, while the fungal-transmitted potyvirus genome is distributed across two molecules of RNA. The potyvirus RNA is covalently bound to a genome-linked protein, VPg, at the 5'-end and contains a poly(A) tail at the 3'-end. The RNA is expressed as a single polyprotein which is subsequently cleaved by proteases to yield several functional proteins, including a conserved, ordered gene set of nonstructural proteins that are involved in RNA replication. The order of these products in the potyvirus polyprotein is : first protein (P1), helper component (HC), third protein (P3), cylindrical inclusion protein (CI), small nuclear inclusion protein (NIa) which includes VPg at its N terminus, large nuclear inclusion protein (NIb) and coat protein (CP). Only two of these, VPg and the coat protein, are present in virus particles (Shukla et al., 1991). Comparisons of the nucleotide sequences of potyvirus genomes show that the degree of identity between the equivalent regions of strains is greater than 96%, while between distinct potyviruses the identity ranges from 30% (3' non-coding region) to 60% (NIb coding region) (Frenkel et al., 1991). This suggests that similar taxonomic assignments can be obtained irrespective of the part of the genome or gene product of potyviruses used. However, for a variety of reasons (Shukla and Ward 1988, 1989a; Frenkel et al., 1989, 1991) coat proteins and the 3' non-coding regions are the most convenient targets for developing general identification and detection strategies for potyviruses.

Structure of Potyvirus Coat Proteins

Particles of potyviruses transmitted by aphids, mites and whiteflies are flexuous rods, 680-900 nm long and 11-15 nm wide made up of 2000 copies of a single protein species of 30 kDa to 37 kDa (Hollings and Brunt, 1981). The potyviruses with fungal vectors have particles of two modal lengths, 200-300 nm and 500-600 nm, and a single coat protein of 30 kDa to 33 kDa (Usugi et al., 1989; Kashiwazaki et al., 1989).

Sequence comparisons and biochemical analysis show that the N termini of coat proteins of distinct potyviruses vary considerably in length and sequence, whereas the C-terminal two-thirds of the proteins are highly homologous (Shukla and Ward 1988; Ward and Shukla, 1991). In fact the N terminus is the only large region in the entire coat protein that is unique to a potyvirus (Shukla and Ward, 1989a, b), is immunodominant (Shukla et al., 1988b, 1989d) and contains virus-specific epitopes (Shukla et al., 1989b, d). Mild proteolysis by trypsin of purified potyvirus particles revealed that the N- and C- terminal regions of coat proteins are exposed on the particle surface (Shukla et al., 1988b). The surface exposed N-terminal regions can vary in length from 30-95 amino acids depending on the virus, whereas the length of the surface-exposed C-terminus is relatively constant (18-20 amino acids) in different potyviruses (Shukla et al., 1988b; Ward and

Shukla, 1991). Removal of the N and C termini leaves a fully assembled virus particle composed of coat protein cores consisting of 215-227 amino acids. These core particles appear indistinguishable from untreated native particles by electron microscopy and are still infectious following mechanical inoculation suggesting that the N and C termini are not required in particle assembly or for infectivity (Shukla et al., 1988b).

These observations suggested that antibodies targeted to the unique amino terminus of the potyvirus coat proteins should be virus-specific whereas those generated to the coat protein core should be excellent broad spectrum probes capable of detecting most, if not all, potyviruses. Much of the contradictory information on serological relationships among potyviruses can be attributed to the presence in antisera of variable proportions of cross-reacting antibodies that are targeted to the coat protein core of potyviruses. Substantial variation in specificity was observed when 11 potyvirus antisera produced in different laboratories were tested with 12 distinct potyviruses (Shukla et al., 1989b). A majority of the antisera recognised all or most of the potyviruses tested whereas two antisera reacted only with their homologous viruses. Such variation in the specificity of the antisera may be due to two factors. Firstly, the state of the purified virus preparations used for immunization may have contributed to this situation. It is known that the N and C termini of coat proteins of potyviruses are degraded during purification and storage by enzymes of plant or microbial origin which co-sediment with the virus particles (Shukla et al., 1988b). The usual practice in different laboratories is to use the same preparation of purified virus for successive immunizations. Since the N terminus contains the virus-specific epitopes, its removal from virus particles in situ would gradually result in virus particles containing only non-specific core epitopes. Secondly, immunization procedure may have had an influence. There is considerable variation in the literature on the number, interval and route of injections, and the amount of antigen administered when producing an antiserum to plant viruses (Van Regenmortal, 1982). Although there is little reliable information available concerning the relative merits of different immunization procedures, these are very likely to affect the reactivities of the antibodies produced. Large differences in the reactivities of antisera taken at different stages of immunization of the same animal have been reported (Koenig and Bercks, 1968); antisera from early bleedings contain virus-specific antibodies whereas cross-reacting antibodies begin to appear in later stages of immunizations (Van Regenmortal and Von Wechmer, 1970). Our investigation of potyviruses gave similar results (Shukla et al., 1989c, d).

Analysis of the 136 possible pairings of the complete coat protein amino acid sequences from 17 strains of eight distinct potyviruses revealed a bimodal distribution of sequence homology (Shukla and Ward, 1988). In this analysis the sequence homology between distinct members ranged from 38-71% (average 54%) while that between strains of the one virus ranged from 90-99% (average 95%). These findings are not consistent with the 'continuum' hypothesis (Bos, 1970; Lana et al., 1988) proposed to explain the unsatisfactory taxonomy of potyviruses, and show a clear demarcation of sequence homology between distinct potyviruses and strains. Since that report coat protein sequences of a further twenty three aphid-transmitted, one mite-transmitted and one fungal-transmitted potyvirus have been determined. Analysis of the new sequences showed a bimodal distribution of sequence homologies between aphid-transmitted distinct viruses and strains as previously observed (Shukla and Ward, 1988), and a trimodal distribution of sequence homologies between aphid-, fungal- and mite-transmitted potyviruses. The third level of sequence diversity correlates with vector transmission. If the potyvirus group is elevated to family status, then vector transmission is the character that defines genus, distinct potyviruses correspond to species and their variants to strains (Ward and Shukla, 1991).

3' Non-Coding Region of Potyvirus RNA

The 3' non-coding regions of distinct potyviruses examined so far vary in length from 147 to 499 nucleotides whereas this region in different strains of individual viruses is generally of the similar length (Ward and Shukla, 1991). A comparison of 3' non-coding regions of RNA from 15 strains of eight distinct potyviruses revealed no significant sequence homology between distinct viruses. The degree of homology, ranging from 39-53%, was comparable to that obtained when the 3' non-coding regions of unrelated viruses from other plant virus groups were compared with potyviruses and were probably close to the value expected for coincidental matching. In contrast, the 3' non-coding regions of related strains showed sequence homologies ranging from 83-99% (Frenkel et al., 1991). These results suggested that the sequence of the 3' non-coding region of the potyvirus genome may be an accurate marker of genetic relatedness and could serve as an aid to the identification and classification of potyviruses. Due to its convenient location, sequence data for the 3' non-coding region can be readily obtained from cDNA clones generated by oligo(dT)-primed synthesis on a viral RNA template. In addition, synthetic oligonucleotides, fragments amplified by the polymerase chain reaction or cloned cDNA from the 3' non-coding region should find ready application as sensitive probes for potyvirus detection and classification (Frenkel et al., 1989).

Immunological and Molecular Approaches to the Detection and Identification of Potyviruses

On the basis of the above molecular studies, four novel, simple approaches have been developed to aid the identification and classification of potyviruses (Shukla and Ward, 1989a, b; Shukla et al., 1991; Frenkel et al., 1991; Ward and Shukla, 1991).

Potyvirus Group-Specific Antibody Probes

Due to the large size of the potyvirus group and the fact that several potyviruses can infect one crop, for example maize, sorghum, peanut, beans, potato, watermelons etc, it would be highly desirable to have a potyvirus group-specific probe that can detect most, if not all, potyviruses. Such a probe would be extremely useful for quarantine purposes and in screening diseased plants for the presence of viruses to ascertain if a potyvirus is the cause. On the basis of the information that the coat protein cores of potyviruses are highly homologous, the N and C termini were removed from particles of one aphid-transmitted virus using trypsin, the core particles were dissociated using formic acid, formic acid removed by dialysis in water and the core protein preparation injected into a rabbit. The resulting antiserum was found to react not only with all the aphid-transmitted potyviruses tested (Shukla et al., 1988b, 1989b, d) but also with potyviruses transmitted by mites and whiteflies (Shukla et al., 1989a) viruses which share only a very low sequence identity with aphid-transmitted potyviruses (Ward and Shukla, 1991). The epitopes recognised by this polyclonal antiserum have been mapped by immunochemical analysis of overlapping peptide fragments (Shukla et al., 1989d). Monoclonal antibodies with broad specificity have also been produced against intact virus particles and have been shown to react with variable numbers of potyviruses. One monoclonal antibody, Mab 1/16, reacted with all 15 potyviruses tested and recognised a surface-located peptide sequence that includes the trypsin cleavage site at the junction of the N-terminal region and the trypsin resistant core (Shukla et al., 1989d). Another monoclonal (PTY 1) has been shown to detect 32 distinct potyviruses so far (Jordon, 1989) and is commercially available from Agdia Inc., USA.

Potyvirus-Specific Antibody Probes Targeted to the N-terminus of Coat Proteins

Since the surface-exposed N terminus is the only large region in the entire potyvirus coat protein that is variable and virus-specific, epitopes contained in this region should generate virus-specific antibodies. On the basis of this information Shukla et al. (1989b) developed a simple chromatographic procedure to obtain virus-specific antibodies from polyclonal antisera by selective removal of the core-targeted cross-reacting antibodies. The method involved : (1) removal of the surface-located, virus-specific N-terminal region of the coat protein from particles of one potyvirus using lysyl endopeptidase, (2) coupling the truncated coat protein to cyanogen bromide-activated Sepharose and (3) passing antisera to different potyviruses through the column. Antibodies that did not bind to the column were found to be directed to the N terminus of coat proteins and were highly specific. Such an approach was employed to show that 17 potyvirus strains infecting maize, sorghum and sugarcane in Australia and the United States were not all closely related strains of sugarcane mosaic virus as previously believed, but represented four distinct potyviruses, namely Johnsongrass mosaic virus, maize dwarf mosaic virus, sorghum mosaic virus and sugarcane mosaic virus (Shukla et al., 1989c). This problem of taxonomy of the sugarcane mosaic virus sub-group of potyviruses had frustrated plant virologists for the past 25 years.

High Performance Liquid Chromatographic (HPLC) Peptide Profiling of Coat Protein Digests

Although amino acid sequence homology between coat proteins of potyviruses can differentiate between distinct members and strains, its value as a taxonomic criterion would be greatly enhanced if this property could be exploited by simpler techniques than sequencing, which is a formidable task in the case of potyviruses given the large number of viruses and strains in the group. One approach is the use of HPLC peptide profiling of coat protein digests (Shukla et al., 1988a). This approach is based on the different degrees of coat protein sequence homology found between the distinct members of the potyvirus group on the one hand and between strains of the one potyvirus on the other. Depending upon the availability of facilities, specific differences between the proteins can be determined by this method as most of the peaks in the profiles represent single peptides which can be directly sequenced. The method involves : (1) overnight digestion of potyvirus coat proteins with trypsin, and (2) separation of peptides on a reverse phase HPLC column. A visual comparison of the peptide profiles with a reference isolate can provide exact identity of the unknown isolate in most cases. The peptide profiles can be stored in a computer data bank and used for comparison with the new isolates. A comparison of 12 strains from six distinct potyviruses showed that the peptide patterns of strains from the same virus were very similar, almost superimposable, but those from distinct potyviruses were different (Shukla et al., 1988a). This approach has recently been used : (1) to confirm the existence of four distinct viruses in the sugarcane mosaic virus subgroup (McKern et al., 1991a), and (2) to reveal that peanut stripe virus, azuki bean mosaic virus and blackeye cowpea mosaic virus are all strains of the same virus (McKern et al., 1991b). The method is currently being used to resolve the taxonomy of cowpea aphid-borne mosaic virus and the large number of strains of bean common mosaic virus which at this stage appear to fall into three distinct viruses (D.D. Shukla, N.M. McKern and C.W. Ward, unpublished results).

Molecular Hybridization Involving the 3' the Non-coding region of the Potyviral Genome

A comparison of nucleotide sequences from the 3' non-coding regions of distinct potyviruses and strains suggested that this region may have great value as a marker for distinguishing viruses and strains in the potyvirus group (Frenkel et al., 1989). On the basis of this information polymerase chain reaction-amplified DNA probes corresponding to the 3' non-coding regions of two potyviruses, namely sugarcane mosaic virus and watermelon mosaic virus 2, were produced. The probes were hybridized with recombinant clones, purified potyviral RNA, partially purified total infected plant RNA and crude extract from infected plant tissue from homologous and heterologous potyviruses. In each case hybridization occurred only with the strains of the homologous viruses (Frenkel et al., 1991). This approach was used : (1) to support the proposal (Yu et al., 1989) that watermelon mosaic virus 2 and soybean mosaic virus-N are strains of the same virus, (2) to confirm that maize dwarf mosaic virus-B is a strain of sugarcane mosaic virus (Frenkel et al., 1991) and (3) to show that bean yellow mosaic, clover yellow vein and pea mosaic are distinct potyviruses (Tracy S.L., Frenkel M.J., Gough K.H., Hanna P.J. and Shukla D.D., unpublished results).

Concluding Remarks

The forgoing discussion shows that the recent use of coat protein amino acid sequences and the sequences from the 3' non-coding regions can differentiate between distinct potyviruses and strains, in contrast to the use of conventional approaches which revealed the presence of a 'continuum' between strains of distinct potyviruses. In addition, the sequence data, in combination with information on the structural organisation of the potyvirus particles, have been used to develop simple techniques (such as potyvirus group-specific and virus-specific antibody probes, HPLC peptide profiling and molecular probes involving 3' non-coding regions) that are more successfully applied than those used previously for potyvirus identification and classification.

Immunochemical analysis of native virus particles, trypsin-treated virus particles, dissociated core proteins and overlapping synthetic octapeptides have established the molecular basis for potyvirus serology and have explained many of the problems associated with the application of conventional serology to this group of plant viruses. As pointed out previously (Shukla et al., 1988b), the critera and approaches developed with potyviruses may well be equally applicable to members of the other plant virus groups with rod-shaped morphology.

Acknowledgements

D.D. Shukla is highly grateful to the Indo-American Hybrid Seeds, Bangalore, India for funding his visit to the "International Seminar on New Frontiers in Horticulture". This project was supported by grants from the Rural Credits Development Fund of the Reserve Bank of Australia and the Australian Wool Corporation.

References

Bos L. (1970) The identification of three new viruses isolated from Wisteria and Pisum in the Netherlands, and the problem of variation within the potato virus Y group. Neth. J. Plant Pathol. 76: 8-46.

Edwardson J.R. (1974) Some properties of the potato virus Y-group. Monograph Series, Florida Agr. Exp. Sta., No. 4.

Francki R.I.B. (1983) Current problems in plant virus taxonomy. In: Matthews R.E.F. ed., A critical appraisal of viral taxonomy, pp. 63-104. CRC Press, Boca Raton, U.S.A.

Francki R.I.B., Milne R.G. and Hatta T. (1985) Atlas of plant viruses, vol. 2, CRC Press, Boca Raton, U.S.A.

Frenkel M.J., Jilka J.M., Shukla D.D. and Ward C.W. (1991) Differentiation of potyviruses and their strains by hybridization with the 3' non-coding region of the viral genome. J. Gen. virol. (submitted).

Frenkel M.J., Ward C.W. and Shukla D.D. (1989) The use of 3' non-coding nucleotide sequences in the taxonomy of potyviruses: application to watermelon mosaic virus 2 and soybean mosaic virus-N. J. Gen. Virol. 70 : 2775-2783.

Harrison B.D. (1985) Usefulness and limitations of the species concept for plant viruses. Intervirology 25: 71-78.

Hollings M., and Brunt A.A. (1981) Potyviruses. In: Kurstac E. ed., Handbook of plant virus infections: comparative diagnosis, pp. 731-807. Elsevier/North Holland, Amsterdam, The Netherlands.

Jordon R. (1989) Mapping of potyvirus-specific and group-common antigenic determinants with monoclonal antibodies by Western-blot analysis and coat protein amino acid sequence comparisons. Phytopathology 79: 1157.

Kashiwazaki S., Hayano Y., Minobe Y., Omura T., Hibino H. and Tsuchizaki T. (1989) Nucleotide sequence of the capsid protein gene of barley yellow mosaic virus. J. Gen. Virol 70: 3015-3023.

Koenig R. and Bercks R. (1968) Anderungen in heterologen Reaktionsvermogen von Antiseren gegen Vertreter der potato virus-X-Gruppe im Laufe des Immunosierungs-prozesses. Phytopathol. Z. 61: 382-398.

Lana A.F., Lohuis H., Bos L. and Dijkstra J. (1988) Relationships among strains of bean common mosaic virus and blackeye cowpea mosaic virus - members of the potyvirus group. Ann. Appl. Biol. 113: 593-505.

McKern N.M., Shukla D.D., Barnett O.W., Vetten H.J., Dijkstra J., Whittaker L.A. and Ward C.W. (1991a) Coat protein properties suggest that azuki bean mosaic virus, blackeye cowpea mosaic virus, peanut stripe virus and three potyvirus isolates from soybean are all strains of the same potyvirus. Intervirology (submitted).

McKern N.M., Shukla D.D., Toler R.W., Jensen S.G., Tosic M., Ford R.E., and Ward C.W. (1991b) Peptide profiles of coat proteins confirm that sugarcane mosaic virus subgroup consists of four distinct potyviruses. Phytopathology (submitted).

Milne R.G. (1988) The plant viruses, vol. 4: the filamentous plant viruses. Plenum Press, New York, U.S.A.

Shukla D.D., and Ward C.W. (1988) Amino acid sequence homology of coat proteins as a basis for identification and classification of the potyvirus group. J. Gen. Virol. 69: 2703-2710.

Shukla D.D., and Ward C.W. (1989a) Structure of potyvirus coat proteins and its application in the taxonomy of the potyvirus group. Adv. Virus Res. 36: 273-314.

Shukla D.D., and Ward C.W. (1989b) Identification and classification of potyviruses on the basis of coat protein sequence data and serology. Arch. Virol. 106: 171-200.

Shukla D.D., Frenkel M.J., and Ward C.W. (1991) Structure and function of the potyvirus genome with special reference to the coat protein coding region. Can. J. Plant Pathol. (in press).

Shukla D.D., Ford R.E., Tosic M., Jilka J., and Ward C.W. (1989a) Possible members of the potyvirus group transmitted by mites or whiteflies share epitopes with aphid-transmitted definitive members of the groups. Arch. Virol. 105: 143-151.

Shukla D.D., Jilka J., Tosic M., and Ford R.E. (1989b) A novel approach to the serology of potyviruses involving affinity purified polyclonal antibodies directed towards virus-specific N termini of coat proteins. J. Gen. Virol. 70: 13-23.

Shukla D.D., McKern N.M., Gough K.H., Tracy S.L. and Letho S.G. (1988a) Differentiation of potyviruses and their strains by high-performance liquid chromatographic peptide profiling of coat proteins. J. Gen. Virol. 69: 493-502.

Shukla D.D., Strike P.M., Tracy S.L., Gough K.H. and Ward C.W. (1988b) The N and C termini of the coat proteins of potyviruses are surface-located and the N terminus contains the major virus-specific epitopes. J. Gen. Virol. 69: 1497-1508.

Shukla D.D., Tosic M., Jilka J., Ford R.E., Toler R.W. and Langham M.A.C. (1989c) Taxonomy of potyviruses infecting maize, sorghum and sugarcane in Australia and the United States as determined by reactivities of polyclonal antibodies directed towards virus-specific N-termini of coat proteins. Phytopathology 79: 223-229.

Shukla D.D., Tribbick G., Mason T.J., Hewish D.R., Geyson H.M. and Ward C.W., (1989d) Localization of virus-specific and group-specific epitopes of plant potyviruses by systematic immunochemical analysis of overlapping peptide fragments. Proc. Natl. Acad. Sci. USA 86: 8192-8196.

Usugi T., Kashiwazaki S., Omura T. and Tsuchizaki T. (1989) Some properties of nucleic acids and coat proteins of soil-borne filamentous viruses. Ann. Phytopathol. Soc. Japan 55 : 26-31.

Van Regenmortal M.H.V. (1982) Serology and immunochemistry of plant viruses. Academic Press, New York, U.S.A.

Van Regenmortal M.H.V. and Von Wechmar M.B., (1970) A reexamination of the serological relationship between tobacco mosaic virus and cucumber virus 4. Virology 41: 330-338.

Yu M., Frenkel M.J., McKern N.M., Shukla D.D., Strike P.M. and Ward C.W. (1989) Coat protein of potyviruses 6. Amino acid sequences suggest watermelon mosaic virus 2 and soybean mosaic virus-N are strains of the same potyvirus. Arch. Virol. 105: 55-64.

Ward C.W., and Shukla D.D. (1991) Taxonomy of potyviruses: current problems and some solutions. Intervirology. (in press).

Table 1 Sensitivity of serological and molecular techniques for plant virus detection

Technique	Detection Range [a]
Double immunodiffusion	2-20 lg/ml
Liquid precipitin tests	1-10 lg/ml
Radial immunodiffusion	0.5-10 lg/ml
Rocket electrophoresis	0.2-1.0 lg/ml
Complement fixation	50-100 ng/ml
Immunoosmophoresis	50-100 ng/ml
Passive hemaglutination	20-50 ng/ml
Latex test	5-20 ng/ml
ELISA	1-10 ng/ml
Immunoelectron microscopy	1-10 ng/ml
Western blotting	1-10 ng/ml
Molecular hybridization	Less than 1 pg of RNA

[a] sources: Van Regenmortal (1982), Frenkel et al (1991) and references therein.

VIRUSES OF ORNAMENTAL - IDENTIFICATION AND DIAGNOSIS

B.P. SINGH, K.M. SRIVASTAVA AND R.K. RAIZADA
PLANT VIRUS LABORATORY, NATIONAL BOTANICAL RESEARCH INSTITUTE, LUCKNOW-226 001, INDIA.

Key Words: Gladiolus, Chrysanthemum, Petunia, Hippeastrum, Amaryllis, hollyhock, BYMV, CMV, Carlavirus, Geminivirus, ELISA, DIBA, ISEM, cDNA.

INTRODUCTION

Advancement in mass clonal propagation and its impact on floriculture industry is pressing for elite propagules. Besides the emphasis on cut flowers, the market demand is getting higher for pot plants, miniculture and foliage plants. The twin demand of high quality propagating material and agrotechniques for specific target has made conspicuous the role of diagnosis of viral diseases in ornamental plants. The ornamental plants are conducive to the multiplication and spread of viruses. Mostly ornamental plants are vegetatively propagated and the viruses and virus like pathogens transmitted mechanically. A small member of infected plants in a production area can adversely affect the quality of the net product. Cutting for mass multiplication may be obtained either from growers own stock or specialist propagators. The problem is intensified by the fact that most field grown plants are virus-infected and also that symptoms may be masked or disappeared due to environmental conditions. The methodologies for detection of virus in ornamentals has become essential to the ornamental industrial as no practical treatments exist to cure virus-infected material in the field. The uses of virus free propagating material thus has been used in controlling virus diseases of ornamental crops.

Floriculture trade is well established in United States of America and European countries. It is a developing industry, in India, for loose flowers, cut flowers and pot cultures of rose, chrysanthemum, carnation, tuberose, gerbera and jasmine. The major limitation in floriculture industry is the availability of clean propagating quality material to produce the flowers/plants of international standards. This task gains more importance as most of the ornamentals of economic value are vegetatively propagated.

The rapid, sensitive and reliable diagnostic tools have been developed to detect and identify viruses. These methods have been employed for mass screening of ornamentals either exhibiting symptoms or as symptomless carrier. The following text deals with the identification of virus infect-

J. Prakash and R. L. M. Pierik (eds.), Horticulture – New Technologies and Applications, 321–328.

ing ornamental plants and the diagnostic methods used in floriculture for virus detection in different areas.

IDENTIFICATION OF VIRUS

Proper characterization of the causal agent is a prerequisite for developing effective diagnostic methods and feasible management practices. In general the virus infected ornamentals frequently exhibit almost common symptoms viz. various types of mosaic, curling of leaf lamina and reduction in plant growth. These abnormalities influence the quality of blooms adversely unmarketable which tender them.

The criteria for identification of viruses, include the mode of transmission virus-vector relationship, morphology of virus particle, serological characteristics, and chemical characterization of viral genome.

A few viruses belong to Potyvirus, Cucumovirus and Carlavirus groups have been identified at NBRI. They infect the commercial floral crops like gladiolus, hippeastrum, petunia, chrysanthemum and mesembryanthemum. A virus causing light and dark green to yellow mosaic in gladiolus has been identified as Bean yellow mosaic potyvirus (BYMV). The method of the purification has been standardized (Srivastava et al. 1983). The 750x 12 nm particles causing the disease is transmitted by several aphid species in non-persistent manner. The isolate is serologically related with BYMV gladiolus strain reported in Florida. The disease incidence was approximately 90-100%.

Hippeastrum mosaic virus which is common on Amaryllis and Hippeastrum has been identified as a Potyvirus measuring 750x12 nm in size (Raizada et al. 1983). Unlike other potyvirus it could not be transmitted by aphids. Purification procedure commonly used for potyviruses could be improved for isolation of HMV by using urea during initial steps of purification. The antibodies of this virus has been raised and ELISA techniques for its detection in infected plants has been standardized (Raizada et al. 1983).

Carlaviruses are not known to cause major diseases in plants. However, a severe virus disease in Petunia was associated with a Carlavirus of 660x 13 nm in size. It was not transmitted by aphids in either persistent or non-persistent manner. The virus causes extreme effect on vegetative growth and flowering characteristics but disease incidence is very low. The isolate has been identified as a serotype of potato virus S (Raizada et al. 1984).

Three viruses are prevalent on chrysanthemum namely, a strain of cucumber mosaic virus (CMV), Chrysanthemum aspermy virus (CAV) and Chrysanthemum virus B (Raizada et al. 1989; Srivastava et al. 1990). These isolates have been characterized on the basis of particle morphology, size, molecular weight of coat protein subunits and serological characteristics. A satellite RNA has been found associated with CMV as well as CAV strains. The role of this satellite RNA, if any, in symptom production is currently under investigation.

Infection of iceplant (*Dorotheanthus* species) with CMV is of high incidence (Aslam et al. 1989). The virus causes severe mosaic symptom and stunting of the plant which has been reported by several workers. They have indicated the serological relationship with other CMV strains (Aslam et al. 1989).

Whitefly transmitted viruses affecting *Zinnia elegans*, *Althea rosea*, sunflower and *Ageratum* sp. have been studied with reference to virus-vector relationship (Srivastava et al. 1977; Dwadash Shreni et al. 1979,). Purification of these viruses, however, was not satisfactory due to extremely fragile nature of the disease agents though, we always observed a few geminate virus particles present in the preparation in the infected preparation. The methods to stabilize the virus are still under investigation.

Virus complex in narcissus plant is also under study. Though we have isolated a potyvirus from the complex, but at present we are not aware whether it is distinct from those reported from other countries. Various properties of some viruses identified from our laboratories has been summarized in Table- 1.

Table 1. Viruses affecting ornamental plants at National Botanical Research Institute.

Viruses	Host	Particle size(nm)	Serological relationship	Nature of genome	Uni-or multi-partite	MW of coat protein
Cucumber mosaic	Ice plant	28	CMV strains	RNA	Tripartite	24.5 KD
	Chrysanthemum	29	CMV strains	RNA	Tripartite	24.5 KD
	Zinnia	28	CMV strains	RNA	Tripartite	24.5 KD
	Petunia	28	CMV strains	RNA	Tripartite +satellite RNA	24.5 KD
Chrysanthemum aspermy	Chrysanthemum	29	Chrysanthemum aspermy & tomato aspermy	RNA	Tripartite +satellite RNA	24.5 KD
Bean yellow mosaic	Gladiolus	750x12	BYMV-Gladiolus strain	RNA	Unipartite	32 KD
Hippeastrum mosaic	Hippeastrum	750x12	Not known	RNA	Unipartite	30 KD
Unidentified Potyvirus	Narcissus	740x13	Not known	RNA	Unipartite	31 KD
Unidentified Carlavirus	Petunia	660x13	PVS-serotype	RNA	Unipartite	
Unidentified Ilarvirus	Catharanthus	35, 32, 28	Not known	RNA	Tripartite	-
Unidentified Gemini	Zinnia	40x70	Not known	DNA*	-	-
	Hollyhock	38x20	Not known	DNA*		

*–Suspected

NATURAL SPREAD

Aphids, whiteflies and thrips are the major vectors, responsible for spread of these viruses in ornamentals. The relationship of the aphid vector with CMV strains affecting Zinnia, Petunia, Iceplant and Chrysanthemum has been studied in detail. The aphid acquires these viruses within a short span of 15-30 sec probing on the infecting plants. The time required for transmission is very short and none of the cases exceeded more than 30 min. Virus in aphids lose its infectivity within 20-30 min after acquisition. Moreover, viruliferous aphids lose infective potential after moulting.

BYMV is also transmitted in non-persistent manner by aphid from Gladiolus to Melilotus alba. The virus-vector studies related to BYMV were not carried out in gladiolus both as donor and recipient host, because of the meager availability of the healthy plant material (Srivastava et al. 1983) due to very high incidence of BYMV.

The relationship of Zinnia yellow net disease agent, hollyhock and sunflower yellow mosaic viruses was found to be strictly circulative. The whitefly acquired the virus from the infected plants within 30 min but the transmission percentage was more when this period was raised to 24 hr (Srivastava et al. 1977; Dwadash Shreni et al. 1979). Similarly, transmission of disease agent by viruliferous whitefly was possible within 30 min to an hour but number of plant infected were much more when period of feeding test plants was extended upto 48 hr. After acquiring the virus/ whitefly requires a latent period between 8-18 hr.

Recently thrips have been suspected to cause major diseases in Impatians balsamina and Catharanthus roseus, but exact relationship of these viruses with their thrips-vector has not been worked out.

Role of weeds in spread of viruses is well documented (Ramakrishna et al. 1981). We have recorded Physalis minima as a major weed host for CMV strains. M. alba, Vicia hirsuta as reservoir of BYMV and Ageratum conyzoides as an alternative host of Zinnia yellow net disease agent (Table 2).

Table-2. Role of weeds in natural spread of viruses.

Name of ornamental plants	Virus	Weed host	Vector & virus-vector relationship
Chrysanthemum	CMV	Physalis minima	Aphis gossypii +
		Solanum nigrum	A. fabesolanella +
Gladiolus	BYMV	Melilotus alba	Myzus persicae +
		Trifolium sp.	A. gossypii +
		Vicia hirsuta	
Petunia	CMV	P. minima	M. persicae +
		S. nigrum	
Zinnia	CMV	P. minima	A. gossypii +
		S. nigrum	A. fabesolanella +
	Unidentified gemini	Ageratum	Bemisia tabaci ++
Hollyhock	"		B. tabaci ++
Sunflower	"		B. tabaci ++
Acalypha sp.	"		B. tabaci ++

+ Non-circulative
++Circulative

PRODUCTION OF VIRUS FREE PLANTS *IN VITRO*

Tissue culture is being used to produce virus free propagules and their mass-multiplication. Needless to say that sensitive detection methods have proved useful to achieve the objective. At NBRI the production of virus free plantlets has been attempted in Dorotheanthus (Aminuddin *et al.* 1988), Petunia and Gladiolus (Aminuddin and Singh, 1985). The efforts are being concentrated on chrysanthemum and gladiolus. Bean yellow mosaic virus (BYMV) has been successfully eliminated from gladiolus mother stock through tissue culture by screening of plantlets grown using apical bud through sero-diagnostic methods i.e., ELISA and Dot immunobinding assay. Selection of the clean stock was made using ELISA, DIBA and ISEM. Similar approach is being made to obtain virus free stock of chrysanthemums.

MAINTENANCE OF CIRUS CULTURE UNDER ASEPTIC CONDITIONS

NBRI has attempted the culture collection of virus/es in plants under aseptic conditions in case of herbaceous and annual ornamental plants as these crops cannot be grown in field throughout the year (Table 3). Detection method is an essential step to maintain the protocol of keeping pure culture collection.

Diagnostic tools play an important role in indexing of ornamental plants for viruses. This decade has seen enormous improvement in the existing techniques of detection and devicing new methods of diagnosis of virus which are sensitive, quick, reliable and permit mass screening of plants for presence or absence of virus or virus like pathogens.

Table 3. Culture of viruses *in vitu* at NBRI

Ornamental Plants	Virus	Detection methods
Gladiolus	BYMV*	ELISA, DIBA
Chrysanthemum	Chrysanthemum Virus B	ELISA, DIBA
Solanum melongena/ S. capsicastrum	EMCV**	Double diffusion

* BYMV = Bean yellow mosaic virus
** EMCV = Eggplant mottled crinkle virus.

DIAGNOSTIC TECHNIQUES

The most widely used techniques for the detection of plant viruses are based either on antibodies which recognise surface structure of viral coat protein or on complementary nucleic acid (cDNA or cRNA).

A) IMMUNO PROBES

The antibody based probes have been used in plant virus work since long and techniques viz., precipitin test, ring interface test, ISEM, double diffusion test of various kind have been used since long for diagnosis of viruses. The recent technique based on immunoprobing like different forms of enzyme linked immunosorbent assay (ELISA) and dot immunobinding assay (DIBA) have permitted quick detection of important viral pathogens with reliability and feasibility of handling several thousand samples in short time. These techniques have become more popular by development of automated machines like ELISA readers, computerization, etc. The basic principle is the specific reaction in between the antigen and its homologous antibody (Mowat and Dawson 1987). The tests have been made in liquid medium, in agarose gels or on solid matrices. Polyclonal antisera has been mostly used for mass detection but monoclonal antibodies are currently being employed for differentiating among strains of virus. Enzyme assisted immuno assay (ELISA, DIBA) is the immobilization of antigen or antibody on polysterine polyvinyl/nitrocellulose surfaces and sequential addition of various reactants at the reaction site (Raizada et al. 1986, 1988, Srivastava et al. 1988). Removal of nonspecific substances is carried out by extensive washing after each step of the procedure leaving only specific immobilised reactors.

B) NUCLEIC ACID PROBES

Immunoprobes, however, are not applicable in some cases. Diffective viruses which have lost their protein coat thus cannot be detected by antibody raised to protein coat of the wild type virus. Similar problems are encountered in the etection of viroids which exist as free nucleic acid in their host (Bar Joseph et al., 1986). The detection of such agents necessitated development of new techniques using cDNA or cRNA probes. The basic principles underlying the detection based on nucleic acid hybridization technique is that a positive strand nucleic acid will specifically bind with its complementary DNA or RNA. These double stranded (ds) hybridized forms can be detected if such probes are labelled with radioactive material. The detection system raised on this approach is very sensitive and permits handling of large amount of samples (Baulcombe et al. 1984; Symons, 1984).

cDNA can be obtained enzymatically within a few hours without a biological propagation system. Thus, any viral RNA or DNA in the presence of suitable buffering conditions and enzyme like, reverse transcriptase or DNA polymerase can be used to make a complementary nucleic acid probe which can be subsequently labelled (Hames and Higgins, 1985). This will continously require availability of viral RNA and DNA propagated in biological system. The system has been further simplified by cloning and propagation cDNAs. The ds cDNA is synthesized in the routine way and inserted into plasmids/phages. Such chimeric plasmids/phages are then introduced into bacterium. Further isolation of cDNA from the bacteria is simple and the whole process can be completed within 24 hr. The advantage of using bacterial culture is the rapidity of the process and simplified procedures. Nevertheless, it requires more qualified labour than that required for production of antiserum.

C)COMPARISON OF IMMUNOPROBES AND NUCLEIC ACID PROBES

A comparison of the immunoprobes and nucleic-acid probes for detecting virus clearly reveals that virion specific antibody can recognise the genetic information which is expressed in the form of surface structure of viral coat protein only. Coat protein gene comprises only a small part of the viral genome which normally never exceeds more than 10% of the total viral genome. The nucleic-acid probes, such as those generated by random primer cDNA, on the other hand, have been considered to be representative for the entire genome of the virus. However, this may not be always entirely true because of sequence dependent premature termination of cDNA chain induced by reverse transcriptase. It is needless to say that cDNAs are more likely to detect differences between closely related and similarities between distantly related viruses as compared to antibody based probes.

Use of nucleic-acid as a probe for detection, requires labelling of complementary nucleic acids. Conversely when antibodies are used as probes, labelling is not always a prerequisite for diagnosis because the surface structure of coat protein are highly repetitive in virions. A single virus particle provides 180 (in isometric virus) to several thousand (in rod-shaped virus) coat protein subunits. In contrast, only one or a few nucleic acid strands with few or no repetitive sequence are present in a virus particle. Nucleic acid hybridization test, thus, essentially requires a sensitive labelling of a probe which is not required in serological tests. The labelling of antibodies is, however, required when techniques like, ELISA and DIBA are being used for mass detection. The labelling in these systems is generally done with enzymes (alkaline phosphatase, urease, horse radish peroxidase, Pencillinase) and gold colloids in case of electronmicroscopy. Radioactive labelling of nucleic acid probes (Lakshman et al. 1986), is however, not a method of choice due to inherent potent danger of radiation hazards. To overcome this hazard, progress has been made in the development of non-radioactive labells such as biotin/streptavadine, photobiotin system etc. (Leary et al. 1983; Habili et al. 1987). Sandwich hybridization techniques on overlapping cDNA probes, ELISA like testing procedures for capturing and detection of the viral/viroid nucleic acid seems to be extremely promising.

It is apparent that choice can be made between immunoprobe and nucleic acid probe for detection of a virus. However, nucleic acid probes remain the only method for detection of viroids and certain viruses which preclude the reliable detection in serological methods. Detection of mutant strains of tobacco rattle virus which fail to produce coat protein can only be detected with cDNA probes. Also, some of the viruses present in very less concentration in their host cannot be detected through antibody probes.

DISCUSSION

To conclude, it is to emphasise that the virus detection methods are of immense use in trade of healthy propagating material for quality control. The successful application of these diagnostic methods will also help in certification schemes for plant introduction, plant quarantine regulations at regiona, national, and international levels. In countries like the Netherlands, England, United States of America, Switzerland, Denmark, etc. virus testing is carried out at Governmental Institutes as well as by provate organizations. In India, commercial organizations have many opportunities and challenges in this area. The availability of low cost antisera against common viruses within the country will help several fold in developing the certification and quality control programmes by Government and private organisations. Need for development of low cost virus detection techniques is another big challenge for the organizations involved in propagation and scale of ornamental plants. The diagnostic tools play further key role in producing virus-free planting material either by conventional methods or *in vitro* mass multiplication of reservoirs and vectors to keep at possible minimum level. These diagnostic methods also act as check points in every protocol of different activity of floral crop management.

ACKNOWLEDGEMENTS

Thanks are due to Drs. Zaidi, Vinod, Gupta, Aslam, Raj, Aminuddin, Govind Chandra, Late Ram Krishna for their investigation towards the valuable studies. Dr. P.V. Sane, Director for allowing such independent work.

REFERENCES

Aminuddin and Singh, B.P. (1985) Indian Phytopath. 38: 375-377.
Aminuddin and Singh, B.P. (1985a) Indian Phytopath. 38: 692-694.
Aminuddin, Aslam, M. and Singh, B.P. (1988) Indian J. Exp. Biol. 26: 815-816.
Aslam, M., Srivastava, K.M. and Singh, B.P. (1989) Z. Schrift Pflanz. Pflan. 96:478-485.
Bar-Joseph, M., Seger, D., Blickle, W. Yesodi, V., Franck, A. and Rasner, A. (1986) In: Jones, R.A.C. and Torrance, C. eds., Developments and applications in virus testing, pp. 13-23, Association of Applied Biologists.
Baulcombe, D., Flavell, R.B., Boulton, R.E. and Jellis, G.J. (1987). Plant Pathology 33: 361-371.
Dwadash Shreni, V.C., Srivastava, K.M. and Singh, B.P. (1979). Source and spread of Zinnia yellow net disease, Indian J. Hort. 37: 190-194.
Habili, N., Mc Innes, J.L. and Symons, R.H. (1987) J. Virol. Methods 16: 225-237.
Hames, B.D. and Higgins, S.J. (1985). Nucleic acid hybridization. IRL Press Ltd., England 245 pp.
Lakshman, D.K., Hiruki, C., Wu X.N. and Leung, W.C. (1986). J. Virol. Methods 14: 307-319.
Leary, J.J., Brigati, D.J. and Ward, D.C. (1983). Proc. Natl. Acad. Sci., USA 80:4045-4049.
Mowat, W.P. and Dawson, S. (1987) J. Virol. Methods 15: 233-247.
Ram Krishna, Srivastava, K.M. and Singh, B.P. (1981). New Botanist 7: 105-109.
Raizada, R.k., Srivastava, K.M. and Singh, B.P. (1983). Tropical Plant Sci. Res. 1: 157-163.
Raizada, R.K., Srivastava, K.M. and Singh, B.P. (1984). Internat. J. Tropical Pl. Dis. 2: 105-116.
Raizada, R.K. (1986). A Report Danish Government Institute Seed Path. Develop. Countries, Denmark pp. 28.
Raizada, R.K., Govind Chandra and Singh, B.P. (1988). Indian Phytopath Soc. Meeting, Jaipur (Abstract).
Raizada, R.K., Srivastava, K.M., Govind Chandra and Singh, B.P. (1989). Indian J. Exp. Biol. 27: 1094-1096.
Srivastava, K.M., Dwadash Shreni, V.C., Srivastava, B.N. and Singh, B.P. (1977). Plant Dis. Reptr. 61: 550-554.
Srivastava, K.M., Raizada, R.K. and Singh, B.P. (1983). Indian J. Plant Path., 1: 83-88.
Srivastava, K.M., Raizada, R.K. and Singh, B.P. (1988). Indian Phytopath. Soc. Meeting, Jaipur (Abstract).
Srivastava, K.M., Raizada, R.K., Raj, S.K., Govind Chandra, and Singh, B.P. (1990). VII Internat. Cong. of Virology, Berlin, Germany (Abstract).
Symons, R.H. (1984). In: Kosuge, T. and Nester, E.W. eds., Plant Microbe Interactions. Vol. I. Molecular and Genetic Perspectives, pp. 93-124, Mac Millan, New York.

Protein-A supplemented immune electron microscopy for diagnosis of potato viruses X, S, Y and leafroll

I. D. GARG AND S. M. PAUL KHURANA
Division of Plant Pathology
Central Potato Research Institute
Shimla-171001, India

Key words: Protein-A, Immune electron microscopy, Potato viruses

Introduction

Immune electron microscopy (IEM) is a well-documented and a highly sensitive virus detection technique (Milne and Luisoni, 1977; Khurana and Garg, 1989). The technique has also been found to be highly effective for potato viruses (Roberts and Harrison, 1979; Garg and Khurana, 1988).Pre-treatment of grids with Protein-A was reported to trap more virions than with the standard IEM (Shukla and Gough, 1979). But it did not increase trapping of virions when 1000-fold diluted antiserum was used (Pares and Whitecross, 1985).

The potato leafroll virus (PLRV) is a small, phloem-restricted luteovirus occurring in very low concentration, poses problems, in its elcetron microscopic detection. IEM for its detection according to the method described by Roberts and Harrison (1979) and Singh *et al.* (1990) has overcome many of the difficulties, yet recognition of virions mainly due to their low concentration and small size is rather difficult. A variant of IEM called Solid Phase Immune Electron Microscopy (SPIEM) has been effectively used for the detection and diagnosis of certain human viruses (Gerna *et al.*, 1984; Lewis *et al.*, 1988). This paper reports the advantage of Protein-A in the detection and identification of the common potato viruses by the variant of IEM which we have termed as Protein-A Supplemented Immune Electron Microscopy (PAS-IEM) and found highly useful for PLRV.

Methods

Virus extracts

Pure cultures of PVX, PVS, PVY and PLRV maintained in *Nicotiana glutinosa*, potato cv. Craig's Defiance, *N. tabacum* cv. Havana and *Physalis floridana*, respectively at Central Potato Research Institute, Shimla were used. Crude virus extracts for PVX, PVS and PVY were prepared by homogenizing 0.1 g of infected leaf tissue with 1 ml of the buffer in a mortar and pestle, the macerate was diluted with the

J. Prakash and R. L. M. Pierik (eds.), Horticulture – New Technologies and Applications, 329–336.

extraction buffer (phosphate buffer, 0.1 M, pH 7.2) to 50 ml in case of PVX and to 10 ml in case of PVS, while in case of PVY the macerate was prepared in 0.1 M phosphate buffer (pH 7.2) plus 0.1 M EDTA and diluted to 10 ml with the buffer. A clarified virus extract of PLRV was prepared by thorough homogenization of 0.1 g of infected leaf veins in 500 ul of 0.05 M phosphate buffer (pH 6.5) by adding a little of 400 mesh carborundum, followed by centrifugation in an Eppendorf tube at 12000 g for 10 min and collecting the supernatant which served as the virus extract.

Antisera

Polyclonal antisera to purified PVX, PVS and PVY raised in rabbits (Khurana et al., 1987, 1990) had initial microprecipitin titres of 1 : 2000, 1 : 512 and 1 : 512 respectively. PVX antiserum was first diluted 4-folds with 0.85% saline to bring it down to a level at par with PVS and PVY and then all these were diluted 1000-fold with saline. Polyclonal antiserum to PLRV (Singh et al., 1990) had a micro-precipitin titre of 1 : 128 and it was diluted 200-fold with saline.

Grid preparation

Collodion-copper grids were used and prepared as follows:

Trapping of virions on Protein-A + antiserum (IgG) coated grids: Grids were coated with Protein-A by floating on 50 ul drops of Protein-A (5 ug/ml) solution in 0.1 M phosphate buffer (pH 7.2) for 1 h at room temperature (20-22 C). They were washed with 30 drops of distilled water, drained and then immediately transferred onto 50 ul drops of the diluted antiserum for 1 h at 37 C followed by washing and draining as above. They were then floated on 50 ul drops of the virus extract for 1 h at 37 C in case of PVX, PVS and PVY. But, in case of PLRV, incubation was initially 1 h at room temperature and then 72 h at 4 C. All incubations were given in humid petri dishes. After incubation, grids were washed, drained and then negatively stained with 2% aqueous uranyl acetate.

Trapping of virions + IgG (clumps) on Protein-A coated grids: Grids were coated with Protein-A as described in the previous paragraph. The clumps were prepared by incubating equal volumes of the virus extract and diluted homologous antisera for 1 h at 37 C in case of PVX, PVS and PVY but, for PLRV 24 h at 4 C. Protein-A coated grids were immersed in 50 ul drops of the mixture containing the clumps for 1 h at 37 C followed by washing, draining and staining.

Trapping of clumps on Protein-A + IgG coated grids: Grids were coated with Protein-A + IgG as described above and the clumps were also prepared as mentioned. Coated grids were immersed in 50 ul drops of the mixture containing clumps and incubated for 1 h at 37 C followed by washing, draining and staining.

Trapping without Protien-A: In case of control, the virions as well as the clumps were trapped on grids coated only with the diluted antisera. Grids were also prepared by trapping and decorating the virions on the grids.

Three grids were prepared in each case. Virions were counted at a magnification of 21,000 through 10x binocular. Mean of counts from 10 fields per hole was found out and 21 such mean values were recorded from three grids per treatment. Mean counts did not differ significantly from the grand mean.

Results

Data on effect of pre-treatment of the grid with Protein-A on trapping of potato virus particles are given in Table 1. There was slight enhancement in trapping of virions. But contrast was rather poor except in PLRV, where treatment caused aggregation of virions in groups of 3-15 and also gave better contrast (Fig. 1) as compared to standard IEM (Fig. 2). Effect of Protein-A treatment on trapping of clumps is illustrated in Figs. 3 and 4. Trapping of clumps on grids coated with Protein-A alone gave good contrast in case of PVX and PVY but a poor contrast in PVS. However, trapping of clumps on grids coated with protein-A and IgG resulted in very good contrast in all the four potato viruses tried.

Table 1 Effect of Protein-A on trapping of particles of potato viruses on a grid

Grid coated with	Trapping of potato virus			
	X[a]	S[a]	Y[a]	Leafroll[b]
Protein-A +IgG	5.5 (1.3)	1.8 (0.4)	2.3 (0.5)	15,000[c] (2,600)
IgG	4.1 (1.2)	1.5 (0.3)	1.7 (0.4)	12,500[d] (2,200)

[a] Average number of virus particles per field of view at 21,000 x through 10x binocular

[b] Average number of virus particles per 1000 um^{2} grid area.

[c] Particles occurred in groups of 3-15.

[d] Particles occurred as singles and twos.

Values in parenthesis are standard deviations

In case of flexuous potato viruses (i.e., X, S and Y) equally good or slightly better results were obtained when virions were trapped and decorated according to standard IEM (Fig. 5). Trapping of clumps of

PLRV on Protein-A + IgG coated grids resulted in larger aggregates having 4 - 30 virions (Fig. 6) which were readily identified due to antibodies attached to them (antibody staining) (Fig. 5, 7).

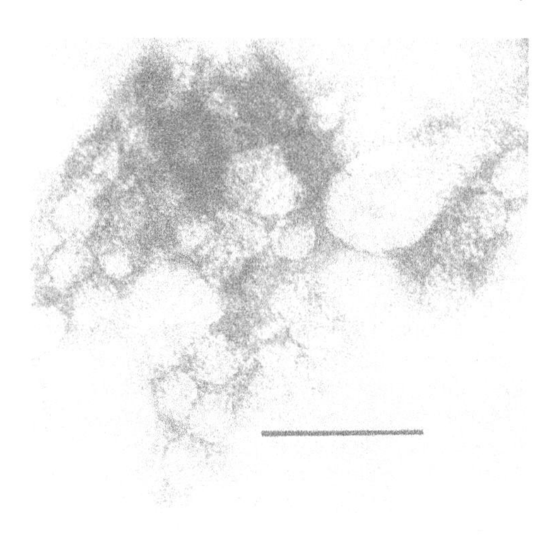

Fig. 1. PLRV trapped on a grid coated with Protein-A and 1:200 diluted PLRV antiserum. Bar = 100 nm.

Fig. 2. PLRV trapped on a grid coated with 1:200 diluted PLRV antiserum. Bar = 250 nm.

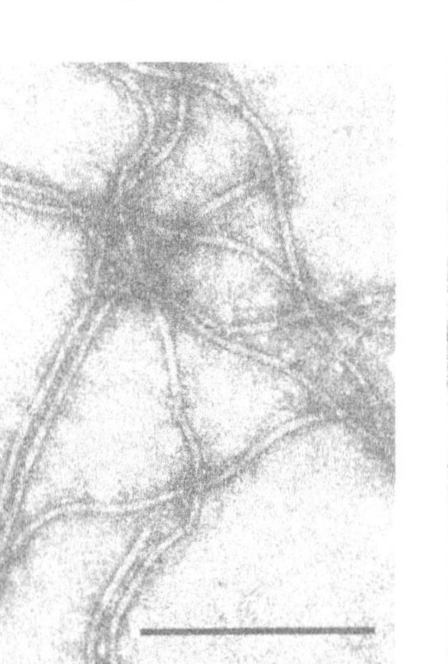

Fig. 3. PVY clumps formed with 1:2000 diluted PVY antiserum and trapped on a Protein-A coated grid. Bar = 250 nm.

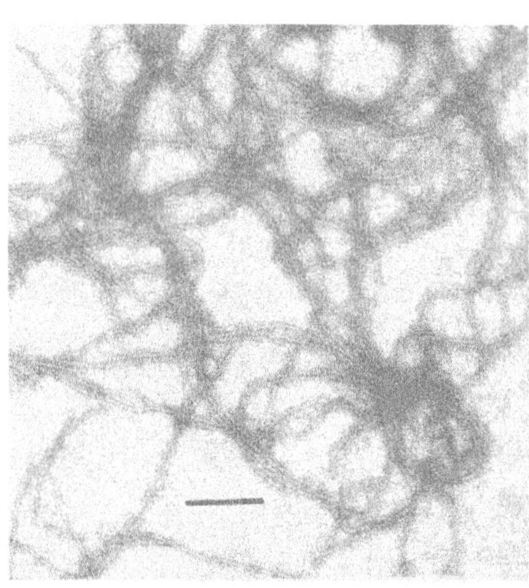

Fig. 4. PVY clumps formed with 1:2000 diluted PVY antiserum and trapped on a grid coated with Protein-A and 1:1000 diluted antiserum. Bar = 250 nm.

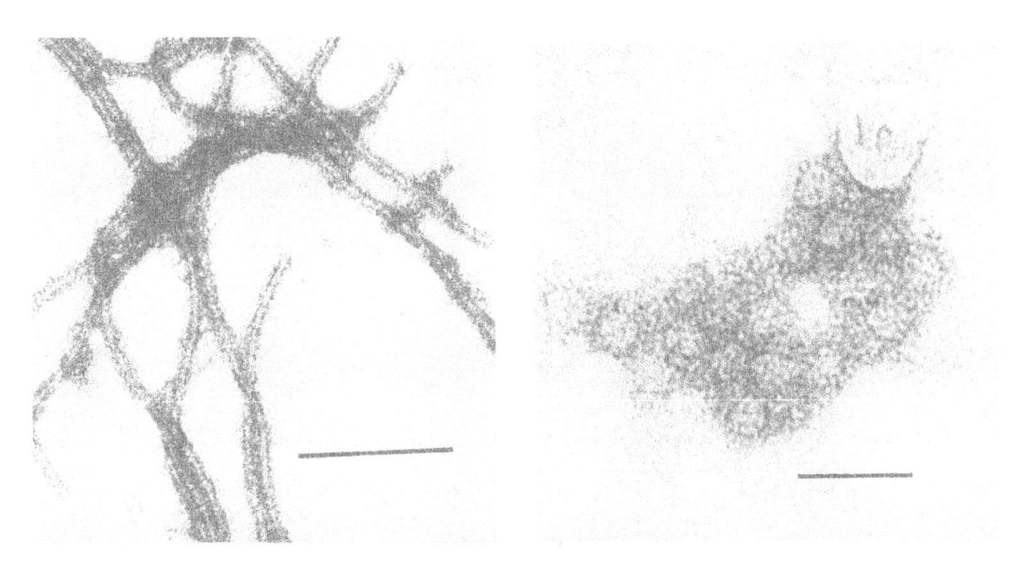

Fig. 5. PVY trapped on a grid coated with 1:1000 diluted antiserum and decorated with the same antiserum. Bar = 250 nm.

Fig. 6. PLRV clumps formed with 1:400 diluted antiserum and trapped on a grid coated with Protein-A and 1:200 diluted antiserum. Bar= 100 nm.

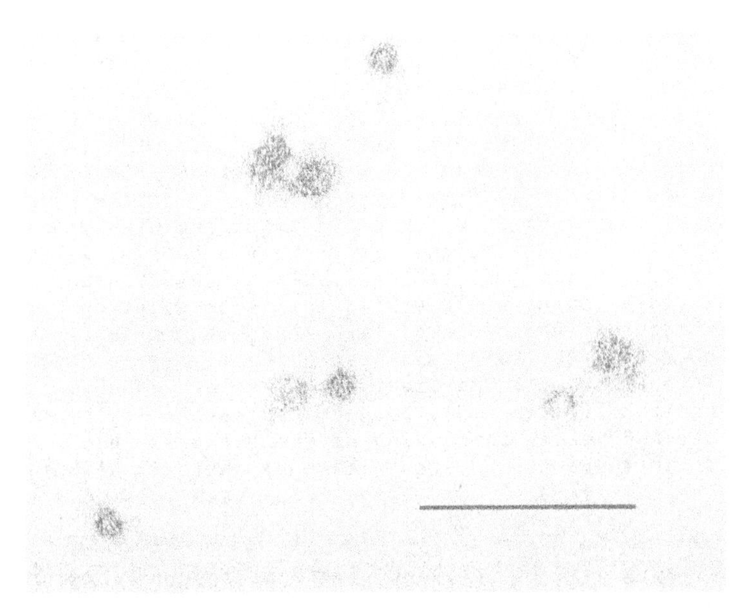

Fig. 7. PLRV trapped on a grid coated with 1:200 diluted PLRV antiserum and decorated with the same antiserum. Bar = 250 nm.

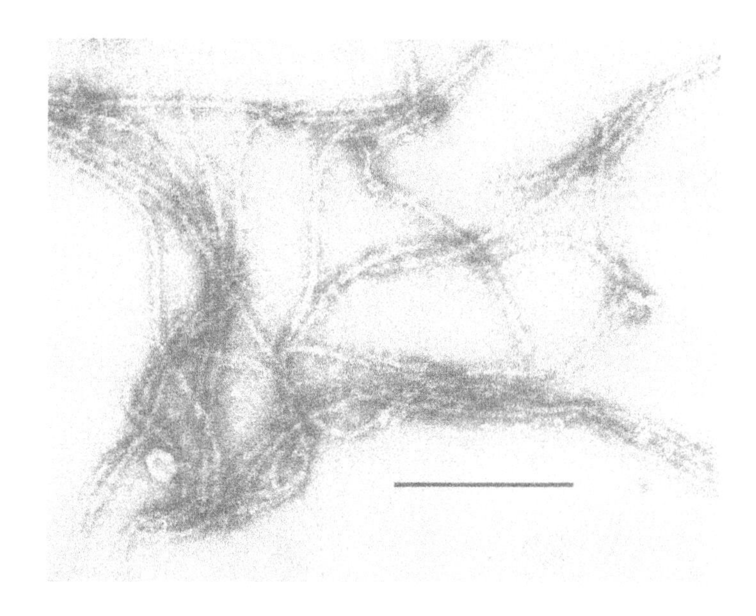

Fig. 8. PVY clumps formed by 1:400 diluted antisaerum and trapped on a grid coated with 1:200 diluted antiserum. Bar = 250 nm.

Discussion

Slight enhancement of trapping of flexuous potato virus particles occurred when trapped on grids coated with Protein-A and IgG, though the contrast was rendered poor in case of PVS. Our results differ from those reported by Pares and Whitecross (1985) for tobacco mosaic virus. They observed marked enhancement in trapping of virions with Protein-A pre-treatment of the grid when antiserum was diluted only 20-fold but, recorded a marked decline in trapping with Protein-A pre-treatment when antiserum was diluted 1000-folds. They explained the results by presuming that at 1000-fold dilution other serum proteins did not interfere with the trapping and that sufficient amount of antibodies was still present to occupy all the available sites. Consequently, there was enhanced trapping at 1000-fold antiserum dilution without Protein-A pre-treatment which caused less attachment of anti-bodies to the grid due to some sites being occupied by Protein-A. Enhancement in trapping of virions with the pre-treatment under our conditions may probably be due to a limiting anti-body concentration in the antisera used where Protein-A helped in binding more antibodies to the grid. This assumption is supported by the pattern of antibody attachment observed presently (Fig. 3, 5, 8) where virions are less covered with antibodies at lower antiserum titre.

Enhanced contrast in case of trapping of the clumps on Protein-A and antiserum coated grids may be due to grainy contrast caused by Protein -A and IgG. The reason for aggregation of the PLRV particles in groups during trapping on Protein-A and antiserum coated grid is not clear.

But formation of the large clumps in liquid phase is obviously due to cross linking of virions by antibodies. This phenomenon helps in quick detection/diagnosis of even small virions like that of PLRV.

Acknowledgements

Sincere thanks are due to Dr. R. P. Singh (Canada), Prof. E. E. Banttari (USA) and Dr. J. A. deBokx (Netherlands) for initial supply of reference antisera, especially for PLRV. Authors are also grateful to Dr. J. S. Grewal, Director, CPRI, Shimla for facilities and encouragement.

References

Garg I. D. and Khurana S. M. Paul (1988) Sensitivity of immunoelectron microscopy for the detection and diagnosis of potato viruses. Proceedings of IInd International Symposium on Electron Microscopy and Biophysics, pp. 275 - 278, Panjab University, Chandigarh, India.

Gerna G., Passarni N., Battaglia M. and Percivalle E. (1984) Rapid serotyping of human rotavirus strains by solid phase immune electron microscopy. J. Clinical Microbiol. 19: 273-278.

Khurana S. M. Paul and Garg I. D. (1989) Plant virus, viroid and mycoplasma-like organisms detection for quarantine. Rev. Tropical Pl. Path. 6: 251-277.

Khurana S. M. Paul, Singh M. N. and Shivkumar (1987) Purification of potato virus Y and the production of antiserum. Current Science 56:420-422.

Khurana S. M. Paul, Singh M. N., Garg I. D. and Shivkumar (1990) Uniform purification scheme for potato viruses. Asian Potato J. 1: 46-48.

Lewis D. C., Lightfoot N. F. and Pether J. V. S. (1988) Solid phase immune electron microscopy with human immunoglobulin M for serotyping of Norwalk-like viruses. J. Clinical Microbiol. 26: 938-942.

Milne R. G. and Luisoni E. (1977) Rapid immune electron microscopy of virus preparations. In : Maramorosch K. and Koprowski H. Eds., Methods in Virology VI, pp. 179-191. Academic Press, New York.

Pares R. D. and Whitecross M. I. (1985) An evaluation of some factors important for maximizing sensitivity of plant virus detection by immune electron microscopy. J. Virol. Methods 11: 339-346.

Roberts I. M. and Harrison B. D. (1979) Detection of potato leafroll and potato mop top viruses by immunosorbent electron microscopy. Ann. Appl. Biol. 93: 289-297.

Shukla D. D. and Gough K. K. (1979) The use of Protein-A, from Staphylococcus aureus, in immune electron microscopy for detecting plant virus particles. J. Gen. Virol. 45: 533-536.

Singh M. N., Khurana S. M. Paul and Garg I. D. (1990) Antiserum production and immunoelectron microscopy of potato leafroll virus. Indian Phytopath. 43: 13-19.

Production of virus-tested seed potatoes

M.J. FOXE,
National Agricultural and Veterinary Biotechnology Centre,
University College Dublin, Belfield, Dublin 4, Ireland

Introduction

Several systems for the production of virus-free potatoes (*Solanum tuberosum*) through tissue culture have been described (Mellor and Stace-Smith, 1977; Westcott, 1983; Foxe, Macken and Wilson, 1983a, 1983b). The purpose of the present investigation was to establish conditions for the *in vitro* propagation of virus tested seed potatoes of commercial importance in Ireland. Experiments were designed to select an appropriate medium for the growth of excised meristem-tips, their multiplication by nodal cuttings and subsequent rooting. Virus elimination by meristem-tip size alone or in combination with thermotherapy was assessed. Virus indexing systems using enzyme-linked immunosorbent assay (ELISA) (Clark and Adams, 1977) and immunosorbent electron microscopy (ISEM) (Derrick, 1973) were incorporated into the overall system.

Methods

The media used in this investigation were all based on Murashige and Skoog Plant Salt Mixture (Flow Laboratories, Catalogue No. 26-330-20), supplemented with $30gl^{-1}$ sucrose and for the solid media $7.2g\ l^{-1}$ (or $6.0g\ l^{-1}$ in medium MSW_1) Oxoid Agar No. 1. Additional supplements are listed in Table 24.1. After autoclaving, 100 ml aliquots of liquid media were dispensed aseptically into 250 ml flasks, and 25 ml aliquots of solid media were dispensed aseptically inst sterile plastic 100ml jars, which were allowed to set on a slope of approximately 45°.

Meristem-tips (0.1, 0.25, 0.5, 0.75 mm) were excised aseptically from shoot tips of potato tubers sprouted in the dark at 24^oC or from the terminal or axillary shoot buds of plants grown in pots at 24^oC in the glasshouse under a 16 h photoperiod. Prior to the excision of the meristem-tips, the source material was immersed in 1% household disinfectant 'Domestos' for 20 min and then rinsed three times in sterile distilled water. Immediately after excision, the meristem-tips were placed (10 meristem-tips per flask), in the liquid MSC_1, MSW_1 or MSI media or on the same media in solid formulation (two meristem-tips per jar). Flasks were incubated on an orbital shaker ($80r\ min^{-1}$) at 24^oC under a 16 h photoperiod with $70\ mEm^{-2}s^{-1}$ of photosynthetically active radiation. Jars containing the solid media were incubated on their sides under the same conditions of lighting and temperature.

J. Prakash and R. L. M. Pierik (eds.), Horticulture – New Technologies and Applications, 337–344.

Table 1. Supplements for the media used for meristem-tip culture and propagation by nodal cuttings of potato cultivars.

Supplement			Median		
(mg l-1)	MSC_1	MSC_2	MSW_1	MSW_2	MSI
Indoleacetic acid	1.0	-	0.5	-	-
Kinetin	1.0	-	-	-	-
Zeatin	-	-	0.5	-	-
Gibberellic acid	-	-	0.2	-	0.1
Inositol	-	100.0	100.0	100.0	-
Thiamine HCl	0.4	0.4	0.5	0.1	-
Pyridoxine HCl	-	-	0.5	0.1	-
Nicotinic acid	-	-	0.5	0.1	-
Glycine	-	-	2.0	0.4	-
NaH2PO4	-	170.0 -	-	-	
CaCl2.2H2O	-	-	-	440.0	-
KH2PO4	-	-	-	170.0	-
pH	5.2	5.2	5.6-6.0	5.6-6.0	5.2-5.8

Nodal cuttings approximately 50 mm in length were excised from shoots derived from the meristem-tip cultures, placed onto MCS_2, MSW_2 or MSI solid media and incubated as above. For transfer to pots in the glasshouse, the nodal cuttings were removed from the solid medium, planted in soil-less potting compost in 3 inch pots and covered with transparent polythene to maintain humidity. A 10 h photoperiod was provided with supplementary lighting.

Both source material and shoots developed from meristem-tip cultures were tested for the presence of the potato viruses X (PVX), Y (PVY), S (PVS), A (PVA), M (PVM) and Potato Leafroll Virus (PLRV) using ELISA and ISEM techniques. Meristem-tips varying in length from 0.1-1.0mm were excised from virus-infected plants which had been subjected to high temperature treatment. Tubers were selected and planted in pots which were placed in a growth chamber at 24°C and 16 h photoperiod. When the plants reach a height of approximately 20 cm, stem tips are clipped off. These clipped plants were placed in a growth chamber at 34°C under continuous light for 3 weeks. Meristem-tips were cultured on MSI medium and indexed for the presence of virus.

Results

The performance of meristem-tips from potato cultivars 'Clada', Kerr's Pink' and 'Pentland Crown' was assessed after 45 days of culture in liquid and solid MSC_1, MSW_1 and MSI media. In liquid culture, 100% survival of the meristem-tips of all sizes of all three cultivars occurred only on MSI, (Table 2), which permitted very prolific and healthy shoot and root growth. Although the percentage survival of meristem-tips was comparably high on MSW_1 medium the mean fresh weight was only 1% of that for MSI and was entirely due to callus formation. Only 'Kerr's Pink' meristem-tips survived (at 30%) on MSC_1 medium (Table 2); again growth was entirely due to callusing.

Table 2. The percentage survival and mean fresh weight of 20 meristems from potato cultivars, 'Clada', 'Kerr's Pink' and 'Pentland Crown' after 45 days in either MSC_1, MSW_1 or MSI liquid and solid media.

Measurement		Medium								
		MSC_1			MSW_1			MSI		
		Clada	Kerr's Pink	Pentland Crown	Clada	Kerr's Pink	Pentland Crown	Clada	Kerr's Pink	Pentland Crown
Liquid Media	% Survival	0	30	0	90	70	85	100	100	100
	Fresh weight (mg)	-	2.03	-	13.23	21.70	15.45	1.20×10^3	1.80×10^3	0.7×10^3
Solid Media	% Survival	65	50	0	85	55	50	90	60	75
	Fresh weight (mg)	15.03	10.33	-	35.94	68.87	63.62	1.82	1.59	1.28

On solid media, the percentage survival of meristem-tips from all three cultivars was highest on MSI medium (Table 2), although the greatest amount of growth occurred on MSW_1. This growth was by callus formation alone as was that of 'Clada" and 'Kerr's Pink' on MSC_1 medium. Only on MSI was there no callus formation. However, although shoots did start to develop on this medium their development was very poor and they did not form.

Table 3. Mean number of nodes per nodal cutting on 20 nodal cuttings of potato cultivars 'Clada', Kerr's Pink' and 'Pentland Crown' on MSD_2, MSW_2 or MSI solid media

Cultivar(a)	Medium (b)		
	MSC_2	MSW_2	MSI
Clada	1.0	1.6	2.1
Kerr's Pink	2.1	2.1	3.1
Pentland Crown	1.6	1.1	2.5

(a) Difference between cultivars significant at P = 0.01.
(b) Difference between media significant at P = 0.01.
Interaction between cultivars and media significant at P = 0.5.

Multiplication of the meristem-tip cultures was assessed on the three solid media MSC_2, MSW_2 and MSI in terms of the possible number of nodal cuttings (i.e. the number of nodes) which developed from a single cutting after 10 days in culture (Table 3). For all three cultivars, more nodes developed on MSI medium than on either MSC_2 or NSW_2 (Table 3) which supported a similar amount of growth but visual observation showed that its quality on MSW_2 was inferior to the other two media.

Almost 100% of the nodal cuttings of all three cultivars on MSI showed shoot and root development after 17 days (Table 4). A high percentage (>80%) of the nodal cuttings had developed shoots and roots on MSC_2, with the exception of 'Clada", the performance of which was very poor both in comparison to the other cultivars and its performance on MSW_2 and MSI. Overall, MSW_2 was the least successful medium for promoting shoot and root development in nodal cuttings.

Table 4. The percentage of 20 nodal cuttings from potato cultivars "Clada, 'Kerr's Pink' and 'Pentland Crown' showing shoot and root development after 17 days on MSC_2, MSW_2 or MSI solid media.

Cultivar	Medium		
	MSC_2	MSW_2	MSI
Clada	30	50	100
Kerr's Pink	85	50	95
Pentland Crown	80	15	100

For all three cultivars, nodal cuttings grown on MSC_2 medium for either 7 or 28 days performed very poorly when transferred to potting compost (Table 5). Most of them had not survived after the longer period, whereas survival of nodal cuttings grown for 28 days on either MSW_1 or MSI medium was significantly better than those grown for only 7 days. In 'Pentland Crown', the number of fully expanded leaves was significantly greater on plants grown from cuttings transferred from MSI than MSW_2 medium. Although no significant differences were observed in 'Clada' and 'Kerr's Pink', the plants developed from cuttings on MSW_2 were relatively of very poor quality. Thus, overall the best plants were obtained from nodal cuttings grown on MSI medium for 28 days before transfer to potting compost.

Virus elimination could not be accomplished by meristem-tip size alone, but required additional thermotherapy. Using 0.1 mm meristem-tip provided elimination of PVY but not of PVX. The use of larger sized meristem tips resulted in virus infected plantlets.

Discussion

These results show that the MSI medium in liquid formulation in a shake flask was best for establishing meristem-tip cultures of the potato cultivars 'Clada', 'Kerr's Pink' and 'Pentland Crown'. Not only was there 100% survival but the fresh weight of tissue produced was, on average, 100 times greater than the next best medium MSW_1. Furthermore, unlike the latter, MSI medium did not induce callus formation at any stage. This advantage is particularly important in the propagation of commercial cultivars through tissue culture as callus formation is considered to lead to genetic instability (Hussey, 1980) and plants regenerated from callus may differ genetically from plants of origin (D'Amato, 1975).

For propagation through nodal cuttings, MSI medium in solid formulation was most suitable in terms of both the rate of propagation and the quality of resultant in vivo plants. The 28-day culture period prior to potting rooted cuttings has now been successfully reduced to 14 days for routine work, providing a further advantage in terms of time and costs.

Fast, reliable, accurate and sensitive virus indexing methods represent important components of this tissue culture system. Virus indexing of meristem-tip derived shoots by ELISA and ISEM prior to taking nodal cuttings ensures that only virus-free material is propagated. ISEM is particularly advantageous when only a small amount of tissue culture material is available and is valuable as a check for some ELISA results. When meristem-tip cultures are derived from virus-free mother tubers (as tested by ELISA), indexing prior to propagation by nodal cuttings provides a further check. Occasionally, cultures derived from infected mother tubers are found to be free of virus. However, using small size meristem-tips (0.1 mm) would not particularly ensure freedom from virus and also resulted in lower survival rates. When a meristem-tip culture is shown to be infected it is eliminated and fresh cultures are initiated either from smaller meristems or heat-treated mother plants (Foxe, Macken and Wilson, 1983,b).

Table 5. The percentage survival, mean numbers of fully expanded leaves formed, and percentage top quality plants from 20 nodal cuttings of potato cultivars 'Clada', 'Kerr's Pink' and 'Pentland Crown' grown on MSC_2, MSW_2 or MSI solid media for either 7 or 28 days before transferring to pots in the glasshouse. The potted plants were assessed after 28 days.

Medium	Time (days)	Clada			Kerr's Pink			Pentland Crown		
		% Survival (%)	Number of Leaves(a)	Top Quality Plants	% Survival (%)	Number of Leaves (b)	Top Quality Plants	% Survival (%)	Number of Leaves(c)	Top Quality Plants
MSC_2	7	0	-	-	15	13.0	0	5	7	0
	28	0	-	-	0	-	-	0	-	-
MSW_2	7	40	6.5	0	30	7.8	0	65	7.2	0
	28	100	14.7	0	95	10.1	0	95	8.2	0
MSI	7	75	10.6	30	70	8.7	50	95	8.8	50
	28	100	15.7	90	100	11.8	45	100	11.1	85

(a) Difference between transfer age significant at P = 0.01
Difference between media MSW_2 and MSI significant at P = 0.5

(b) Difference between transfer age significant at P = 0.1
Difference between mediaMSW_2 and MSI not significant

(c) Difference between transfer age significant at P = 0.01
Difference between media MSW_2 and MSI significant at P = 0.01

An advantage of the tissue culture system described here is that both initiation and propagation of cultures can successfully be carried out on the same medium (MSI) reducing both time and expense. Gibberellic acid (0.1 mg l-1 in MSI) appears to be a particularly important medium component and has been shown by other workers to improve the growth of meristems (Gregorini and Lorenzi, 1974) and the rooting of meristem-tips (Pennazio and Redolfi, 1973), and to be a most effective growth regulator for promoting shoot and root development (Novak *et al.*, 1980).

Significant differences have been reported between cultivars in terms of their media requirements (Quak, 1961; Mellor and Stace-Smith, 1977; Novak *et al.*, 1980). However, although cultivar-media interactions were obtained here MSI medium was suitable for the three cultivars. Furthermore MSI medium has now been used successfully for meristem-tip culture and micropropagation by nodal cuttings of over 100 commercial potato cultivars in Ireland with an average time of 8 weeks from excision of meristem-tips to production of rooted virus-free nodal cuttings suitable for potting.

References

Clarke M.F. and Adams A.N. (1977) Characteristics of the microplate method of enzyme-linked immunosorbent assay for the detection of plant viruses. J. gen. Virol., 34, 475-483.

D'Amato F. (1975) The problem of genetic stability in plant tissue and cell culture. In: O.H. Trankel and J.G. Hawkes, eds, pp 333-348. Crop Genetic Resources for Today and Tomorrow. Cambridge University Press.

Derrick K.S. (1973) Quantitative assay for plant viruses using serologically specific electron microscopy. Virology, 56, 652-653.

Foxe M.J., Macken S. and Wilson U.E. (1983a) Initial investigations on production of virus-tested seed potatoes by tissue culture. In: A.C. Cassells and J.A. Kavanagh, eds., pp. 66-71, Plant Tissue Culture in Relation to Biotechnology. Dublin, Royal Irish Academy.

Foxe M.J., Macken S. and Wilson U.E. (1983b) Tissue culture methods for the production of virus-tested seed poatoes. Indian J. Plant Pathol., 1, 5-9.

Gregorini G. and Lorenzi R. (1974) Meristem-tip culture of potato plants as a method of improving productivity. Potato Res., 17, 24-33.

Hussey G. (1980) In vitro propagation. In: D.S. Ingram and J.P. Helgeson, eds., pp. 51-61. Tissue Culture Methods for Plant Pathologists. Blackwell Scientific Publications, Oxford,

Mellor F.C. and Stave-Smith R. (1977) Virus-free potatoes by tissue culture. In: J. Reinert and Y.P.S. Bajaj, eds., pp. 616-635, Applied and Fundamental Aspects of Plant Cell, Tissue and Organ Culture. Springer-Verlag. Berlin, Heidelberg, New York,

Novak F.J., Zadina J., Horackova V. and Maskova I. (1980) The effect of growth regulators on meristem tip development and in vitro multiplication of Solanum tuberosum L. plants. Potato Res., 23, 155-166.

Pennazio S. and Redolfi P. (1973) Factors affecting the culture in vitro of potato meristem-tips. Potato Res., 16, 20-29.

Quak F. (1961) Heat treatment and substances inhibiting virus multiplication in meristem culture to obain virus-free plants. Adv. Hort. Sci. Appl., 1, 144-148.

Westcott R.J. (1983) Micropropagation of virus-free potatoes. In: A.C. Cassels and J.A. Kavanagh eds., Plant Tissue Culture in Relation to Biotechnology, pp. 61-65. Dublin, Royal Irish Academy.

Pesticide decontamination of fruits and vegetables

S.K. MAJUMDER, Ex-additional Director and Coordinator UN University Center,
Central Food Technological Research Institute,
Mysore 570 013, India.

1. INTRODUCTION

The pesticides and their terminal residues remain on harvested fruits and vegetables. Sources of the residues are due to the direct application for control of pests and diseases during production of the crop, or due to translocation from the contaminated adjacent and surrounding areas where high volume spray applications are carried out (Dahm, 1957). These xenobiotics remain in the parent form and also as the oxidised or biotransformed or photodegraded compounds. They are metabolised and persist as terminal residues in the tissues of the fruits and vegetables. Often these terminal residues are more toxic than parent compounds (Klein, 1976). The problems of consumer safety, quality of products and legal tolerances of the compound and the marketability of such produce become important. Hence, abatement of residues by the redeeming technologies becomes imperative, until approaches to new leads for insecticides are successful (Elder & Keyserlingk, 1985) which would be ideal and safe to environment.

The 'pesticide cycle' and 'global chromatography' as natural phenomena have been postulated to explain the episodes of pesticide residues in the biosphere including on fruits and vegetables (Majumder, 1968, 1979). The over application of pesticide to the environment is due to the inefficient devices like the normal spraying techniques. Only 2% of spray is normally deposited on target and thus contaminate non-target areas. Stack emissions from chemical plants also lead to migration of foreign chemicals and contamination on crops. The pesticides and other xenobiotics migrate from the point of application to water, soil, air and the biosphere. During the migration to man, animal and crop, they pass through the phenomenon of biomagnification and the food chain (Majumder, 1979, 1983). In this paper some approaches to decontamination of Soil, Water Fruits & Vegetables are discussed.

2. APPROACHES TO DECONTAMINATION

For plant protection it is time to replace the recalcitrant pesticides with the compounds of high Pesticide Suitability Rating (PSR) to avoid the problems of residues (Deo et al. 1988).

J. Prakash and R. L. M. Pierik (eds.), Horticulture – New Technologies and Applications, 345–351.

Isomerization of beta-isomer of HCCH:

Although BHC (HCCH) and DDT are recalcitrant and still persisting in the environment, the isomerisation and biodegradation are the redeeming phenomena in nature. A significant discovery was made on the phenomenon of non-biological isomerisation of beta-isomer of HCCH to other isomers in water as this isomer of BHC is regarded as most persistent. The GLC and other analysis revealed its isomerisation to alpha and gamma isomer within a few minutes while delta-isomer was also detected after 25 days. This phenomenon opens up the need for investigating chronic and acute toxicities likely to be implicated in the cumulative effects of beta-isomer in biological system (Krishnakumari 1982). Evidence was also obtained on the conversion of beta isomer to more insecticidal forms by bioassay and GLC analysis, of extract of beta-HCCH in aqueous filtrates. Deo, et al (1980) studied interconversion of Hexachlorocyclohexane isomers. In view of large differences in the toxicity of individual HCCH isomers, the phenomenon of their interconversion appears to be quite important especially so, from the stand point of environmental toxicology. Available evidences indicate that interconversion of HCCH isomers do occur in nature and can also be carried out in the Laboratory by employing conditions like high temperature, pressure and UV irradiation. Although much work has been done to study the transformation of HCCH isomers in the environment and biota, the subject of physical interconversion has not received due attention. Currently available literature shows that interconversions of HCCH isomers do occur in nature (Benezet & Matsumura, 1973). It is possible that two types of mechanisms may exist-biological and non-biological.

Processing

Lockwood et al (1974) observed that the pesticide residue losses during processing ranged from 30 to 75%, during the preparation of chapaties from wheat and sorghum flour, total residues losses range from 59 to 93% for whole grain (wheat and sorghum) processed to chapaties. With rice product, wet cooked, all the protectants were totally degraded, except their metabolites, undetected with the equipment used, may have formed. In India, steam cooking of rice has been reported to almost completely (95%) destroy malathion. Similarly fruit and vegetables during thermal processing and cooking processes undergo volatilization and chemical changes (Visweswariah and Jayaram, 1972). Calcium hydroxide treatment of maize as is used in traditional Mexican corn processing was found to reduce the pesticides from 33.6ppm to 1.8ppm, particularly of Fenitrothion and Methyl parathion. The final concentration of Fenitrothion in flat chapati when baked at a temperature of 180°C, the concentration came down to 0.68ppm (Annual Report CFTRI, 1988 p176).

Decontamination of Soils

Crops were grown on different soils containing pesticides residues. The preferential uptake by different crops from the soils were determined (Lichtenstein, 1959; Majumder, 1969). Most of the crops were less tolerant at 50ppm level of beta HCCH in soil as compared to other isomers. The test insecticides were detected in the edible crop of all the crops studied and concentrations varied depending on the crop and insecticidal level of residue, when coriander, amaranthus, carrot, knolkhol, chillies, cucumber and tomato were compared. The residues were, however, did not exceed the tolerance limit (to 3ppm) established for human consumption for BHC (Karanth, et al, 1983).

Growth of tomato was retarded by chemicals resulting from their phyto-toxicity, while french bean, brinjal and ragi were not adversely affected (Srimathi et al 1987). Paddy exhibited reduced growth only in the presence of X-factor. Tomato plants contained considerable amount of test chemicals and other had either no or negligible amount present in them. Possibility of using french bean, paddy, ragi and brinjal as rotation crops in soil containing such chlorinated pesticides is suggested (Karanth, et al. 1981, 1982).

Residues in the soil could be minimised through microbial biodegradation (Majumder, 1968). Soil inoculation with the active organisms such as Pseudomonas spp. was required. Growing scavenger crops to remove residues of pesticide through selective translocation processes are promising. The surface residues of pesticides from leaves and fruits can be removed by biosurfactants, nontoxic detergents and solvents. The pesticides residues that are present within the tissues are difficult to remove but can be degraded by enzymatic and alkaline treatments. Cooking with permissible alkalies or incubation with enzymes provide the breakdown of complex molecules to smaller molecules and help in their leaching with water in ionised and soluble forms.

Monitoring these xenobiotic residues by instrumental analysis, bioassay techniques and enzymatic measurements are applicable under the existing processing and quality control conditions of industry (Visweswaraiah & Jayaram, 1971). Majumder et al, (1965) reported on a very elegant and rapid Paper Chromatographic technique for screening volatile chemicals for their reactivity with the constituents of foods. In this study, extremely low quantities (5 micrograms) of amino acids spotted on paper and exposed to the volatile chemicals indicated high reactivity of methyl bromide, methyl iodide, ethyl bromide, vapona and Beta-Propiolactone with methionine, after development of the chromatogram. This Chromatographic technique offers good scope for picking out the non-reactive fumigants with ease, rapidity and reliability. It can be used as a tool for predicting the reactivity of the pesticide or any chemical with the constituents of the food particularly dry fruits and oilseeds.

Decontamination of Fruits and Vegetables:

A Redeeming Technology, attempted to remove the surface residues of fruits and vegetables at the market, in the kitchen or in a processing plant, incorporates in it a detergent effect and also solubilisation of the xenobiotics (Raju, et al, 1983). The solutions based on non-toxic organic acids. The crops raised with modern agricultural system, grapes, tomatoes, cabbages, potatoes, brinjal and green leafy vegetables, have been decontaminated. A comparative biotechnology for decontamination of Fruits and vegetables is presented in FIGURE-1.

Decontamination of water:

The other redeeming technology deals with production of potable water (Visweswariah, et al, 1977). Water is often the carrier of pesticides. Both the irrigation water as well as the potable water are the sources of bio-translocation and biomagnification of pesticide residues. The aquatic flora and fauna also mobilise these residues. The extent of pesticide residues in the intensive agricultural areas, green houses and also in the vicinity of organic chemicals synthetic plants is of great concern. The decontamination of water by passing through the four pot system with the sorptive properties of the pore dimensions of the activated carbon, silica gel, activated kaolin along with the filtration through sterile sand could provide potable water free from organic pesticides. This water can be used in sprinkler irrigation or drip system. But, this will need the activation plant for production of active carbon, sterile sand and activated kaolin of the mineralogical nature of meta-Hydrogen-halloysite (Venugopal and Majumder, 1968, Venugopal et al, 1981).

Non-toxic protectants: for Poultry Houses & Green Houses

The activated clay developed for seed protection containing meta-Hydrogen-halloysite as active ingredient (Majumder et al, 1959; Majumder & Venugopal, 1968). It has been found to be of unique usage in the low grade poultry feeds, otherwise such deteriorated product with mycotoxins and heavy metal can be used as manure only. The diets pretreated with 0.5% to 1% of meta-Hydrogen halloysite, the low grade feed materials become safe for poultry. Equally novel approach is in reducing the breeding of houseflies and mosquitoes with the secondary metabolites of Bacillus thuringiensis and related species. The feeding of bacterial preparations, along with the fodder or feed to the poultry, dairy animals and cattle, reduces the capacity for insect breeding on the animal excreta, particularly houseflies (Majumder, 1990). A built-in sanitation with regard to such excreta is achieved in green houses, compost heaps, poultry houses and cattle farms.

3. ANALYSIS

Analytical kits, that are required for measuring the residues prior to treatment and residue after the treatment, have been developed based on TLC, Chromogenic agents and enzyme assay (Visweswariah & Jayaram, 1971; Muthu, et al, 1978; Karanth, et al, 1982, 1983). Venugopal et al, (1980) carried out studies on the characteristics of the isomers of Hexachlorocyclohexane (HCCH) by thermogravimetry. Such analytical kits for water, soil, fruits and vegetables for their comprehensive residue monitoring have been developed. Standard portable instruments like photoionisation gas chromatograph, olfactometry, antennography, electron scanning topography, oscilloscopy are becoming tools for monitoring residues. A finger-printing technique for pesticide residue mapping from fruits and vegetables using chromogenic paper has been developed (Karanth et al, 1983). A flexible TLC plate, pre-coated and pre-activated, has also been developed for utilisation by the analysts (CFTRI Annual Report 1988 p.196).

4. FUTURE MONITORING METHODS

Finally there appears to be need for non-invasive and non-destructive monitoring of pesticides residues in environment and animals to detect and provide insight into probable toxicological effects (Krishnakumari, et al, 1978; Salimath, et al, 1988; Karanth, et al, 1981). Bacteria, protozoa, algae, cell culture and even an enzyme have biosensors (Majumder, 1990). They can detect molecular species even below parts per trillion. Instrumental analyses have developed quite rapidly and sensitivity to parts per billion levels are used for detection, and quantification. The GC-MS and HPLC-MS with sensitive detector systems have minimised the time-effort for analysis of xenobiotics in food, air, water, soil and in plants and animals. Chronic effects of pollutants on plant and animal are discernible. There are the biosensors of insects, dogs, birds, rodents and in aqueous system for fishes, protozoa, bacteria and algae.

The primordial neuroreceptors, antenna or chlorophyll, even antenal structure of DNA are the most sensitive, selective detector, sensors, and recorders. Insects find their food crop from long distances with sensillar configuration. They not only detect the friendly molecules floating in the air but even decipher their stereo-isomers. It is now understood that behaviour changes and chemotactile responses in test organism are proportional to their exposures. Even it is quantitative. Olfactometry, antennography, soft x-ray real time analysers, DTG, EAG, immunoassay, photomigratory response analysis, and many bioassay methods, locomotion recording including xerography have opened new possibilities of bioassay of toxicants in the environment. Future autoanalysers and monitors would be functional or effect measuring instruments, in contrast to to-day's approaches to the separation and identification of molecular species.

The foregoing discussions are pointing towards the need of future studies for using insects, their sensory properties and communication systems for their control with semiochemicals. They could also provide very sensitive methods for detection and analysis. The prospects of developing Industrial Entomology as a biotechnology towards the goal of pest control, pest monitoring and environmental protection have been described (Majumder, 1990). Future investigations to provide tools for fruits and vegetables, free from residues of pesticides and reducing green house contamination are needed. In the meantime decontamination of soil, water, fruits and vegetables from xenobiotics are to be followed as described above.

REFERENCES

1. Dahm, Pant A. (1957), Uses of radioisotopes in pesticide research. In: Metcalf R.L. ed., Pest Control Research 1: 81--146, Interscience Publishers, New York, U.S.A.
2. Deo,P.G., *et al*, (1980) Isomerization of beta-HCH in aqueous solution. J. Environ. Sci.Health, B15: 147--164.
3. Deo, P.G. *et al*, (1988) Toxicity and suitability of some insecticides for household use. International Pest Control, 22: 118--129.
4. Elder, U. and Von Keyserlingk, H.C. (1985) The Challenge of finding new insecticides for a mature market. In: Von Keyserlingk ed. Approaches of New Leads for Insecticides, pp.1--18, Springer-Verlag, Heidelberg.
5. Karanth, N.G.K., *et al*, (1981) Phytotoxicity of X-factor. Pesticides, 15: 7--10.
6. Karanth, N.G.K., *et al*, (1982) A Chromogenic paper for ultrarapid detection of organochlorine insecticide residues in vegetables. Bull, Environm.Contam. Toxicol., 28: 221--224.
7. Karanth, N.G.K., *et al*, Insecticidal residue in vegetables obtained from soil treated with hexachlorocylohexane. J.Fd. Sci. Technol., **19**: 14--19.
8. Karanth, N.G.K., *et al*, (1983), Observations on the growth and residue levels of insecticide in vegetable plant raised on Hexachlorocyclohexane treated soil. Comp. Physiol. Ecol., **8**: 357--361.
9. Karanth, N.G.K., *et al*, (1983) Insecticide finger printing technique for detection and location of organochlorine insecticide residues in foods. J.Environ.Sci.Health, **B18**: 745--755.
10. Klein, W. (1976) Environmental pollution by insecticides. In: Metcalf R.L. & Mekelevey J.J. ed., Future for insecticides-needs and prospects. pp. 65-95.
11. Krishnakumari, M.K. *et al*, (1978) Culturing *Ariophanta madraspatana* (gray) in the laboratory for pesticide bioassay. Proc. Indian Acad.Sci., **87B**: 137--144.

12. Krishnakumari, M.K. et al, (1982) Mammation toxicity of technical X-factor I. An acute oral study, J.Environ.Sci.health B-17: 241--252.
13. Lichetenstein, E.P. (1959) Absorption of some chlorinated hydrocarbon insecticides from soil into various crops. J. Agric. Fd. Chem., **7**: 430--434.
14. Lockwood, L.M. et al, (1974) Degradation of organophosphate pesticides in cereal grains during milling and cooking in India. Cereal Sci. Today, **19**: 330--346.
15. Majumder, S.K. et al, (1959) insecticidal effects of activated carbon and clays. Nature (London) 184: 1165--1166.
16. Majumder, S.K., (1960) Pesticides and health hazards. Annual Review of Food Tech. **1**: 36--72.
17. Majumder, S.K. et al, (1965) A paper chromatographic technique for screening volatile chemicals for their reactivity with the constituents of foods. J.Chromatg., **17**: 373--381.
18. Majumder, S.K., (1968), Soil as a source of microorganisms for industrial fermentations. Proc. of the First All-India Symp. on Agrl. Microbiology. pp. 186--189. Publ. Univ. Agrl. Sciences, Bangalore, India.
19. Majumder, S.K., (1979) Possible impact of Wiesner and Cook Committees; Reports on Future trends of pesticide research In Pesticides. Acad. Pest Control Sciences, India. pp. 256--269.
20. Majumder, S.K., (1979) Storage in relation to tropical developing countries. Pesticides, Ann. Number, pp. 56--78.
21. Majumder, S.K., (1983) Problems of pesticide residues in foods. Pesticides Information. PAI, 2: 50--54.
22. Majumder, S.K., (1990) Integrated pest management for post-harvest technology. In:Proc. National Symp. Newer Dimensions in Integrated Pest Management. Pest Control Sciences, India, pp. 13--51.
23. Muthu, M., et al, (1978) A simple method of determining the sorption affinity of food stuffs to phosphine. Chemistry and Industry, **18**: 120--131.
24. Salimath, B.P., et al, (1988), Hexachlorocyclohexane inhibits calmodulin dependent Ca^{2}- ATPase activity in rice shoot membranes. Pesticide Biochem and Physiol., **31**: 146--154.
25. Srimathi, M.S. et al, (1986), Influence of Bromophos on some biological activities of the soil. J.Soil. Ecol., **6**: 9-14.
26. Venugopal, J.S. & Majumder, S.K. (1968) Active mineral in insecticidal clays. In: Majumder, S.K. ed. Pesticides. Academy of Pest Control Sciences, India. pp. 200-209.
27. Venugopal, et al, (1980) Studies on the characteristics of the isomers of hexachlorocyclohexane (HCCH) by thermogravimetry. J. Thermal Anal. **18**: 15--19.
28. Venugopal, J.S. et al, (1981) Differential moisture absorption as an indicator of insecticidal quality in halloysites. Research and Industry, **26**: 220--222.
29. Visweswariah,K and Jayaram, M. (1971), Detection and quantitative determination of certain chlorinated pesticides by microthin layer chromatography. J. Chromatogr. **62**: 479--484.
30. Visweswariah, K. and Jayaram, M. (1972), The effect of processing on lowering the BHC residue in green leafy vegetables. Pesticides Sci., **3**: 345--347.
31. Visweswariah, K. et al, (1977) Wood charcoal as a decontaminating agent for the removal of insecticides in water. Indian J. Environ. Health, **19**: 30--37.

FIGURE-1

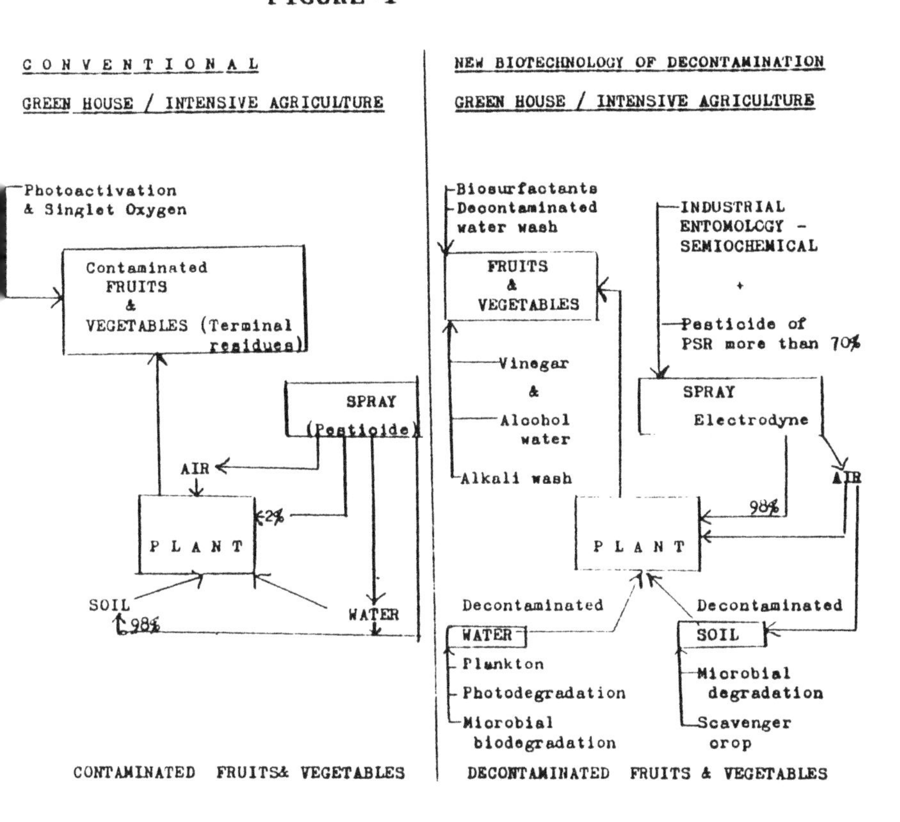
C O N V E N T I O N A L
GREEN HOUSE / INTENSIVE AGRICULTURE
Photoactivation
& Singlet Oxygen
Contaminated
FRUITS
&
VEGETABLES (Terminal
residues)
SPRAY
(Pesticide)
AIR
2%
P L A N T
SOIL
98%
WATER
CONTAMINATED FRUITS& VEGETABLES
NEW BIOTECHNOLOGY OF DECONTAMINATION
GREEN HOUSE / INTENSIVE AGRICULTURE
Biosurfactants
Decontaminated
water wash
FRUITS
&
VEGETABLES
INDUSTRIAL
ENTOMOLOGY -
SEMIOCHEMICAL
+
Pesticide of
PSR more than 70%
Vinegar
&
Alcohol
water
Alkali wash
SPRAY
Electrodyne
AIR
98%
P L A N T
Decontaminated
WATER
Plankton
Photodegradation
Microbial
biodegradation
Decontaminated
SOIL
Microbial
degradation
Scavenger
crop
DECONTAMINATED FRUITS & VEGETABLES

Pathogenesis - related proteins of tomato : comparative studies **on protein profiles of resistant and susceptible lines of tomato following infection with** *Alternaria solani*

LALITHA ANAND
Molecular Biology Laboratory,
Division of Plant Physiology and Biochemistry,
Indian Institute of Horticultural Research,
Hessaraghatta, Bangalore - 560 089, India.

Key words : PR-proteins; tomato chitinase; *Alternaria solani*; protein profile; disease resistance.

Introduction

Plants elaborate a number of inducible defence responses following microbial attack. These include synthesis of phytoalexins, reinforcement of cell wall and production of the so-called "PR-proteins" or pathogenesis-related proteins (Lamb *et al.*, 1989). These proteins have been detected in various plant species (Van Loon, 1985). Some of these proteins have been shown to be proteinase inhibitors or lytic enzymes like chitinases and Beta-1, 3-glucanases (Kombrink *et al.*, 1988; Legrand *et al.*, 1987). Chitinases and glucanases have attracted considerable attention as defence weapons in recent years (Boller, 1987), owing to the fact that these enzymes have the ability to directly attack specific structures of pathogens. The enzyme chitinase has been shown to be induced in tomato plants following infection with *Verticillium albo-atrum* (Pegg and Young, 1982). Preliminary investigations carried out in our laboratory have demonstrated the induction of chitinase in tomato plants following infection with the fungus *Alternaria solani*, the causal organism of early blight disease of tomato, as well as the ability of partially purified preparations of tomato chitinase to hydrolyse cell walls of *A.solani* and inhibit its mycelial growth. In the present investigation, an attempt has been made to examine the role of chitinase in the defence response exhibited by two susceptible and four resistant lines of tomato following *A.solani* attack. The protein profiles of these lines have also been examined in order to compare the differential response of resistant and susceptible lines of tomato at the level of protein expression.

J. Prakash and R. L. M. Pierik (eds.), Horticulture – New Technologies and Applications, 353–357.

Methods

Extraction of basic and total proteins: Finely-chopped tomato seedlings were ground in a mortar and pestle with cold buffer (0.05M glycine - NaOH, pH 9.5 for basic proteins and 0.05M Tris HC1, pH 7.2 for total proteins). The homogenate was centrifuged at 17000 g for 30 min at 4 deg C. The supernatant was dialyzed against three changes of buffer. The basic protein extract was used as source of crude chitinase.

Enzyme assay:

The reaction mixture (4ml) consisted of 2ml of crude enzyme extract, 100 μg of bovine serum albumin and 0.5 ml of a 1% suspension of colloidal chitin (prepared according to the method of Shimahara and Takiguchi, 1988). Incubation was carried out for 3 hours at 37 deg C and the reaction was stopped by placing the tubes in a boiling water bath. This was followed by centrifugation at 3000 rpm for 10 min. To 0.5 ml of the supernatant was added 50 μg of Beta-1, 4-glucosidase (Sigma, from almonds) and incubated at 37 deg C for 1 hour. N-acetylglucosamine released was estimated according to the method of Reissig et al. (1955). One unit of enzyme activity is defined as the amount of enzyme that liberates 1 μmole of N-acetylglucosamine/ml of enzyme extract/hr at 37 deg C. Specific activity is expressed in terms of units of enzyme activity/mg protein.

Electrophoresis of protein samples:

Protein samples were analysed on 12% polyacrylamide slab gels as described by Gabriel (1971) except that 0.05M Tris-borate buffer, pH 9.5 was used as both running and resolving gel buffer. SDS-PAGE was carried out on 12% polyacrylamide gels according to the method of Laemmili (1970). After electrophoresis, proteins were stained with Coomassie Brilliant Blue G.

Disease induction:

One month old nursery grown tomato plants were sprayed with spores of Alternaria solani, isolated in our laboratory from infected tomato leaves, kept covered for disease development and subsequently analysed after 0, 2, 4 and 9 days of treatment.

Results:

Chitinase activity in resistant and susceptible lines of tomato:

Levels of enzyme activity after 0, 2, 4 and 9 days of treatment with fungal spores are shown in Table-1. It can be seen that induction of chitinase activity is observed only in the case of susceptible lines. However, in three of the resistant lines, constitutive levels of the enzyme are higher than in the susceptible lines.

Table-1. Chitinase activity in resistant and susceptible lines of tomato.

Line		Chitinase activity, units/mg protein			
		Days after infection			
		0	2	4	9
Sel-13	(S)	0.017	0.026	0.069	0.035
AL-1	(R)	0.034	0.032	0.022	0.024
AL-50	(R)	0.033	0.026	0.016	0.018
1816	(R)	0.030	0.018	0.019	0.015
1817	(R)	0.015	0.019	0.019	0.016
Pusa Ruby	(S)	0.017	0.021	0.043	0.030

S = Susceptible R = Resistant

Polyacrylamide gel electrophoresis of protein samples:

Fig. 1 depicts the protein profiles obtained following electrophoresis under the conditions specified. 100 μg of protein from susceptible lines Sel-13 and Pusa Ruby (PR) were loaded in lanes 1 & 7 and lanes 6 & 12 respectively. 100 μg each of proteins from AL-1 was loaded in lanes 2 & 8, AL-50 in lanes 3 & 9, 1816 in lanes 4 & 10 and 1817 in lanes 5 & 11.

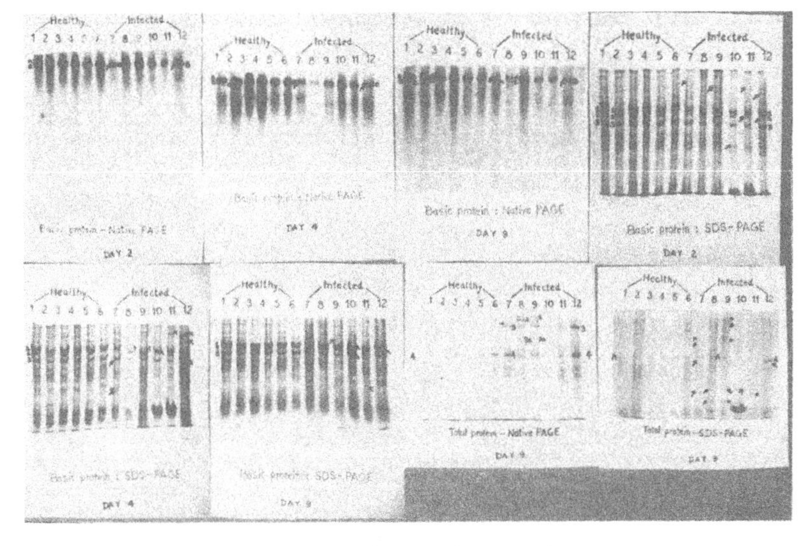

Fig. 1. Polyacrylamide gel patterns of protein samples.

The salient features of the gel patterns can be summarised as follows:

i) In the case of basic proteins of susceptible lines, following native PAGE, two major bands, A and B, of equal intensity are present before infection. After infection, intensity of B increases in both the cases. In the case of resistant lines, three bands A, B and C are obtained, A and B being the major bands. After infection, there is not much change, except in the case of AL-1 on Day 9 when Band C becomes prominent.
ii) After SDS-PAGE, about 15-18 bands can be seen in all the lines before infection, of which three viz. 1, 2 and 3 (Fig. 1) are prominent. After infection, changes are observed in terms of appearance of new bands in some cases, increase in intensity of some bands in certain other cases (indicated by arrows in Fig. 1) and disappearance of certain bands in some other cases (indicated by question marks in Fig. 1).
iii) When total proteins were analysed by native PAGE, 6-7 prominent bands were observed before infection and on SDS-PAGE these resolved into about 15 bands. Following infection, a number of changes occur in all the lines and these have been indicated by arrows (Fig. 1).

Discussion

Following infection, there is an induction of chitinase activity only in susceptible lines and a concomitant increase in the intensity of Band B obtained following PAGE of basic proteins under non-denaturing conditions. No such observation is made in the case of resistant lines. It would, therefore, appear that isoforms of chitinase may be involved in defence response of tomato against A.solani attack. Another significant observation is that the protein profiles of all the four resistant lines analysed under denaturing and nondenaturing conditions are more or less similar before infection. After infection, however, quite a number of differences, both qualitative and quantitative are observed among the four lines examined. Some of these changes are manifested as the infection progresses. Two aspects need further study. One is the characterisation of the new polypeptides that are synthesized in some of the lines analysed and to ascertain their biological functions. The other aspect is the isoform status of chitinase in susceptible and resistant lines in order to establish in a more definite manner whether chitinase has any role at all in the in vivo defence responses of tomato.

Acknowledgements:

The author is grateful to Director, I.I.H.R., for the facilities and Dr N. Anand, Division of Vegetable Crops, I.I.H.R., for providing the susceptible and resistant lines of tomato used in this study.

References

Boller T (1987) Hydrolytic enzymes in plant disease resistance. In: Kosuge T. and Nester E.W. eds., Plant Microbe Interactions, 2: pp 385--413. Macmillan, New York.

Gabriel O. (1971) Analytical Disc Gel Electrophoresis. In: Jacoby W.B. ed., Methods in Enzymology, 22: pp 565--578. Academic Press, San Diego.

Kombrink E., Schroder M. and Hahlbrock K. (1988) Several "pathogenesis - related" proteins in potato are 1, 3 - Beta-glucanases and chitinases. Proc. Natl. Acad Sci. USA 85:782--786.

Laemmili U.K. (1970) Cleavage of structural proteins during the assembly of the head of bacteriophage T-4. Nature 227:680--85.

Lamb C.J., Lawton M.A., Dron M. and Dixon R.A. (1989) Signals and transduction mechanisms for activation of plant defenses against microbial attack. Cell 56 : 215 - 224.

Legrand M., Kauffman S., Geoffroy P. and Fritig B. (1987) Biological function of pathogenesis - related proteins: four tobacco pathogenesis-related proteins are chitinases. Proc. Natl. Acad. Sci. USA 84 : 6750--6754.

Lowry O.H., Rosebrough N.J., Farr A.C. and Randall R.J. (1951) Protein measurement with the Folin phenol reagent. J. Biol.Chem. 193: 265--275.

Pegg G.F. and Young D.H. (1982) Purification and characterization of chitinase enzymes from healthy and *Verticillium albo-atrum* infected tomato plants and from *Verticillium albo-atrum*. Physiol. Plant Pathol 21 : 389--409.

Reissig J.L., Strominger L.J. and Leloir L.F. (1955) A modified colorimetric method for the estimation of N-acetyl amino sugars. J.Biol.Chem. 217 : 959--966.

Shimahara K. and Takiguchi Y. (1988) Preparation of crustacean chitin. In: Wood W.A. and Kellog T.S. eds., Methods in Enzymology, 161 : pp 417--423. Academic Press, San Diego.

Van Loon L.C. (1985) Pathogenesis-related proteins. Plant Molec. Biol. 4 : 111--116.

Recent advances in diagnosis of nematode diseases

P. PARVATHA REDDY, R.M. KHAN AND M.S. RAO
Indian Institute of Horticultural Research
Hessaraghatta, Bangalore-560 089, India.

Key words: Restriction enzyme analysis, DNA probing, nucleotide sequencing, gel electrophoresis, serology.

Introduction

Accurate and reliable identification of plant parasitic nematodes is fundamental to many aspects of their effective management. Control programmes using either resistant cultivars or crop rotation are usually directed at specific species of parasitic nematodes. Since different species of a particular genus are important pathogens on different crops, the species present in a field must be accurately and often rapidly identified. The same speed and reliability in the identification of races would be helpful. Biochemical analyses offers accurate, reliable and rapid identification of nematode species and races.

Biochemical analyses

Proteins, nucleic acids and lipids have all been used to charecterize nematode species, host races and pathotypes.

Proteins

Proteins are a manifestation of the sequence of nucleotides in a gene and analysis of these macromolecules by gel electrophoresis and serology provides a reliable approach for comparing the genotypes of organisms.

Gel electrophoresis: Polyacrylamide-gel electrophoresis provides a high resolution technique for separating protein molecules on the basis of size and net charge. Unique nonenzymatic protein profiles were found for females of *Meloidogyne incognita*, *M. arenaria*, *M. hapla* and *M. javanica* (Dickson *et al*., 1970). The sensitivity of polyacrylamidegel electrophoresis is greatly enhanced

J. Prakash and R. L. M. Pierik (eds.), Horticulture – New Technologies and Applications, 359–362.

when specific enzymatic proteins are localized in the gels through staining technique. Electrophoretic patterns of α-glycerophosphate dehydrogenase, malate dehydrogenase, glucose-6-phosphate dehydrogenase and non-specific esterases varied among the four *Meloidogyne* species (Dickson *et al.*, 1971). Hussey *et al.* (1972) reported most striking differences between *M. incognita* and *M. arenaria* in esterase, malate dehydrogenase and α-glycerophosphate dehydrogenase patterns.

Dalmasso and Berge (1978) used the microelectrophoresis to study enzymes and non enzymatic proteins from several populations of the four common *Meloidogyne* species. The b-esterase patterns were the most reliable for identifying the common *Meloidogyne* species. Trudgill and Carpenter (1971) noted that non-enzymatic profiles of *Globodera rostochiensis* and *G. pallida* consistently differed by three bands. The identification of two pathotypes of *G. rostochiensis* has been achieved by means of two dimensional electrophoresis (Bakker and Gommers, 1982). Pathotype PO_5 had one protein that was absent in the protein pattern of pathotype PO_1 of *G. rostochiensis*.

Serology: More recently, serological methods that were once dismissed as too cross reactive are enjoying a revival with the introduction of monoclonal antibody technology which appears to be very discriminating.

Nucleic acids

Recent advances in molecular biology, including restriction enzyme analysis and DNA cloning, allow direct exploitation of DNA sequence polymorphism.

Restriction enzyme analysis: The restriction fragments from a total DNA digest are separated according to size by electrophoresis in an agarose gel. Isolates of *G. pallida* and *G. rostochiensis* were clearly separated using this method (Burrows and Boffey, 1986). This method was also employed to differentiate pathotypes of the potato cyst nematodes, *G. pallida* and *G. rostochiensis* and host specific races within *Meloidogyne* (Curran *et al.*, 1986) and *Heterodera glycines* (Kalinski and Huettel, 1988).

DNA probing: Two DNA fragments that differentiate *G. Pallida* from *G. rostochiensis* have been isolated from a *G. pallida* genomic library (Burrows and Perry, 1988). Burrows (1988) successfully used a biotin-labelled probe to identify *G. pallida* by using a single egg or second-stage juvenile.

Restriction enzyme analysis from mitochandrial DNA (mt DNA)

are readily identified in agarose gels stained with ethidium bromide and these bright bands have been used to separate five root-knot nematode species, *M. incognita*, *M. arenaria*, *M. javanica*, *M. hapla* and *M. chitwoodii* (Powers *et al*., 1986). Likewise, studies of mt DNA can be used to differentiate *Heterodera schachtii* and *H. glycines* (Radice *et al*., 1988).

Conclusions

The use of biochemical analyses though in its infancy, has considerable potential for identification of plant parasitic nematodes. The comparison of enzymes of adult females with microdisc electrophoresis seems to be quite promising. Two dimentional electrophoresis can seperate more than 1000 proteins and may allow detection of protein differences among races of plant parasitic nematode species. The monoclonal antibody technology should facilitate the production of species-specific antisera and enzyme-linked immunosorbent assays provide a highly sensitive method for detection of phytoparasitic nematode antigens.

Direct analysis of genetic variation using restriction enzymes and DNA cloning will have great impact on routine identification of species or races or pathogypes of plant parasitic nematodes. Using simple diagnostic kits based on biotin-labelled DNA probes, individual nematodes or eggs may be detected. This level of sensitivity is particularly important for statutory control and quarantine regulations where the presence of even a single nematode is significant.

References

Bakker J. and Gommers F.J. (1982) Differentiation of the potato cyst nematodes *Globodera rostochiensis* and *G. pallida* and of two *G. rostochiensis* pathotypes by means of two-dimensional electrophoresis. Proc. K. Ned. Akad. Wet. Ser. C, Biol. Med. Sci. 85: 309-314.

Burrows P.R. (1988) The differentiation of *Globodera pallida* from *G. rostochiensis* using species specific DNA probes. Nematologica 34: 260.

Burrows P.R. and Boffey S.A. (1986) A technique for the extraction and restriction endonuclease digestion of total DNA from *Globodera rostochiensis* and *G. pallida* second stage juveniles. Revue de Nematologie 9: 199-200.

Burrows P.R. and Perry R.N. (1988) Two cloned DNA fragments which differentiate *Globodera pallida* from *G.*

rostochiensis. Revue de Nematologie 11: 441-445.

Curran J. McClure M.A. and Webster J.M. (1986) Genotypic differentiation of Meloidogyne populations by detection of restriction fragment length differences in total DNA. J. Nematol. 18: 83-86.

Dalmasso A. and Berge J.B. (1978) Molecular polymorphism and phylogenetic relationship in some Meloidogyne spp. Application to the taxonomy of Meloidogyne. J. Nematol. 10: 323-332.

Dickson D.W. Sasser J.N. and Huisingh D. (1970) Comparative disc-electrophoretic pattern analysis of selected Meloidogyne, Ditylenchus, Heterodera, and Aphelenchus spp. J. Nematol. 2: 286-293.

Dickson D.W. Huisingh D. and Sasser J.N. (1971) Dehydrogenases, acid and alkaline phosphatases and esterases for chemotaxonomy of selected Meloidogyne, Ditylenchus, Heterodera, and Aphelenchus spp. J. Nematol. 3: 1-16.

Hussey R.S. Sasser J.N. and Huisingh D. (1972) Disc-electrophoresis studies of soluble proteins and enzymes of Meloidogyne incognita and M. arenaria. J. Nematol. 4: 183-189.

Kalinski A. and Heuttel R.N. (1988) DNA restriction fragment length polymorphism in races of the soybean cyst nematode, Heterodera glycines. J. Nematol. 20: 532-538.

Powers T.O. Platzer E.G. and Hyman B.C. (1986) Species specific restriction site polymorphism in root-knot nematode mitochondrial DNA. J. Nematol. 18: 288-296.

Radice A.D. Powers T.O. Sandall L.J. and Riggs R.D. (1988) Comparison of mitochondrial DNA from the sibling species Heterodera glycines and H. schachtii. J. Nematol. 20: 443-450.

Trudgill D.L. and Carpenter J.M. (1971) Disc electrophoresis of proteins of Heterodera species and pathotypes of Heterodera rostochiensis. Ann. Appl. Biol. 69: 35-41.

Diagnosis and management of bean flies in beans

P.N. Krishna Moorthy and K. Srinivasan
Indian Institute of Horticultural Research
Hessaraghatta Lake, Bangalore-560 089, India

Key words: beans, bean fly, diagnosis, Ophiomyia phaseoli

Introduction

Bean fly Ophiomyia phaseoli (Tryon) is a major pest of beans (Phaseolus vulgaris L.) in tropics (Spencer, 1973). Soil application of granular insecticides like phorate or foliar application of endosulfan are effective against this pest (Krishna Moorthy and Tewari, 1987). These treatments are given as a fixed schedule irrespective of the pest population. However, a sound management programme requires understsanding of the biology of the pest. The biology of the pest is given by Goot (1930). The eggs are laid in ovipositional cavities, larvae mine the leaf, petiole and enter the stem . The pest is generally diagnosed in the stem. The early diagnostic characters of the pest incidence are puncture marks, vein-mining and petiole mining. Of these, petiole mining thresholds (PMT) has been used for managing the pest (Krishna Moorthy and Srinivasan, 1989). However, one application of insecticide based on PMT was not consistent in controlling the pest. The present paper reports the use of puncture marks and vein-mining in addition to PMT for the effective management of the pest in beans.

Methods

Two field experiments were conducted in 1989 and 1990 at the experimental farm of Indian Institute of Horticultural Research, Bangalore. 'Arka Komal' beans was sown on 31 June, 1989 and 11 July, 1990. Endosulfan treatments were given at unifoliate plant

J. Prakash and R. L. M. Pierik (eds.), Horticulture – New Technologies and Applications, 363–366.

stage based on diagnostic characters of bean fly incidence, i.e. appearance of puncture marks, vein-mining, petiole-mining and petiole -mining thresholds of 5 (when on an average of 5 leaves had petiole-mining/10 leaves i.e. 5 PMT). In 1989, treatments based on puncture marks are not given. These treatments were compared with scheduled application of endosulfan given at 7,14 and 21 days after sowing (DAS). Endosulfan (0.07% at .5 Kg a.i./ha) treatments were given using a high volume foot sprayer delivering 700 litres of spray fluid per hactare.

Monitoring the plants for diagnostic characteristics of bean fly incidence was done one day after germination. Fifty plant in untreated plots at random were observed every day. Appearance of puncture marks is followed by vein-mining (VM) and petiole-mining (PM) and treatments were given on the day of appearance of the symptoms in a few leaves. For giving treatments based on 5 PMT, monitoring was done using the method given by Krishna Moorthy and Srinivasan, 1989. Egg and larval counts at the time of vein-mining was taken by staining 50 leaves collected at random; method given by Krishna Moorthy et.al., 1988 was followed. The data of green pod yield per plot was also computed and statistically analysed.

Results and Discussion

Adults of bean fly pierce the lamina by ovipositor and lay eggs. The punctures thus formed are visible after a day or two. These puncture marks appeared at 8 DAS in 1990 (Table 1). The eggs hatch and larvae mine the lamina and veins. There was a gap of two days before vein-minings were visible after the appearance of punctures. In 1989 it might have occured at 9 DAS, as VM occured at 11 DAS in that year. Treataments based on initiation of petiole-mining were given at 14 DAS in 1989 and at 12 DAS in 1990 (Table 1). The 5 PMT threshold also occured early in 1990 due to early vein-mining i.e. early pest incidence. The rate of increase in PMT per day from PM to 5 PMT was also more in 1990. This is due to the higher pest population occured at VM. The rapid rise in PMT also reduced the gap between sprays given at VM/PM and at 5 PMT. However, the 5 PMT treatment was given after a gap of 5 days after the treatment at punture mark.

When yields were compared, treatments in which the second

application of endosulfan at 5 PMT was given were very effective and were at par with schedule application of endosulfan.

An ovedrall analysis of the data revealed that the initial treatment at puncture mark suppressed the pest population for a long time compared to other treatments given at VM/PM. Therefore this method of diagnosing and treating the crop against bean fly at puncture mark appeareance and again at petiole mining thresholds can be effectively used for managing the pest in beans.

Acknowledgement: The technical assistance given by M.B. Munirathnaiah and S. Venkateshaiah is gratefully acknowledged. The help given by V.R. Srinivasan in computerising this research paper is also gratefully acknowledged.

References

Goot P. van der. (1930) Agromyzid flies of some native legume crops in Java. Tropical Vegetable Information Service, AVRDC, Shanhua, Taiwan (English Translation from Dutch). 98 pp.

Krishna Moorthy P.N., Alexander M.P. and Tewari G.C. (1988) A versatile method for staining insect eggs and larvae in leaves. J. Econ. Entomol. 81: 403-405.

Krishna Moorthy P.N. and Srinivasan K. (1989) Petiole-mining thresholds and potassiun chloride sprays for management of bean fly (Diptera: Agromyzidae) in snap beans and cowpea. J.Econ. Entomol. 82: 246-250.

Krishna Moorthy P.N. and Tewari G.C. (1987) Management of stemfly Ophiomyia phaseoli (Tryon) on french beans with reduced insecticidal doses Entomon. 12: 363-366.

Spencer K.A. (1973) Agronyzidae (Diptera) of economic importance. Junk, The Hague, 418 pp.

Table 1. Parameters of bean fly incidence in beans under different treatments.

Treatments of endosulfan o.5 kg a.i./ha at	Day of application 1989	1990	Yield (t/ha) 1989	1990
Weekly interval	7,14,21	7,14,21	4.8	7.9
Puncture mark	-	8	-	7.7
Puncture mark+ 5 PMT	-	8,13	-	8.3
Vein-mining (VM)	11	10	4.2	6.3
VM + 5 PMT	11,15	10,13	4.8	7.7
Petiole-mining (PM)	14	12	3.7	7.0
PM + 5 PMT	14,18	12,14	3.8	8.1
Control	-	-	2.3	3.9
CD at 5%	-	-	1.0	1.2

DAS - days of sowing.
Mean egg + larva per leaf at VM was 2.07 in 1989 and 2.7 in 1990.

POST-HARVEST OF GERBERA CUTFLOWERS

HERMAN SCHOLTEN, export salesmanager of
FLORIST DE KWAKEL B.V.

Florist is a company that supplies youngplants of Gerbera. Florist produces 5 till 6 million youngplants per year of which 60% is exported throughout the whole world.
We work with 3 salesmen, one for our home-market and two for export.

All youngplants we produce are propagated by way of tissue-culture in laboratories. We receive the plants from about ten laboratories in Europe. The plants are supplied to us as ex-agar plants and immediately after they arrive we transplant them in rockwool or in a peat/perlite mixture. The plants are grown in fully conditioned greenhouses. Because we use artificial lighting we can produce top-quality plants year-round.
The plants are available in several different kinds of growing-media :

- In Jiffy pot 6 cm.
- In Jiffy pot 4 cm.
- Transplanted
- In rockwool plugs
- On rockwool blocks

After planting the youngplants for cutflower production it takes 2 till 3 months to get the first flowers.
Soon after this period we reach the full production and from there on we can produce Gerbera-cutflowers for 1 till 2 years. One plant produces about 25 till 40 flowers per year, depending on the variety, the climate and the cares of the grower.

J. Prakash and R. L. M. Pierik (eds.), Horticulture – New Technologies and Applications, 367–370.

Actually Gerbera is not a cutflower : we do not use a knife, but we break the flower out of the plant with a rotatory motion.
The right moment to pick the flower is when it reaches the stage in which two till three rings of male flowers are blooming. It is very important not to harvest the flowers too young or too old because in both cases the vase-life will be shorter.
In single type flowers with a dark centre it is very easy to recognize the right picking-stage. Specially in yellow double type flowers it is more difficult to recognize the optimum picking-day. Someone who is picking Gerbera flowers every day will learn very fast to see whether a flower is ready to pick or not.

The flowers are picked per bunch of 15 till 20 flowers, depending on the flower size. After picking the bunches must be cut within a short time. It is best to cut about three cm. from the stems.
It is most important to put the flowers on water with chlorine immediately after cutting the stems.
The Gerbera is very sensitive for dirty water.
The bacteria in the water will block the water-transport system in the stem, starting at the top of the stem just underneath the flower-head, and the flower will die to soon.
Therefor we use chlorine in the water. A pre-treatment right after picking for 3 till 5 hours is the best method to enlarge the keeping-quality of the Gerbera flower.
The concentration chlorine in the water must be in between 100 and 250 ppm.

If we do not put the flowers on water right after cutting the stems, another problem occurs : At the bottom of the stem some air comes in. At the moment we put the flower on water it starts sucking the water. The air that came in at the time the flower was kept dry, will be pushed upwards by the water.
Eventually the air will come at the top of the stem, just underneath the flower-head. At that point the air can not escape end it will block the water-transport.
The result we see is the same as a blockade of the water-transport by bacteria.

At our nursery we are using transport-cars with water to bring the flowers to the grading and packing room.
To the water we add the chlorine so that the concentration will be in between 100 and 250 ppm.
The flowers stay in the cars for 3 till 5 hours before we start the grading.
We use cartons with 25 holes to put in the flowers.

After filling the cartons we have two options :

1. If we want to store the flowers for a longer period before they can be transported we hang the cartons in a water-basin. In the basin we use water with chlorine. The concentration of the chlorine must be lower than in the transport-cars because else we can get damage at the stems.
   The concentration must be in between 10 and 50 ppm.

2. If the flowers will be transported within a short time we will pack them right away.
   We pack two cartons per box, one on each side, the stems to the middle of the box.
   In this way each box contains 50 flowers.
   We staple the cartons to the side of the box so that they can not move. After closing the box we put the box upside down for transport.
   This is because the flowers will grow upwards.
   If we do not put the box upside down the flowers will grow curved, now the stems will remain straight.

This is the way we are packing the Gerbera in Holland.
The way of packing depends on the market the flowers are sold on.
In Holland almost all Gerbera's are for export and the way of packing we use is most suitable for this purpose.

If the flowers are to be sold on the local market other ways of packing can be more suitable.
The Gerbera's can be packed in bunches, but in this way we get a lot of damaged flowers. It would be better to pack them in a kind of display that can be used in the flower-shops as well.
We can pack the flowers in a display meant for a mass-consumption market.
To make the gerbera a more exclusive product, like for sales in hotel shops, we need another kind of display.

In Holland we sell the flowers at the auction.
The grower brings his flowers over there for sale and exporters and retailers are buying.
The selling-system works with clocks. The clock starts at a certain price and runs down to a minimum price.
All buyers have their own seat with buttons to stop the clock at the price they want to pay and they have to tell how many of the product they want.
The starting-price and minimum price are different per product. If the product reaches the minimum price it will be destroyed.

After the products are sold they are distributed and the buyers can collect their goods.
The retailer will sell the flowers in his shop.
The exporter will arrange the transport to his clients all over the world.

EFFECT OF PREPACKING TREATMENTS ON STORABILITY AND QUALITY OF SAPOTA (Achras sapots L.)

S.K.GAUTAM & B.S.CHUNDAWAT
Aspee College of Forestry & Horticulture,
Gujarat Agricultural University,
Navsari-396 450. (Gujarat), INDIA

ABSTRACT

Sapota fruits of cvs. 'Kalipatti' and 'Cricket Ball' were subjected to various prepacking treatments and their ripening behaviour and final quality estimates were made. It was found that treatment of fruits with gibberellic acid 300 ppm significantly increased the time required for ripening thereby increased the storability. Quality of ripe fruits in terms of higher level of total soluble solids, vitamin 'C', total sugar,and carotenoid and lower level of acidity was also highest in this treatment. This was followed by kinetin 100 ppm and silver nitrate 40 ppm. All these treatments slowed down the process of ripening by retarding pre-climacteric activities of catalase and PME (pectin methyl esterase) enzymes and rate of repiration and ethylene production and postponed their peaks of activities compared to control.

Prepacking treatment of sapota fruits in GA 300 ppm can be commercially exploited for extending the storability and marketing of this fruit.

INTRODUCTION

The sapota (Achras sapota L.) is one of the important tropical fruits cultivated in warm humid tropical region of India especially in the states of Maharashtra, Gujarat, Karnataka, Tamil Nadu, Andhra Pradesh and Kerala. However, its cultivation is highly concertrated in the region between Bombay in Maharashtra and Surat in Gujarat with an estimated area of 20,000 ha and production around 3,20,000 t/annum. The total

J. Prakash and R. L. M. Pierik (eds.), Horticulture – New Technologies and Applications, 371–377.

produce is used for table purpose and handled at ambient condition to reach to market in North India. The sapota fruits are highly perishable and begin to ripe within 4 or 5 days and become over ripe within 5 to 7 days. Therefore, post harvest losses are enormous during transit and ripening and at the same time the fruit is sensitive to cold temperature. Therefore, to minimize the heavy post harvest losses and to extend the shelf life of sapota fruits various ripening retardants which had shown promise in tropical fruits (Chundawat and Rao, 1981; Wade and Bradey, 1971; Mehta et al. 1986 and Gautam and Chundawat,1989) have been given as prepacking treatments and their effects have been evaluated in this investigation.

MATERIAL AND METHODS

This investigation was carried out at the laboratories of the Department of Horticulture, Gujarat Agricultural University, Anand Campus during the year 1988 on CVS. 'Kalipatti' and 'Cricket Ball'. In this area crop is harvested during summer. Fruits for experimentation were harvested at colour change stage with slightly yellowish surface colour denoting optimum maturity. Following post harvest treatments were affected as prepacking treatments : T1-GA 300 ppm dip for 20 minutes and ripened at ambient conditions; T2-Kinetin 100 ppm dip for 20minutes and packed; T3-Silver nitrate 40 ppm dip for 20 minutes and packed and T4-No treatment and packed (Control). The fruits were packed in ventillated card board boxes. Treatments were replicated five times in a SPD (Split Plot Design). Each replicate consisted of 20 fruits. Ripening changes were evaluated periodically. Non destructive parameters were : average days to ripening and physiological loss in weight (PLW). Destructive parameters estimated were : TSS, acidity, vitamin 'C', total sugars (Ranganna, 1976), carotene (Roy, 1973), relative activity of catalase and pectin methyl esterase (Colowick and Kalpan, 1968), rate of repiration (Umbriet et al.1957) and ethylene production (Das, 1964). Ambient condition for storage consisted of minimum and maximum temperature in the range of 28.2-39.2 C with RH between 72-100 per cent during investigation. Qualitataive status of fruits was compareds by using C.D. for interaction (Treatment x Date) as fruits ripened on various dates under different treatments.

RESULTS

Days taken to ripening and percentage physiological loss in weight (PLW %)

Treatments significantly affected the days taken for ripening and PLW per cent (Table 1). GA 300 ppm, Kinetin 100 ppm and AgNO3 40 ppm dip treatments in that order significantly and effectively extended the shelf life/storability and reduced PLW compared to control.

Effect of various treatments on days to attainment of climateric peak and average activity of various biochemical components

Periodic observations on enzymatic activities, rate of repiration and ethylene production during ripening exhibited significant variation due to treatment The most significant effect of these prepacking treatment was on the shift of climacteric peaks which were delayed in treat ed fruits compared to control (T4) (Table-2). The average activities of various biochemical components were also significantly lower in treated fruits compared to control (Table-3). The effect was more discernible on catalase and ethylene production with GA 300 ppm being most effective.

Quality of ripe fruits

Ripe fruits of 'Kalipatti' cultivar when given prepacking dip in GA 300 ppm, kinetin 100 ppm and silver nitrate 40 ppm showed significantly higher level of total sugars, carotenoid and vitamin 'C' compared to control. However, the TSS level was higher in control compared to treated fruits. Almost identical trend was recorded in CV. 'Cricket Ball' but the level varied being a varietal character (Table 4).

DISCUSSION

As expected ripening/storability of fruits in prepacking dip treatment with GA 300 ppm, kinetin 100 ppm and silver nitrate 40 ppm was significantly extended and reduced PLW. This has been due to slow rate of catabolic activities and postponement in days to attainment of peaks of catalase, PME, respiration and ethylene production evinced through record of observation in this investigation. Due to controlled and slow rate of metabolic changes under different prepacking treatments vis : GA 300 ppm and kinetin 100 ppm being antisenescence and silver nitrate being anti-ethylene compound, the final quality of ripe fruits was also higher as fruits contained high level of total sugars, vitamin 'C' and carotenoid compared to untreated fruits (control). However, the level of TSS was higher in control fruits probably due to complete ripening as compared to treated fruits. Identical reports are available on sapota (Gautam and Chundawat, 1989; Chundawat and Rao,1981) mango (Parmar and Chundawat, 1985) and papaya (Mehta et al. 1986).

Based on these results it can be conclusively be brought out that if the fruits are given dip treatment with GA 300 ppm before prepacking in containers, their shelf life can be extended by about two days with no detrimental effect on ripe fruit quality. Thus the market zone can be increased and more period is available for selling in the market. Kinetin and AgNO3 are also effective in extending shelf life of sapota fruits. The organisation bulk handling in these fruits can make use of these chemicals.

LITERATURE

Chundawat, B.S. and D.V.Raghava Rao.(1981) Effect of certain plant regulators and waxol on ripening of sapota (Manilkara achras (Mill) Forsberg). Paper presented at "Symp. On Recent Advances in fruit Development" Ludhiyana (India).

Colowick, S.P. and Kalpan N.O. (1968) Methods in enzymology. Academic press Inc. New York pp.159.

Das, M.N. (1964) Estimation of ethylene with mercuric acetate by non-aqueous titration. Anal.Chem. 26 (6): 1086-87.

Gautam, S.K. and Chundawat, B.S.(1989) Postharvest changes in sapota CV. 'Kalipatti' I - Effect of various postharvest treatments on biochemica changes. Indian J. Hort. 46 (3):310-315.

Mehta, P.M., Raj, S.S. and Raju, P.S.(1986) Influence of fruit ripening retardents on succinate and malate dehydrogenases in papaya fruits with emphasis on preservation. Indian J.Hort. 32 : 166-173.

Parmar, P.B. and Chundawat, B.S.(1985) Effect of various postharvest treatments on the physiology of mango fruits. IInd International Symp. on Mango. Bangalore (India), May, 1985.Abstract Sr.No. 8-9, pp.82.

Ranganna, S.(1979) Manual of analysis of fruit and vegetable products. Tata Mc Graw Hill Pub.Co. Ltd. New Delhi.

Roy, S.K. (1973) A simple and rapid method for estimation of total carotenoid pigments in mango. J.Food Sci. Technol. 10 : 45.

Umbriet, W.W., Barris, R.H. and Stauffer, J.P.(1957) Manometric technique and related methods for the study of tissue metabolism. Burgess Publ.Co., Mi nnesota.

Table 1.

Effect of prepacking treatments on keeping quality of sapota fruits cv. 'Kalipatti'

Treatment	Days to ripening	PLW (%)
T1	7.7	2.48
T2	6.7	2.10
T3	6.2	2.30
T4	5.2	2.95
C.D. at 5 %	0.4	0.07

Table 2.

Effect of prepacking treatments on days to attainment of climacteric peak of various biochemical components.

Treatment	Catalase	PME	Respiration	Ethylene Production
CV 'Kalipatti'				
T1	8	3	6	6
T2	8	5	6	6
T3	8	5	6	6
T4	6	3	5	6
CV. 'Cricket Ball'				
T1	8	4	6	4
T2	8	5	6	4
T3	8	4	6	6
T4	6	3	5	4

Table 3.

Effect of prepacking treatments on average activity
CV. 'Kalipatti'

Treatment	Catalase (μl H2O2 Oxidised/ min./ml. of enzyme)	PME (PME units/ g/min.)	Respiration μl. of O2 consumed/ hr./g/pulp)	Ethylene production μg/kg fr.wt/ hr.)
T1	2685.9	4.7	270.3	4.6
T2	2754.2	3.9	242.4	4.9
T3	3285.3	4.9	238.8	6.7
T4	5278.9	5.4	296.2	9.1
C.D. at 5 %	72.8	0.1	2.8	1.5
CV. 'Cricket Ball'				
T1	1782.6	2.7	244.0	4.9
T2	1973.7	2.8	217.0	6.8
T3	2516.3	3.3	215.5	6.3
T4	3332.4	4.2	272.1	11.7
C.D. at 5 %	70.0	0.04	4.5	NS

Table 4.

Effect of prepacking treatments on quality of ripe sapota fruits.
CV.'Kalipatti'

Treatment	TSS (%)	Acidity (%)	Vitamin 'C' (mg/100g)	Total sugar (%)	Carotene (mg/100g)
T1	25.0	0.06	6.1	20.6	1.15
T2	24.5	0.08	10.6	23.0	1.04
T3	24.5	0.14	10.0	18.5	1.08
T4	27.0	0.14	4.0	19.9	0.73
C.D. at 5 %	0.81	0.005	0.68	0.40	0.04
CV. 'Cricket Ball'					
T1	23.5	0.10	5.1	19.5	1.18
T2	23.0	0.08	7.9	18.2	1.24
T3	22.6	0.12	8.8	17.1	0.96
T4	25.0	0.15	3.1	15.7	0.57
C.D. at 5 %	0.72	0.015	0.96	0.41	0.21

POTENTIAL OF INDIAN WILD PLANTS AS ORNAMENTALS

S.C.SHARMA & ANIL K.GOEL, BOTANIC GARDEN, NATIONAL BOTANICAL RESEARCH INSTITUTE, LUCKNOW-226001 (INDIA)

KEY WORDS : WILD PLANTS; EX-SITU CONSERVATION; ORNAMENTALS; PHYTOGEOGRAPHICAL REGIONS

1. INTRODUCTION

Flora of Indian subcontinent is very rich in diversity and endemism. It is represented by over 15,000 species of flowering plants distributed in nine phytogeographical regions of the country (Jain, 1983). Climatically and geographically it exhibits a great degree of diversity ranging from tropical evergreen, subtropical, desert, temperate, alpine and arctic vegetations. Hence, the floristic composition depicts a great variety of plants from coastal regions to the alpine Himalayas. Out of over 15,000 species, there are nearly 4,000 species, endemic to the subcontinent.

Due to depletion or loss of plant species through over exploitation, rapid and ruthless deforestation and unplanned developmental activities, the natural habitat s of plants are under great pressure and serious threat. In India, it is estimated that nearly 15-20% of total vascular flora i.e. 2,500 species of plants are under various degrees of endangerment (Jain, 1987). There are about 2,50,000 species of plants on the earth, the lone biosphere known so far (Heywood, 1978). As per the data available, the global deforestation is estimated to be 16 to 20 million hect. every year. During the next 30-40 years, above 60,000 plants will face threats to their survival or even become extinct. Most of them could have been of great value to the mankind for various purposes such as economic, medicinal or ornamental.

All the plant species which are presently under cultivation for economic, medicinal or ornamental purposes, have come from wild germ-plasm resources during medievial period. The wild relatives of these plants contribute to the vast genetic resources for their varietal improvements and genetic manipulations.

2. BRIEF HISTORY OF EXPLORATIONS IN INDIA FROM HORTICULTURAL VIEW POINT

A glimpse of the history of botanical explorations and plant collections in India reveals that besides from professional botanists, several other British horticulturists and amateur botanists have extensively explored the subcontinent from time to time in search of not only the established ornamentals like Rosa, Primula, Senecio,

J. Prakash and R. L. M. Pierik (eds.), Horticulture – New Technologies and Applications, 379–382.

Rhododendron and orchids etc. but also for many other beautiful herbs, shrubs and trees for introduction in their gardens. The famous 'Assam Tea delegation' consisting of Wallich, Griffith and mc Cleand brought to knowledge, numerous interesting plants from North-East India, apart from establishing the Wild Tea plant (Burkill, 1965).

The renowned British botanists, J.D.Hooker and T. Thomson spent nearly six months for exploring the rich flora of Meghalaya. These botanists not only established the richness and specialities of the floristic wealth, but also collected head loads of wild plants for introduction in the British Gardens as ornamentals specially for Royal Botanic Gardens, Kew, United Kingdom. J.F.Royle and J.F. Duthie, the famous plant collectors in North-West Himalaya also collected many beautiful wild plants of ornamental value for Royal Botanic Gardens, Kew.

While hundreds of Indian wild plants from North-East India have found their way into many European Botanic gardens, where they have been much appreciated. A few such remarkable examples from North-West Himalaya are: Impatiens glandulifera, Morina longifolia, Potentilla nepalensis, Meconopsis robusta, Primula denticulata, Paris polyphylla, Polygonatum verticillatum, Eremurus himalaicus, Rosa brunonii, Lloydia serotina, Tulipa stellata and some genera of orchids like Bulbophyllum, Coelogyne, Cymbidium, Cypripedium, Dendrobium, Paphiopedilum etc.

3. SCOPE OF NEW ORNAMENTALS FROM INDIAN WILD PLANTS

Indian subcontinent possesses the vast potential of new ornamentals from wild sources. The ornamental plants which are under usage in floriculture and landscaping presently hail from Europe and have endless list. The most prominent genera under cultivation as ornamentals are Rosa, Chrysanthemum, Gladiolus, Orchids, Dianthus, Gerbera etc.

The indigenous floristic wealth in our country proclaims its own significance and is deeply involved in our culture, literature, socio-economic life, romance and poetry. It would be very interesting but a bit precarious task to incorporate such ornamental wild plants and flowers in which we may have natural lead, in the floriculture trade. The prime source of introduction of these plants are Botanic Gardens which can very well coordinate exchange of seeds and plant materials within the country and abroad. Organised expenditions and more important individual botanists, gardners and connoisseurs should collect such plants which are not available commercially and found in nature only. It will necessitate a network of Botanic Gardens and good liaison within the subcontinent as well as abroad. Such plants can be collected from the wild, introduced, acclimatised at different altitudinal zones, multiplied and made available for distribution and sale to the nurserymen. Studies on their phenology and various floricultural parameters will be of utmost importance for these plants. Some of the explorations by Rao & Haridasan (1985) and Sharma (1987), have been made in Meghalaya and Punjab and they have listed a fair number of wild plants of horticultural importance.

India has been divided in the following nine phytogeographical regions (Jain 1989-90). A few important plants having potential for floricultural trade and occuring in these regions have been discussed. They can be utilized as ornamentals for landscaping, rockery plants, climbers, arboriculature and for various other similar purposes.

3.1. NORTH-WESTERN HIMALAYA

This region includes the hills of Uttar Pradesh, Himachal Pradesh and Jammu & Kashmir. This area consists of several plants possessing potential for ornamental exploitation. Few of them are: Anemone rivularis, Geum roylei, Potentilla atrisanguinea, Rosa macrophylla, R. brunoni, Hydrangea anomala, Anaphalis busua, Erigeron alpinus, Primula denticulata, Cyperipedium cordigerum, Iris kumaonensis, Tulipa stellata etc.

3.2. EASTERN HIMALAYA

They include states of Sikkim, North-Bengal, Arunachal Pradesh and Upper Assam. The important taxa worth for introduction in floricultural trade are: Magnolia campbellii, Clerodendrum bracteatum, Ixora acuminata, Anemone narcissifolia, Rosa sericea, Camellia kissi, Pieris formosa, Primula sikkimensis, Jasminum amplexicaule, Paphiopedilum faireanum, Vanda coerulea etc.

3.3. WESTERN DRY REGION

It includes Rajasthan, Gujrat, Haryana, Delhi and parts of Punjab and Maharashtra. Plants which can be utilized as ornamentals are: Hybanthus enneaspermus, Mansonia senegalensis, Oxystelma secamone, Telosma pallida, Melhania hamiltoniana, Waltheria indica, Galactia tenuiflora, HYphaene dichotoma etc.

3.4. GANGETIC PLAINS

The Gangetic plains stretching from eastern Rajasthan through Uttar Pradesh to Bihar and West Bengal, are mostly agricultural regions. The wild plants of ornamental importance are: Inula cuspidata, Butea monosperma, Erythrina resupinata, Rosa clinophylla, R. moschata, Primula floribunda, Jasminum arborescens, Trachelospermum lucidum, Eria convallarioides,Dendrobium bicameratum, Asparagus gracilis etc.

3.5. EASTERN INDIA

It includes plains of Arunachal Pradesh, Assam, Meghalaya, Nagaland, Manipur, Mizoram and Tripura. Wild plants of ornamental importance are : Cyathea gigantea, Cycas pectinata, Erythrina fusca, Colquhounia coccinia, Ardisia meghalayensis, Ixora acuminata, Mussaenda roxburghii, Trachelospermum auritum, Bulbophyllum leopardinum, Calanthe angusta, Eria fragrans, Vanda pumila etc.

3.6. THE DECCAN PLATEAU

It is quite a large triangle-shaped plateau with its base in south of Vindhyachal mountains and two sides lying on the east and west of Western and Eastern Ghats. Plants of ornamental value are: Clematis triloba, Costus speciosus, Alpinia malaccensis, Lillium wallichianum, Eria dalzellii, Cymbidium intermedium, Aerides crispum, Cycas beddomei etc.

3.7. WESTERN GHATS

The Western Ghats include the regions of Karnataka, Kerala

and Tamil Nadu, Species of floricultural value are: Clematis smilacifolia , Sonerila rheedii, Ixora polyantha, I. malabarica, Jasminum azoricum, Crossandra indundibuliformis, Mussaenda laxa, Nilgirianthus heyneanus, Dendrobium barbatulum etc.

3.8. EASTERN GHATS

Flora of eastern ghats is not as rich as on the western ghats. It covers parts of Orissa, Andhra Pradesh and Tamil Nadu. Some of the notable wild plants are: Michelia nilagirica, Guettarda speciosa, Ixora finlaysoniana, I. notoniana, Rosa leschenaultiana, Mussaenda tomentosa, Jasminum rottleriamum, Aerides crispa, Coelogyne odoratissima, Nervilia prainiana etc.

3.9. ANDMAN & NICOBAR ISLANDS

The Andman Islands are said to be continental fragments, whereas the Nicobar Islands are volcanic and their flora shows a marked difference. Wild plants of ornamental value are: Magnolia andamanica, Bauhinia nicobarica, Melastoma affine, Gardenia tubifera, Ixora barbata, Mussaenda villosa, Guettarda andamanica, Eria andamanica, Vanilla andamanica, Geodorum densiflorum, Symplocos chengapae.

4. CONCLUSION

In countries like India which possess a rich and diverse vegetal wealth, mostly the exotic plants are preferred in floricultural trade as compared to the use of indigenous beautiful wild plant species. There is a need to take-up R&D work by intertwining the botanists and floriculturists. In our country, a large variety of wild plants of all habitats, which can be grown in the Botanic Gardens and used in landscaping, await the attention of garden lovers as well as specialists, nurserymen, town planners etc. The first step is to bring them in Botanic Gardens and Regional Stations for their conservation, propagation and dissemination. This plant wealth will be also helpful in the improvement and evolution of new ornamental cultivars and will play pivotal role in the floriculture industry.

5. ACKNOWLEDGEMENT

The authors are thankful to the Director, National Botanical Research Institute, Lucknow for facilities provided for R&D work.

6. REFERENCES

Burkill I.H.(1965) Chapters on the History of Botany in India, BSI, Calcutta, India, 245 pp.

Heywood V.H. (1978) Flowering Plants of the World, London,335 pp.

Jain S.K. (1983) Flora and Vegetation of India-An outline, BSI, Howrah, India, 24 pp.

Jain S.K. (1987) The problem of endangered species, its study and solution, Presidential address. National Academy of Sciences, Allahabad, India, 44 pp.

Jain S.K. (1989-90) Botanical Regions and Flora of India, Everym.Sci. 24:213-223.

Rao R.R. and Haridasan K. (1985) A survey of wild plants of horticulture importance from Meghalaya, N.E.India, J. Econ.Tax.Bot. 6: 529-537.

Sharma M. (1987) A survey of wild plants of Horticulture value from Punjab,Bull.Bot.Surv.India.29:120-126 (1990).

POST HARVEST STUDIES ON KINNOW MANDARIN

P.K. JAIN * AND K.S. CHAUHAN
DEPARTMENT OF HORTICULTURE, H.A.U., HISSAR, 125004
INDIA

Key words - **Kinnow, Packages, Transportation and Storage.**

Introduction

Citrus fruits in India are next to banana and mango in acreage and economic importance. Among these, mandarins occupy a prominent position and account for over 50 per cent of the citrus area. 'Kinnow' a recent introduction from California is being cultivated on large scale by the fruit growers of North India because of high yield potential, superior fruit quality, adaptability under varied agroclimates, freedom from granulation etc.

The estimated total loss of fruits and vegetables in India due to inadequate post-harvest, handling, transportation and storage is 20 to 30 per cent. To avoid this glut and to prolong the availability of fruits in the market and for better returns to the farmers, proper packages, transportation and storage aspects need to be standardized. Keeping in view the above facts the present studies were conducted to find out suitable packaging and transportation methods and to know the effect of different fungicides for extending self-life of kinnow fruits at room temperature and in zero energy cooling chambers.

Methods

The following two experiments were conducted separately :

Experiment-I Studies on various packages and mode of transportation for kinnow. Fresh fully ripe uniform fruits were harvested and packed in the various packages (Gunny bag, Bamboo basket, Wooden box, Wire bound wooden box and Corrugated paper box) to their full capacity (8-10 kg) using paper cuttings as cushioning material. The fruits packed in these

* Present Address : Scientist-Horticulture, Jawaharlal Nehru Krishi Vishwa Vidyalaya, Jabalpur-482004 INDIA

J. Prakash and R. L. M. Pierik (eds.), Horticulture – New Technologies and Applications, 383–387.

packages were transported for about 600 Kms by road and by rail. The fruits after transportation were kept at room temperature for 10 days to find out the effect of transportation modes and packages on fruits. Various physiological, qualitative, pathological and economical observations were recorded.
Experiment-II Studies on storage behaviour of kinnow under different storage conditions. The freshly harvested fruits were washed, air dried and dipped in aqueous s olution of fungicides for 2 minutes. The fungicide treatments were (1) Control (water dip) (2) Benomyl (200 ppm) (3) TBZ (200 ppm) and (4) Bavistin (200 ppm). After dipping the fruits were surface dried before packing into perforated wooden boxes, using newspaper as lining material.

There were two storage conditions. In modified storage condition fruits were covered with polyethylene bags and bags were tightened by closing the mouths with 'sutli' just to prevent the exchange of gases and loss of moisture. In case of non-modified storage condition the fruits were kept in wooden boxes without any polyethylene covering. The treated fruits after packaging were kept separately under (i) room temperature (ii) in zero energy cooling chambers. Various physiological, qualitative, pathological and economical observations were recorded at 7 days interval.

Results and Discussion

Experiment-I The physiological loss in weight increased after transportation and in the storage period. However, the minimum loss in weight was observed in the fruits packed in corrugated paper box and maximum in gunny bag. The firmness of the fruits were also maintained in the wire bound box and corrugated paper box as compared to other packages. The maximum mechanical injury (punctures) per fruit was noted in the fruits packed in gunny bags and transported by railway. The corrugated paper box significantly reduced mechanical injury of the fruits under both the transportation modes and minimum injury was observed in the fruits transported by road. This might be due to the better structure and strength of boxes, which resisted the jerks during transportation and providing better handling during loading/unloading of fruits and thus prevented damage during transportation. However, fruits transported by railway showed more mechanical injury than transported by road. This may be due to the carelessness of railway porters while loading and unloading. Secondly, in the gunny bag and bamboo basket the fruits were not packed tightly as compared to wire bound box or corrugated paper box where the packages could not provide chance for movement of the fruits resulting in less injury and loss in weight (Singh and Gupta, 1983).

It was observed that the various packages used significantly affected the juice and pulp content of the fruits after transportation as well as during storage period. The fruits either packed in various packages or transported by railway or road did not differ in organoleptic rating. Chemical composition of fruits was not appreciably affected by various packaging and transportation modes. However, T.S.S. increased slightly after transportation and during storage period and the maximum was observed in the fruits packed in gunny bag and lowest in the corrugated paper box. It was observed that various packages and mode of transportation did not influence the total sugar, reducing sugar and non-reducing sugar contents of the fruits significantly during the course of investigation. Decay losses were recorded maximum after transportation and also during storage in fruits packed in gunny bag. Decay loss was found least in the fruits packed in corrugated paper box may be due to better condition of fruits, when stored after transportation, while in other packages, damages occurred to the fruits were higher which provided a media for the initiation of microbial growth. It was further observed that railway transportation was found cheaper than trnsportation by road for the same distance covered. The differences in market value received by fruits in various packages might be due to visible damage to the fruits, their appearance, shrivelling, glossiness, weight loss and also for attraction of a particular package.

Experiment-II Modified storage conditions significantly reduced the loss in weight during storage at room temperature as well as in zero energy cooling chamber (ZEC champer). The reduction in loss in weight might be due to the retardation in transpiration process. The loss in weight was significantly lower in ZEC chamber, it might be because of lower temperature inside the chambers due to evaporative cooling and higher relative humidity which favours slow evapotranspiration rate. The juice and pulp content, specific gravity and organoleptic rating were also decreased with the increase in storage period but the decreasing rate was higher in control fruits as compared to treated fruits. The losses were minimum in the fruits stored in ZEC chamber with modification due to comparatively low temperature and higher relative humidity. Secondly, the low respiration rate in modified storage could possibly have resulted in slowing down or delaying the senescence process and thereby maintaining the fruit quality acceptable even after 77 days of storage.

The quality parameter viz. total soluble solids, total sugars, reducing sugars, and non-reducing sugars were increased, while acidity and ascorbic acid content decreased during storage at room temperature as well as in ZEC chamber. The storage conditions did not affect the quality parameters though higher increase was observed in fruits under non-modified storage condition. This might be due to more loss of water and concentration of the juice and conversion of non-soluble fraction to soluble fraction. It was also observed that almost all the fungicide treatments were effective in reducing the fruit rot during storage. However, the benomyl was found more effective followed by TBZ. It might be due to effective control of microorganisms present on the fruit surface. It was also observed that the TBZ treated fruits were remained decay free for a longer storage period than other treatments (Gutter, 1985).

It was observed that the fruits with fungicide treatments and stored under modified condition fetches better market value at different sampling dates. It was also noted that the fruits stored in ZEC chamber received more market value than those stored at room temperature. This might be due to better appearance, low decay loss and weight loss after storage under modified condition and in ZEC chamber (Roy and Khurdiya, 1986).

From the proceeding discussion and results it may be inferred that Kinnow mandarin fruit can be trnsported successfully to a longer distance without much inju ry and losses, in woodem box and wire bound box. Transportation of fruits by road was found better than railway transportation. Kinnow mandarin can be stored in acceptable quality for longer period with the help of various fungicide dip treatments. Among the storage conditions, the modified condition was best and under this condition fruits could be retained in acceptable quality upto 77 days and 56 days in ZEC chamber and at room temperature storage respectively.

References

Gutter, Y. (1985) Combined treatment with thiabendazole and 2-aminobutane for control of citrus fruit decay. Crop Prot. 4 (3) : 349-350.

Roy, S.K. and Khurdiya, D.S. (1986) Studies on evaporative cooled zero energy input cooled chambers for the storage of horticultural produce. Indian Food Packer. 40 (6) : 26-31.

Singh, J.P. and Gupta, O.P.(1983) Evaluation of various packings of ber in relation to decay loss caused by various microbes. Haryana Agric. Univ. J.Res. 13 (4) : 593-595.

Influence of Gibberellic Acid on Flowering and Plant Sugar Content of China Aster (Callistephus chinensis (L) Nees) Cv. "Heart of France"

R.JAYANTHI AND J.V.NARAYANA GOWDA
DIVISION OF HORTICULTURAL SCIENCES, UNIVERSITY OF AGRICULTURAL SCIENCES, GANDHI KRISHI VIGNANA KENDRA, BANGALORE - 560 065, KARNATAKA , INDIA.

Introduction

Though China aster is grown as a popular garden annual throuhout the world, it has a special commercial importance in India because of its extensive use in making bouquets, button-holes and garlands. Very little research work has been reported in this flower crop with respect to the use of growth regulators, particularly Gibberellic acid for improving cut flower production (Reddy,1978). The present study was designed to improve flower quantity and quality, by the application of gibberellic acid and to determine the optimum concentration of gibberellic acid spray for china aster.

Methods

The experiment was conducted with Callistephus chinensis Nees Cv. 'Heart of France' in 30cm diameter earthern pots in a completely randomized design with 3 replications. Different concentrations of gibberellic acid were sprayed as foliar spray at 10, 25, 50, 100, 250, 500 and 1000 ppm twice at 30 and 45 days after transplanting. Spraying of distilled water served as control. Morphological observations were recorded on flowering characteristics. Chemical analysis of different plant parts were done both at vegetative and flowering stages for reducing, non-reducing and total sugars using Nelson Somogyi micro copper method (Nelson,1944).

Results and Discussion

It is evident from Table 1, that there is significant and considerable acceleration of anthesis in all the gibberellic acid treated plants compared to control. The best result was obtained at 250 ppm giberellic acid spray where the plants flowered about 52.6 days earlier compared to control. Earliness in flowering may be due to acceleration of flower bud development which is evident from the fact that all the gibberellic acid treated plants took less number of days from flower bud appearance to flower opening. It was noted that gibberellic acid treated plants started flowering at a time when the non-treated control plants

J. Prakash and R. L. M. Pierik (eds.), Horticulture – New Technologies and Applications, 389–392.

were still in the vegetative stage. This possibly indicates that gibberellic acid might have substituted the long day requirement of china aster and responsible for overcoming the juvenility in plants and determined that developmental sequence of sex organs of flowers. Acceleration of flowering, though to a lesser extent was also observed by Reddy (1978), in china aster at concentration of 300 ppm.

The production of non-salable (i.e., small sized, deformed) flowers reduced significantly as a result of Gibberellic acid spray between 50-250 ppm and the maximum reduction of 67% was recorded at 250 ppm compared to control. There are several reports in a number of other flowers where gibberellic acid application improved flower yield (Sen and Maharana, 1971, Shanmugam and Muthuswamy, 1974. El-Shafie and Hassan, 1978). Sen and Sen(1968) reported that Gibberellic acid treatment resulted in more but smaller flowers in Zinnia.

Flower diameter, flower stem length and duration of flowering (longevity of flowers on the plants) increased significantly as a result of gibberellic acid spray. (Table 1). In both these attributes the best results were obtained at 50 ppm. While Tamberg(1963) noted improvement of flower diameter in china aster and Reddy (1978) reported prolonged flowering in china aster as a result of Gibberellic acid treatment. Improvement in vase life of cut flowers was also recorded in gibberellic acid treated flowers compared to non treated flowers (Table 1). Gibberellic acid spray at 50 ppm and above was found to be useful in increasing the fresh weight of flowers and this improvement could be attributed to the increase in flower diameter and individual flower weight as discussed earlier.

Though the results of the chemical analysis of different plant parts at two stages of crop growth were found to be statistically non-significant, a relatively higher sugar content was recorded in the gibberellic acid treated plants (Table 2). It was also observed that gibberellic acid treated plants had more sugar content in the leaves at vegetative stage than at flowering stage. It might be probable that this higher contents of sugar was utilised for flower bud development, flower quality and subsequently yield. Zeislin and Halevy (1976) suggested that in rose, gibberellic acid participates in endogenous control of flower development and possibility is that it acts by directing the translocation of metabolites to the flower buds. El-Shafie (1978) reported an increased total carbhohydrate content of flowers, stem and leaves due to gibberellic acid treatment in roses at concentration from 50-250 ppm compared to control.

Relatively higher amounts of sugars recorded in the flowers treated with gibberelic acid at 25,50,250 and 500 ppm might be responsible for the extended vase life of flowers in these treatments, since the keeping quality of cut flowers depends to a large extent on the stored carbhohydrate in flowers and sugars are known to provide energy for the metabolic activity of the cut flowers.

Conclusion:

The results indicate that gibberelic acid spray at 50-250 ppm applied as foliar spray at 30 and 45 days after transplanting can accelerate anthesis, and improve flower quality and consequently yield in china aster (Callistephus chinesis(L) Nees) Cv. 'Heart of France. The duration of flowering and vase life of cut flowers also improved as a result. Gibberellic acid treatment increased the sugar contents of leaves at vegetative stage.

References:

El-Shafie, S.A., 1978. Effect of gibberelic acid on the growth and flowering of Montezuma roses. Archiv fur Gartenbau, 26(6) : 287-296.

El-Shafie, S.A. and Hassan, H.A., 1978. Effect of gibberelic acid and chloromequat on the growth and flowering of gerbera. Archiv fur Gartenbau, 26(7) : 333-342.

Nelson, N. 1954. A photometric adoptation of the somogys method for the determination of glucose J. Biol. Chem., 153 : 575-638.

Reddy, Y.T.N., 1978. Effects of growth substances on growth and flowering of china aster (Callistephus chinensis Nees). Mysore Journal of Agricultural Sciences. 12(3) : 526

Sen, S.K. and Maharana T., 1971. Growth and flowering response of chrysanthemum to growth regulator treatments. Punjab Horticultural Journal 11(3/4) : 219-224.

Shanmugam A. and Muthuswamy. S 1974 Influence of photoperiod and growth regulators on the nutrient status of chrysanthemum. Indian J. Hort. 31(2) : 186-193.

Tamberg, T.G., 1963. The effect of gibberellin on ornamental plants. Tr. Priklad. Bot. Genet. Selek. 35(2) : 85-93.

Zeislin, N. and Halevy, A.M., 1976 Effect of environmental factors on gibberellin activity and ethylene production in flowering and non-flowering shoots. *Physiologia plant.* 37(4) : 331-335.

TABLE 1

Effect of Gibberellic acid spray on various floral attributes of China aster Cv. 'Heart of France'

Treatments	Days taken for anthesis	Total no. of flowers	No. of salable flowers	Flower diameter (cm)	Flower stem length (cm)	Total fresh weight of flowers/plant	Duration of flowering	Vase life of flowers (days)
GA_3 at								
10 ppm	166.00	23.95	13.8	5.87	16.96	26.9	15.27	5.00
25 ppm	167.47	26.00	15.50	5.98	15.40	30.76	13.27	5.40
50 ppm	153.50	26.40	17.20	6.46	17.18	56.67	18.13	7.40
100 ppm	150.90	25.30	18.45	6.24	18.64	45.56	15.33	7.80
250 ppm	149.77	24.75	18.60	6.16	19.98	40.36	15.26	10.20
500 ppm	150.00	21.80	12.00	6.27	15.04	34.23	15.20	9.00
1000 ppm	150.27	23.35	15.75	6.11	14.04	29.35	13.53	9.00
Control	202.40	30.20	12.65	5.08	13.38	23.85	12.73	4.60
Mean	152.12	25.22	15.53	6.02	16.33	33.96	14.84	7.30
F test	**	NS	**	**	**	**	**	**
CD at 5%	9.48	-	3.28	0.33	3.97	14.97	2.33	2.03

TABLE 2

Total sugar content of leaves, stem and flowers of China aster at two stages of growth as influenced by gibberellic acid. (% sugar on dry wt. basis)

Treatments	Veg. stage	Flowering Stage		
	Leaf	Leaf	Stem	Flower
GA_3 at				
10 ppm	2.56	1.83	1.28	8.00
25 ppm	2.63	1.47	1.46	9.60
50 ppm	3.40	1.56	2.38	9.87
100 ppm	2.94	1.08	2.53	8.92
250 ppm	5.64	3.06	1.41	9.57
500 ppm	4.30	1.55	3.01	9.64
1000 ppm	4.30	2.64	1.81	8.06
Control	1.72	2.08	1.69	8.90
Mean	3.44	1.90	1.94	9.06
F Test	NS	NS	NS	NS

Postharvest physiology of gladiolus flowers as influenced by cobalt and sucrose

T.P.MURALI and T.VENKATESH REDDY
Division of Horticultural Science, Univ. Agri. Sci., Bangalore - 560 065, INDIA

Key words: cut flower, gladiolus, vase life, water relations

Introduction

The use of flower preservatives to promote the quality and to prolong the life of cut flowers has been known since many years. Cobalt salts have been shown to improve the water balance of cut roses (Venkatarayappa et. al., 1980), and tuberose (Balakrishna et.al., 1989), thereby extending their longevity. It is known that sucrose improves water balance in cut flowers (Borochov et. al., 1976) and maintains the respirable substrates in the flower (Nichols, 1975). Although several attempts have been made to study the effect of different chemicals, sugars and harmones on the longevity of commercial cut flowers (Halevy amd Mayak, 1979, 1981), relatively little work has been done to prolong the vase life of gladiolus, an important commercial flower. The present paper discusses the effects of cobalt and sucrose on the water relations, fresh weight and vase life of gladiolus flowers.

Methods

Gladiolus (cv. Friendship) flowers grown as per standard cultural practices and harvested when the lower most floret started unfolding the petals were cut to an uniform length of 75 cm. All leaves were removed except one or two bract-like leaves below the florets. After recording the fresh weight, each spike was placed in a 500 ml bottle containing 500 ml of distilled water or aqueous solution of 2% or 4% sucrose alone, 0.25 mM, 0.5 mM or 1.0 mM cobalt sulphate alone or their combinations. Each treatment had four flower spikes with each spike representing a replication. The buds were opened under normal laboratory conditions. At 3-day intervals, the weight of the container plus solution with and without the spike and spike alone were recorded. From these data water uptake and transpirational loss of water by the individual spike were computed. The longevity of individual florets was measured by noting the day of opening and fading of the florets. Flower spikes were discarded when 5th floret from the bottom wilted. This stage is considered to be the end of the potential useful longevity of gladiolus spikes (Kofranek and Halevy, 1976).

J. Prakash and R. L. M. Pierik (eds.), Horticulture – New Technologies and Applications, 393–396.

Results and discussion

Sucrose and cobalt either alone or in combination increased the water uptake and the transpirational loss of water by the cut gladioli compared to the control (Table 1). However, water loss : water uptake ratio was reduced by Co 0.5 mM alone or in combination with sucrose 4%. Co and sucrose either alone or in combination increased the fresh weight of the flower, the combination treatments maintaining higher fresh weights (Table 2).

When the rate of water loss is more than that of water uptake, the flower tissue experiences water stress which has a direct effect on wilting and senescence of cut flowers. Spikes held in distilled water were found to have undergone water stress earlier, which may be due to the plugging of

Table 1. Water uptake and water loss by gladiolus spikes as influenced by cobaltous sulphate and sucrose.

Treatments	Water uptake during 9 days (g/spike)	Water loss during 9 days (g/spike)	Water loss : water uptake ratio
Sucrose (%)			
0	64.99	82.18	1.26
2	89.42	98.53	1.10
4	102.92	122.97	1.19
CD at 5% level	10.24	11.41	
Cobaltous sulphate (mM)			
0.0	57.10	65.81	1.15
0.25	82.24	93.48	1.14
0.50	119.27	116.30	0.98
1.00	95.01	115.54	1.22
CD at 5% level	11.83	13.17	
Sucrose (%) + Cobalt (mM)			
0 + 0.0	40.31	40.31	1.00
0 + 0.25	70.49	89.34	1.27
0 + 0.50	84.68	89.98	1.06
0 + 1.00	70.95	89.35	1.26
2 + 0.0	82.32	83.32	1.00
2 + 0.25	86.96	97.02	1.12
2 + 0.50	115.09	117.75	1.02
2 + 1.00	92.53	136.32	1.47
4 + 0.0	144.61	144.61	1.00
4 + 0.25	93.44	108.76	1.16
4 + 0.50	126.24	107.28	0.85
4 + 1.00	99.46	139.53	1.40
CD at 5% level	20.49	22.82	

the xylem either physically by the microorganisms entering through the base or physiologically (Halevy and Mayak, 1979,1981). Cobalt might act to inhibit vascular blockage as has been reported in roses (Reddy, 1988). A deterioration in the water balance of flower leads to a rise in endogenous ABA content (Borochov et al, 1976). As a result of changes in the water balance shown in the present investigation, possible changes in ABA and its implication in the senescence of cut gladioli cannot be ruled out. Since cut gladioli spikes treated with cobalt and sucrose, separately or in combination, maintained better water relations, the increase in ABA levels might have been checked leading to delayed senescence. As a consequence of improved water relations, the fresh weight of the spikes was increased. In addition, cobalt (Lau and Yang, 1976) and

Table 2. Effect of cobalt and sucrose on fresh weight, longevity of individual florets and vase life of gladiolus spikes.

Treatments	Fresh weight on day 6 (% of initial weight)	Longevity of individual florets (Days)	Vase life (Days)
Sucrose(%)			
0	92.58	4.42	7.75
2	113.41	5.25	9.58
4	111.36	5.58	10.83
CD at 5% level		0.42	0.44
Cobaltous sulphate(mM)			
0.0	95.31	4.67	7.73
0.25	104.35	5.11	9.00
0.50	115.18	5.67	10.22
1.00	105.64	5.22	10.00
CD at 5% level		0.48	0.51
Sucrose(%) + Cobalt(mM)			
0 + 0.0	81.80	4.00	7.00
0 + 0.25	101.90	4.67	8.67
0 + 0.50	117.45	5.33	9.33
0 + 1.00	105.51	4.67	8.67
2 + 0.0	89.65	5.00	8.33
2 + 0.25	113.74	5.33	9.67
2 + 0.50	116.92	5.67	10.67
2 + 1.00	97.67	5.33	10.33
4 + 0·0	100.82	5.67	8.67
4 + 0.25	117.90	5.67	10.00
4 + 0.50	130.97	6.00	12.33
4 + 1.00	98.91	5.33	10.67
CD at 5% level		0.84	0.89

sucrose (Dilley and Carpenter, 1976) have been shown to inhibit ethylene biosynthesis.

The longevity of florets was extended by both cobalt and sucrose (Table 2). The longevity of florets was maximuym (6 days) with sucrose 4% + 0.5 mM Co as compared to control (4 days). Vase life of the flower spikes was extended by sucrose and cobalt either separately or in combination (Table 2), combination treatments being better than the individual ones. Maximum vase life (12.33 days) was observed in the flowers placed in sucrose 4% + 0.5 mM Co as compared to the vase life of 7 days observed in flower spikes placed in distilled water.

References

Balakrishna H.V. Reddy T.V. and Rai B.G.M. (1989) Postharvest physiology of cut tuberoses as influenced by some metal salts. Mysore J. agric. Sci. 23 : 344-348.

Borochov A. Tirosh T. and Halevy A.H. (1976) Abscisic acid content of sensecing petals of cut rose flowers as affected by sucrose and water stress. Plant physiol. 158 : 175-178.

Dilley D.R. and Carpenter W.J., (1776) The role of chemical adjuvants and ethylene synthesis on cut flower longevity. Acta Hort. 41: 117-132.

Halevy A.H. and Mayak S. (1979) Senescence and postharvest physiology of cut flowers, Part 1. Hort. Rev. 1:204-236

Halevy A.H. and Mayak S. (1981) Senescence and postharvest physiology of cut flowers, part 2. Hort. Rev. 3:59-143.

Kofranek A.M. and Halevy A.H. (1976) Sucrose pulsing of gladiolus stems before storage to increase spike quality. HortScience 11 : 572-573.

Lau O.L. and Yang S.F. (1776) Inhibition of ethylene production by cobaltous ion. Plant physiol. 58: 114-117

Nichols R. (1975) Senescence and sugar status of the cut flower. Acta Hort. 41 : 21-30.

Reddy T.V. (1988) Mode of action of cobalt extending the vase life of cut roses. Scientia Hort. 36 : 303-313.

Venkatarayappa T., Tsujita M.J. and Murr D.P. (1980) Influence of cobaltous ion (Co^{2+}) on the postharvest behaviour of Samantha roses. J. Amer. Soc. Hort. Sci. 105 : 148-151.

A few simple facts to reveal the stupendous dimensions

NEW FRONTIERS IN WILD FRUIT TREE CROPS

* S. SHYAM SUNDER
** S. PARAMESWARAPPA
* a member of the Indian Forest Service, and retired as P.CCF.
** a member of the IFS,& currently the P.CCF,Karnataka.

Tropical Forests, Rainforests pilinut, commercialnut, Tropical deforestation, grafting, Tissue culture, gene cloning, Hybridisation Bioenginering, Biosphere reserve.

Introduction

A few simple facts to reveal the stupendous dimensions. of the possibilities which await us or, on the contrary, on which we have closed our doors:

Of the 3000 or so of the species of plants that have been cultivated throughout history of mankind only 150 have been farmed on a large scale. Yet there are an estimated 75,000 potential food species and many of these are located in the equatorial forests (1).

Many of the most popular fruits have their origin in tropical forests, yet many species remain unexplored. Of the 2500 rain forest fruits identified, only 50 are well known and a mere 15 commercially traded on a large scale. Most are not yet available outside of their native localities(1).

The accidentally discovered Serendipity berry, Dioscoreophyllum cuminsii is 3000 times sweeter than sucrose(1).

The miracle fruit Synsepalum dulcificium contains the sweetening agent moneltin (1).

The increasingly sought after Macademia nut,Mecademia fermifolia, is native to the rain forest of Queensland, Australia, though it is now being cultivated in latin America and Pacific Islands. Other nuts from the region offering the same commercial potential include the Bush Walnut, the Red bopple and the Daintree nut (1).

J. Prakash and R. L. M. Pierik (eds.), Horticulture – New Technologies and Applications, 397–400.

From the Philippines comes the Pili nut, Canarium ovatum, which can be eaten directly, used for oil extraction or roasted to provide the basis for confections (1).

Another promising addition to the commercial nut range is the Paradise nut, Lecythis Zabucajo, a native of Brazil (1).

Babussa, Orbignya species, yield prolific quantities of high grade colour-less kernel oil suitable for margarine, general edibles, soap, plastic, emulsifiers and detergents (2).

Milpesos, Jessenia polycarpa, produces a clear and golden oil resembling olive oil (2).

The list is endless. These are some of the species of proven value but mostly still confined to the areas in which they occur.

Discussion

The next category consists of a far larger number of species the scope and potential of which are yet to be discovered. This point will be appreciated only when we compare the cultivated varieties of some of our well known fruits with their wild cousins in the forests. The evergreen forests of Karnataka have more than a hundred varieties of mango but none of which hint at the delicasy of the commercial crop. The wild tamarind trees of Sudan have small pods with one or two seeds encased in fibrous pulp while today we have cultivated trees yielding pods 25 cms long and trees yielding as much as 800 Kg of pulp. A tree has been discovered in Karnataka (distant from its original home) the pods of which are seedless. There are trees yielding sour pulp and those yielding pulp that is sweet and less sour. This equally applies to almost all the fruit and nut crops of the world today. The controlled crossing, hybridisation and grafting have produced spectacular results far removed from the original lowly parents.

The potential for uncovering valuable new products is immense with such a vast gene base. But as many as 50 species are becoming extinct each day as a result of tropical deforestation (3).

The vast myriad of genetic base is now threatened with elimination - even before the use of a majority of these is understood. The tragedy is that it is taking place at a time when the developments in the frontiers of biological research, viz., tissue culture, gene cloning, hybridisation and bioengineering are available to shorten nature's evolutionary process of speciation of millions of years, to a short span and to usable bounties. Newton discovered gravity but even without the discovery, the law would have continued to operate. Not so in the case of the rich biological potential, hidden in the tropical forests. If not discovered, brought out and developed, this is liable to be lost, once for all, with the loss of forests or the relentless degradation which is taking place.

The obvious answer would be to save the forests in the tropics - easier said than done. Unlike a decade back, today in almost all the tropical countries there is appreciation of the need to conserve the forests. The problem however is multifaceted and the remedy that is applicable to one country may be totally inapropriate to another. The cause of loss or deterioration of forests could be due to expansion of agriculture, excessive commercial logging, unrestrained removals to meet the fuel needs, uncontrolled grazing, annual incediary fires, etc. A few countries have responded to the call to save the tropical forests by stopping or reducing removal of timber. Some countries have sought to achieve this result by banning export of timber. A few other countries have imposed import ban. But none of these can save the forests if the degrading factors listed above continued to operate. Loss of forest and with it, the elimination of the genetic resource base will continue unless all the biotic pressures are eliminated and investments provided for protection and tending of the natural forests. Simultaneously alternate resources will have to be developed to meet the needs, including of those who cannot pay for meeting it. In most developing countries, the major threat to forests is from the relentless pressure due to the serious gap between demand and supply of firewood - not usually appreciated. Randianasolo, Minister of Livestock, Fisheries and Forests of Madagascar, inaugurating the International Conference on Conservation for Development, in 1985, made it clear that ecology of Madagascar with its forests, of the type and composition not met elsewhere, would be destroyed unless Madagascar gets to be self-sufficient in food and fuelwood.

The problem in Madagascar need not be the same as in Philippines or Honduras, nor be the same even within the same country as in Sabah and Sarawak.

The national and international efforts to save the tropical forests from degradation and destruction should continue. Meanwhile, with time running out on several hundreds and hundreds of species, definite measures aiming solely at conservation of the gene pool are called for. One of these is establishment of biosphere reserves, taken up in many countries but not with the seriousness it deserves nor the financial support it requires. The other would be to organise surveys, list out the species in use or not in use but with hints of possibility of use with genetic enginnering particulary in the forests of the tropical world. Next would be to grade the chances of survival of the different species and list out the endangered ones. This should be followed up by efforts at ex-situ along with in situ conservation of those species in the endangered list, reminiscent of Noah's efforts.

Vast amount of work in gene conservation has already been carried out in the field of agriculture. Not so with tree crops and the field in the case of little known and unknown species likely to yield fruits, pods and nuts of food, nutritive and medicinal values, is as yet untapped. Time is running out, and 'Extinction is for ever'.

Let us not foreclose the gateway to the new frontiers in the field of biotechnology just when it is emerging as an effective tool to achieve mastery over nature in genetics.

References

1. The Rainforest harvest - Agriculture and Food - Friends of the Earth, 1990.

2. The Rain Forest harvest - Industrial products - Friends of the Earth - 1990.

3. The Rain Forest harvest - Sustainable strategies for saving Rain forests - Friends of the Earth -1990.

LANDSCAPING OF INDUSTRIAL REGIONS

S.C.SHARMA, A.N.SHARGA AND R.K.ROY
BOTANIC GARDEN, NATIONAL BOTANICAL RESEARCH INSTITUTE,
LUCKNOW-226001, INDIA

KEY WORDS : Pollution; Environment; Industrial region; Tolerant

1. INTRODUCTION

Since independance rapid industrialisation has taken place in India in order to achieve self sufficiency. As a consequence emissions from industries have also increased manifolds causing pollution of environment in an alarming proportion. According to an estimate, coal based power generation units in India alone contribute 13 million tonnes of fly ash, 8 million tonnes of particulates, 4,80,000 tonnes of SO_2 ; 2,80,000 tonnes of NO_x; 16,000 tonnes of CO and 5000 tonnes of hydrocarbons to the atmosphere annually. In addition, automobiles have further aggravated the problem. There are 10 million vehicles on the road in India at present and the number may increase up to 40 million by the end of the century (Anonymous, 1989). Automobiles contribute 1.8 million tonnes of pollutants in the country annually and worst affected are metropolitan cities which receive 80% of the pollutants.

The ambient air quality standard of industrial, residential and other places is furnished below:-

Ambient Air Quality Standard of India

Categories	Concentration (Micrograms/cu.m.)		
	SPM	SO_2	NO_x
Industrial Area	500	120	120
Residential & Rural Area	200	80	80
Sensitive Areas: Hill Station Sanctuaries, National Parks and Monuments	100	30	30

Several measures have already been initiated to combat this global menace. Landscaping in and around industrial regions with pollution tolerant plant species is an effective way to minimise the level of air pollution.

2. METHODS

2.1 *Selection of plants and pattern of landscaping*

Depending on type of pollutant and its intensity, selection of plants and their number to be planted, will vary. Evergreen plants having long life span with dense foliage will be the ideal choice. For thermal power station and other industries which emit

J. Prakash and R. L. M. Pierik (eds.), Horticulture – New Technologies and Applications, 401–404.

huge quantity of SO_2, SPM, NO_x etc. judicious planning and selection is needed with tolerant plants to the particular pollutant. Studies have revealed that a green area of 500 m. width surrounding a factory is capable of reducing SO_2 concentration by 70%. Therefore, it is recommended that vacant spaces and road sides may be planted with following trees and shrubs.

Trees - Albizzia lebbek, Ailanthus excelsa, Alstonia macrophylla, Alstonia scholaris, Azadirachta indica, Lagerstroemia flosreginae, Lagerstroemia thorelii, Mimusops elengi, Parkinsonia aculeata, Polyalthia longifolia, Terminalia arjuna, Terminalia muelleri.

Shrubs - Caesalpinia pulcherrima, Cassia surattensis, Hibiscus rosa-siensis, Rosa hybrida Gruss an Teplitz, Vinca rosea.

In order to have best effect close planting is necessary. Dwarf trees should be planted in the front followed with medium-tall trees in the back so that all plants are well exposed to the pollutants. Plantation should be done across the direction of the wind blowing from the source of emission. For roadside plantation double row planting is recommended as air passes through green barrier the pollutants are either get absorbed or settled. Cement factories and other industries which release appreciable quantity of dust and SPM in the atmosphere are of serious concern. To overcome this problem, the remedy is green belt by planting trees and shrubs having dust trapping ability. The width of green belt and density of plantation will depend on source of pollutants and their intensity. It has been observed that dust fall is reduced by 2-3 times with the help of green belt of 8 m. width between roads and buildings. A reduction of overall dust fall upto 42% by conifers in urban areas of temperate zone has been reported. The dust trapping ability of plants depends on certain morphological characters viz. branching habit, arrangement of leaf, its size, shape, surface (smooth/striate), presence or absence of trichomes (Ahmad et al., 1989). The more compact is branching and arrangement of leaves, the more will be the dust trapping ability. Some of the plant species which have been found promising are namely - Alstonia macrophylla, Cassia siamea, Cordea sesbestena, Delbergia sissoo, Diospyros citriodora, Ficus benghalensis, Ficus infectoria, Peltophorum ferrugineum, Polyalthia longifolia, Tectona grandis, Terminalia arjuna.

In the industrial and adjoining urban areas automobiles contribute significantly to air pollution. The major pollutants released by automobiles are CO_x, NO_x, SO_2, Pb and unsaturated hydrocarbons. In order to attenuate these pollutants it is recommended to plant pollution tolerant trees and shrubs along the road in double row system keeping shrubs in the front line facing the road. The idea is to develop a green buffer against the vertical and horizontal movement of pollutants including sound. Traffic islands and narrow central verge of two way roads may be planted with hardy plant species for abatement of automobile pollution. The plant species befitting for the purpose are namely -

Trees - Acacia auriculaeformis, Butea frondosa, Cassia

marginata, *Madhuca indica*, *Polyalthia longifolia*, *Putranjiva roxburghii*, *Thespesia populinea*.

Shrubs - *Bougainvillea*, *Lantana depressa*, *Murrya exotica*, *Nerium odorum*, *Tabernaemontana coronaria*, *Thevetia nerifolia*.

3. RESULT AND DISCUSSION

3.1. *Pollutants and Pollution*

Main pollutants, usually released in the atmosphere from industries besides automobiles, are namely Oxides of carbon (CO, CO_2); Oxides of sulphur (SO_2, SO_4); Oxides of nitrogen (NO, NO_2); Lead (Pb); Fly ash (silica-52%, iron Oxide-26%, alumina 16.26%); Suspended Particulate Matters (SPM); Dust; Unsaturated hydrocarbons; Factory waste materials; Water pollutants specially heavy metals and sound nuisance.

3.2. *Types of pollution*:

All the above mentioned pollutants have deleterious effect on human health either directly or indirectly as enumerated below:-

Air pollution - SO_2 is a major constituent of Air pollutants. It is generated in huge quantity as a result of coal burning (coal contains 0.2 to 1.4% sulphur) in thermal power stations, cement and steel plant, pulp and paper industries. All these industries are usually located near the urban areas and SO_2 concentration is high in cities. Calcutta and Bombay are worst affected where SO_2 concentration is 29 and 81 micrograms/cu.m. of air respectively (Gokarn, 1990). It becomes toxic when inhailed at 1 ppm for one hour or 0.3 ppm for 8 hours (Deoras, 1990). Lead (Pb) another pollutant emitted by automobiles once entered in the blood stream may retard mental development of child.

Noise pollution - Sound of high frequency and intensity exceeding the level of 65 decibles (dB) which causes annoyance becomes pollutant. Major sources are factories and automobiles. In Indian cities and industrial areas, noise level is increasing alarmingly and it ranges between 90-120 dB. People who are exposed to high level sound may suffer from insomnia, cardiac diseases, psychiatric and nervous disorders etc. apart from loss of hearing ability (Srivastava & Gupta, 1988).

Water pollution - Industries like paper mill, sugar factories dischange huge quantity of liquid waste materials containing sulphates, chlorides, sodium hydroxide, cellulose etc. to the rivers or open fields causing extensive damage to aquatic flora and fauna.

Dust pollution - Dust and SPM which remain suspended in the atmosphere are released from different sources like factories, automobiles. These particulate matters are inhailed during respiration and the get settled in respiratory tract and lungs. As a result several respiratory diseases occur.

Vegetation of industrial and urban areas also gets adversely affected by pollutants. Some plants are too sensitive and manifest symptoms of injury like chlorosis, necrosis, defoliation, deformities and stunted growth and thus serve as bio-indicator of air-pollutants.

3.3. *Necessity of landscaping in industrial regions*

Earlier, landscaping in industrial areas was mainly for

aesthetic purpose. Gradually the awareness came for checking pollution and its hazardous effects on flora, fauna and mounments. The concept of using tolerant plant species to combat pollution in industrial regions for ameliorating environment is now well known and universally accepted because of their tremendous sink capacity. Plants remove significant quantities of pollutants from air without sustaining serious foliar damage or decline in growth. On the other hand, they provide oxygen to the atmosphere. Annual need of oxygen of one person is met by 150 sq.m. of leaf surface which is obtained from 30-40 sq.m. of greeneries (Moffet & Schiler, 1981). Plant species which are suitable for landscaping as well as tolerant to pollution have been screened. These plants grow well in polluted regions, improve aesthetics and the overall environment.

The time has come to enact legislative measures for compulsory plantation and development of green belts in and around industrial regions with suitable plant species to combat this menace. The plants have got tremendous power to act as pollution sink as a natural process and will remain effective in abatement of pollution and re-vitalization of environment for long period of time.

4. ACKNOWLEDGEMENT

Authors are grateful to the Director, National Botanical Research Institute for providing necessary facilities.

5. REFERENCES

Ahmad, K.J.; Yunus, Mohd.; Singh, S.N.; Srivastava, Kanti; Singh, Nandita; Pandey, Vivek; & Mishra, Jyoti (1989). Study of Plants in Relation to Air Pollution - A report published by Envir. Prot. Lab., NBRI,LUCKNOW.

Anonymous (1989) Indian News & Views. I.J. Envir. Prot. 9(3): 220.

Deoras, P.J. (1990). Some Environmental Problems. I.J. Envir. Prot. 10(1): 4.

Gokarn, A.N. (1990). SO_2 Pollution Abatement. NCL Bull. XV (3): 31-39.

Moffet, A.S. & Schiler, M. (1981). Landscape Design That Saves Energy, W.M.& Co. New York, 223 pp.

Srivastava, A.K. & Gupta, B.N. (1988). Noise Pollution-The Indian Scene.I.J.Envir.Prot. 8(7): 481-486.

Evaporative cool storage of tomato fruits

K P GOPALAKRISHNA RAO
Indian Institute of Horticultural Research
Hessaraghatta Lake Post, Bangalore-560 089. India.

Key words : Tomato, Evaporative cool storage, ripening, lycopene, hue, storage life.

Introduction:

Tomato fruits are commercially harvested at earlier stages of ripeness for fresh market purpose. In the tropical climates like in India, when the temperature rises higher than 30 C especialy in summer, these fruits do not develop optimum red colour, remain yellow and become unmarketable. Surface colour is a major factor in marketing of fresh tomatoes (Garrett et al, 1960). Consumer preference for fresh tomatoes is highly correlated to the surface colour and firmness of tomatoes(Beattie et al, 1983, Jordan et al, 1985, Resurrecion and Shewfelt, 1985). Lycophene synthesis during ripening of tomatoes has been considered to be the principal factor in formation of surface colour (Meredith and Purcell, 1966). But the lycopene synthesis is temperature related. As reported by Koskitalo and Omrod (1972) normal ripening patterns do not occur below 12 C or above 30 C.At high temperatures, β-Carotene accumulates but lycopene synthesis is inhibited resulting in yellow fruit. The objective of this experiment was to obtain normal colour development and prolong the storage life of tomato fruits harvested at earlier stages of ripeness in summer, by storing them in the evaporative cool storage.

Methods

The studies were conducted at Indian Institute of Horticultural Research, Hessaraghatta, Bangalore, (India). Tomato fruits of two cultivars viz. Pusa Ruby and Roma harvested at mature green and breaker stages

J. Prakash and R. L. M. Pierik (eds.), Horticulture – New Technologies and Applications, 405–409.

were used for the study. The study was conducted during April-May of 1986,1987 and 1988. The fruits were stored inside the evaporative cool storage and in the storage room at ambient conditions (Control-temperature 16.5 C to 35.5 C and relative humidity 36.3% to 93.8%). The evaporative cool storage has been developed at the Indian Agricultural Research Institute, New Delhi, using low cost easily available materials viz. brick, sand, bamboo and gunney bags (Roy and Khurdiya, 1986). The cool chamber which is known as zero energy cool chamber,do not require any energy, functions based on the principle of evaporative cooling. The fruits were observed periodically for physiological loss in weight (%) firmness (measured by using EFFEGI fruit pressure tester with 0.7 cm plunger and expressed in kg) lycopene (mg\100 g) ripening rate (days) and storage life (days). Surface colour characteristics were measured using colorgard system 1000 colorimeter of Pacific Scientific U.S.A. and expressed in the Hunter (L, a, b) system.

Results

The evaporative cool chamber maintained a temperature of 20 C to 22.0 C and relative humidity of 90% to 97% throughout the year when the temperature and relative humidity in the storage room (Control) ranged from 16.5 C to 35.5 C and 36.3% to 93.8% respectively.

The physiological loss in weight (%) of the tomato fruits was reduced by about three times by storage in the cool chamber as compared to control. Firmness (Kg) of tomato fruits harvested both at mature green and breaker stage and stored in the cool chamber showed higher values when red ripe, compared to control fruits stored at ambient conditions (Table 1). Storage in the cool chamber also delayed the fruits harvested at both the ripeness stages to reach the red ripe stage and extended the storage life from an average of 6.25 days to 10 days (Table 1). Lycopene synthesis was greatly enhanced in the fruits stored in the cool chamber (Table 2). Higher synthesis of lycopene were seen in the breaker harvested fruits compared to mature green harvested fruits in both the cultivars. Observation of surface colour characteristics showed that fruits stored in cool chambers registered higher 'a' and hue (a\b) values while lower 'L' and 'b' values than ambient stored fruits (Table 3).

Table 1. Firmness, ripening rate and storage life of tomato fruits stored in cool chamber and at ambient conditions

Cultivar	Stg of Hvst.	Firmness(kg) C.C	Firmness(kg) Cntl.	Days to Red-ripe C.C	Days to Red-ripe Cntl.	Stg.Life C.C	Stg.Life Cntl
Pusa Ruby	MG	2.25	2.00	6	3	9	5
	BR	2.25	1.90	4	2	7	4
Roma	MG	2.80	2.40	8	5	14	9
	MR	2.50	2.10	5	4	10	7
Average		2.45	2.20	5.75	3.75	10.00	6.25

Table 2. Lycopene (mg\100g) content of tomato fruits stored in cool chamber and at ambient conditions

Cultivar	Stage of Harvest	Lycopene (mg/100 g) C.C	Lycopene (mg/100 g) Control
Pusa Ruby	MG	4.088	0.687
	BR	4.275	2.684
Roma	MG	3.370	1.435
	BR	4.650	3.027
Average		4.096	1.938

Table 3. Surface colour characteristics of tomato fruits stored in cool chamber and at ambient conditions

Stg hvst	L C.C	L Control	a C.C	a Control	b C.C	b Control	hue(a/b) C.C	hue(a/b) Control
MG	47.26	80.74	30.84	3.89	21.30	40.63	1.45	0.09
BR	44.97	72.88	36.19	11.68	19.57	36.11	1.85	0.32
Avg	46.11	76.81	33.51	7.78	20.43	38.37	1.65	0.20

Discussion

The evaporative cool chamber effectively reduced the temperature and increased the relative humidity inside the chamber. Thompson and Kasmire (1981) found evaporative cooling as an effective means of providing low temperature and high relative humidity for cooling and short term storage. Due to the lowered temperature (20-22 C) and increased relative humidity (90-97%) by the evaporative cool chamber, the physiological loss in weight (%)of the tomato fruits stored inside the chamber was greatly reduced, higher firmness was retained, ripening of the mature green and breaker fruits was delayed and their storage life was prolonged. Roy and Khurdiya (1986) obtained similar performance by zero energy cool chamber for several other vegetables.

Lycopene synthesis and colour development : It was observed that the fruits stored in the cool chamber were bright red ripe with higher content of lycopene, while those fruits stored at ambient temperature (control fruits) remained yellow with very low content of lycopene. This was because of the fact that the cool chamber which maintained a temperature of 20-22 C provided optimum condition for the normal synthesis of lycopene in the fruits stored inside the chamber, while the lycopene synthesis was reduced or inhibited in the fruits stored at ambient temperature (above 30 C). As reported by several workers (Koskitalo and Omrod, 1972; Buescher, 1979) lycopene synthesis is temperature related, which gets inhibited at higher temperature. Hardenburg et al (1986) recommended a temperature of about 18-21 C as ideal for optimum colour development and ripening of tomato fruits. The normal synthesis of lycopene and colour development in fruits stored in the cool chamber were further proved by studying the colour characteristics of the fruits. The fruits stored in the cool chamber showed higher `a` values indicating higher redness. These fruits also recorded higher a/b values (hue) showing higher colour and lower `L` values indicating higher darkness in colour. The fruits stored at ambient conditions showed very high `b` values which proved their high yellowness and the higher `L` values indicated their lightness of colour. Thus, storage of tomato fruits (at mature green or breaker stage) in the evaporative cool storage was proved beneficial in getting optimum lycopene and colour development, delay in ripening and extension of storage life.

Acknowledgement

The author is thankful to the Director for providing the

facilities and to Mr. Srinivasan,V.R. for helping to prepare the manuscript on Word Perfect Software.

References

Beattie, B.B Kavanagh E.E Mc Glasson W.B Adams K.H. Smith E.F and Best D.J (1983) Fresh market tomatoes. A study of consumer attitudes and quality of fruit offered for sale in Sydney, Food Technol. Australia 35: 450-457.

Buescher, R.W. (1979) Influence of high temperature on physiological and compositional characteristics of tomato fruits. Lebensm Wiss. u. Technol. 12: 162-164.

Garrett A.W. Ammerman N.W. Desrosier,N.W. and Fields M.L (1960) Effect of colour on marketing of fresh tomatoes. Amer. Soc. Hort. Sci. 76: 555-559.

Hardenburg R.E. Watada A.E and Wany C.Y (1986) The commercial storage of fruits, vegetables and florist and nursery stocks. Washington,D.C. United States Department of Agriculture, Agriculture Handbook No.66 (revised): 71.

Jordan J.L. Shewfelt R. Prussia S.E. and Hurst W.C (1985) Estimating the price of quality characteristics for tomatoes. Aiding the evaluation of the postharvest system HortScience 20: 203-205.

Koskitalo D.N and Omrod D.P (1972) Effect of sub-optimal ripening temperatures on the colour quality and pigment composition of tomato. J. Food. Sci. 37-56.

Meredith F.L and Purcell A.E. (1966) Changes in the concentration of carotenes of ripening Homstead tomatoes. Proc. Amer. Soc. Hort. Sci. 89: 544.

Resurreccion A.V.A and Shewfelt R.L. (1985) Relationships between sensory attributes and objective measurements of post harvest quality of tomatoes. J. Food Sci. 50 (5): 1242-1246.

Roy S.K and Khurdiya (1986) Studies on the evaporatively cooled zero energy input cool chamber for the storage of horticultural produce. Indian Food Packer 40(6): 26-31.

Thompson J.F. and Kasmire R.F. (1981) An evaporative cooler for vegetable crops. California agric. 35: 20-21.

Effect of containers, on total losses during grapes transportation and improvements for reduction in 3 years

V. SHANKARAIAH[1] and S.K. ROY[2]
Division of Fruits & Horticultural Technology, Indian Agricultural Research Institute, New Delhi-110012

KEYWORDS
Transportation, ideal package, total losses (PLW, shattering, bruising and decay), ventilation and in-package fumigant.

INTRODUCTION

In India, during the period 1980-85, the grape registered an increase by over 9 times in area and 27 times in production (Kaul, 1987). India produces annually 2.37 lakh tonnes with the average production of grapes in India is 25 ton/ha which is the highest in the world. The crop is harvested from December to March and the glut in the market is created in this period in southern states. Whereas in northern states, harvest starts from May and peak period is from June to July. So this difference in the seasons of availability offers great scope for distant marketing.

Traditionally wood and bamboo have been the choice of packaging material to withstand to crude transportation, to resist to high humid climate and excellent staking strength, but there was lot of physical damage, such as bruising, very high tare weight of the package and mostly ecologically undesirable, more over-packaging occupies third position in consumption of wood which is estimated at about 5 million cubic meters per annum (Anon., 1985). So more wooden crates meant more felling of trees. The packaging in isolation cannot solve the problems of post harvest management of grapes and can be an utter failure, if subsequent handling and transport system are not proper. Considering the above points, if an ideal packaging for grapes is developed with proper transport system, then it cannot only boost the grape industry, but also brings sizeable foreign exchange to India (Roy, 1985).

No systematic work in reducing post harvest losses has been carried out, particularly in selecting suitable package material, ventilation, cushioning, material for transportation. Hence the present investigations were carried out in order to screen and improve the package and to use chemicals for reducing transportation losses of grapes.

---

1. Part of his Ph.D. Thesis submitted to the IARI, New Delhi.
2. Project Coordinator (PHT) S.4 Horticulture and Fruit Technology, IARI, New Delhi.

J. Prakash and R. L. M. Pierik (eds.), Horticulture – New Technologies and Applications, 411–414.

MATERIAL AND METHODS

The present investigation on transport of Thompson Seedless and Anab-e-Shahi grapes was undertaken in the Division of Fruits and Horticultural Technology, Indian Agricultural Research Institute, New Delhi. The experimental materials of Thompson Seedless and Anab-e-Shahi grapes were obtained from the Grape Research Station of Andhra Pradesh Agricultural University, Hyderabad. The initial analysis of fruits (Berries) and other studies after harvest were carried out in the Horticulture Division, College of Agriculture, A.P.A.U., Hyderabad. The studies on changes during transport were undertaken in the Division of Fruits and Horticultural Technology, I.A.R.I., New Delhi, during 1986 to 1988.

In the year 1986, investigations were carried out to screen the containers used commercially at Hyderabad for long distance transportation. Five different commonly and locally available containers like Corrugated Fibre Board Box (CFBB), Deal Wood Box (DWB), Date Palm Basket (DPB), Bamboo Basket (BB) and Mud Pot (MP) having approximately 3 kg capacity were used for both varieties of Thomson Seedless and Anab-e-Shahi grapes and were taken in triplicate. Using 60 g of newspaper cutting as cushioning material, in all containers except in mud pot in which bunches were simply placed one above the other.

The CFB box which is more in use and is gradually replacing the wooden box and the newly developed Wire Bound Box (WBB) were used in the subsequent year, i.e. 1987, with four levels of ventilations 1%, 0.75%, 0.5%, 0%. The CFB box was locally made at Hyderabad and WB box was designed and manufactured at Forest Research Institute, Dehradun. The ventilation was provided by punching holes in CFB box and open spacing between the planks in wire bound box. Three kg of grapes was packed in each container by using 60 g of newspaper cuttings as a cushioning material. The WB box was closed by inter-locking its wire clips and it was tied with jute string on all sides. The CFB box was closed and tied on all sides.

In the year 1988, the CFB box and WB box with 0.5% ventilation were used with in-package fumigant (IPF). It consisted of 3.0 g of potassium metabisulphite (KMS) + 0.03 g citric acid placed together in a small muslin cloth bag 4 cm x 4 cm in size. This IPF bag was put at the bottom of the box and 3 kg of grapes was packed in each box and this was replicated four times in each treatment and compared with grapes packed without IPF.

After packing, the grapes were transported from Rajendranagar (Grape Research Station) to Secunderabad by road, from Secunderabad to New Delhi by rail (A.P. Express and

Dakshin Express) and further from New Delhi Railway Station/Nizamuddin to I.A.R.I. campus, by road. During transportation, the packed containers were staked one above the other, except mud pots. Post-transportation observations were recorded within 48 hours after harvest in 1986 and 1987 and within 72 hours after harvest in 1988.

RESULTS

The total losses comprising physiological loss in weight (PLW), shattering, bruising and decay were more in the Anab-e-Shahi grapes than in Thompson Seedless grapes in all three years of study During 48 hours transportation with five different containers , it was observed that the total losses were less in CFB box with Thompson Seedless grapes followed by mud pot, whereas in Bamboo basket, the total losses were maximum. In the case of Anab-e-Shahi grapes, the total losses were least in Deal wood box followed by CFB box and such losses were maximum in Bamboo basket and Date palm basket. It is also evident that the total losses were least with 0.5% vented containers CFB box as well as WB box for both varieties. But such losses were maximum in 1% and 0% vented containers. In further study for 72 hours transportation , such losses were still least in 0.5% vented WB box followed by CFB box with IPF bag for both varieties, but these losses were maximum in same types of containers without IFP usage.

DISCUSSION

Out of 5 different containers , the total losses comprising of PLW, shattering, bruising and decay were found to be less in CFB box packed Thompson Seedless grapes because of its smooth surface and flexibility. But the same CFB box had more losses of PLW and shattering in Anab-e-Shahi variety. The total losses were less in DW box packed Anab-e-Shahi grapes. DW box might have protected the delicate skin of Anab-e-Shahi grapes and had less PLW and shattering. Mud pot also had less total losses in both varieties compared to other containers which had more of PLW and bruising losses due to its much aeration and rough surface. However, the mud pot was not found suitable inspite of having less total losses as it was easily breakable.

Only 0.5% ventilation was found better out of 4 levels of ventilations, in having less total losses in CFB box in Thompson Seedless and WB box and Anab-e-Shahi variety , because without ventilation, the decay and shattering were more in both containers and in both varieties. As ventilation was increased, the PLW losses were increased (Uota, 1957). The difference between 0.5% and 0.75% in having total losses was not much except more PLW in 0.75% as compared to 0.5%. Thus, the 0.5% was found suitable for both the containers.

The total losses were considerably reduced with the use of in-package fumigant (IPF) in both the containers during transportation for both varieties during 72 hours after harvest and transportation as compared to control samples
Because the SO_2 liberated from IPF bag in the container reduced PLW, shattering and decay (Krishna Murthy et al., 1984).

ACKNOWLEDGEMENT

I express my sincere gratitude to the Vice Chancellor, Director of Research and Senior Scientist (Horticulture) of A.P.A.U., Hyderabad; and Head of the Department of Fruit and Horticulare Technology, I.A.R.I., New Delhi, for their continuous encouragement and cooperation for conducting it as part of research work, during my Ph.D. study.

REFERENCES

Anonymous (1985) Role of packaging materials in post harvest management of grapes. Federation of corrugated box manufacturers of India. In Proc. 1st National Workshop on Post Harvest Management of Grapers (India), pp.90-93.

Kaul, G.L. (1987) Role of horticultural crops in crop diversification in India. In ICAR Consultancy Report submitted to the FAO Regional Office for Asia and Pacific Region, Bangkok, Nov. 1987.

Krishna Murthy, G.V., Bererh, O.P., Giridhar, N. and Raghuramaiah, B. (1984) Studies on transportation of Anab-e-Shahi grapes. J. Fd. Sci. Technol., 21: 132-134.

Roy, Susanta K. (1988) Package of fruits and vegetables. In a souvenir on Packaging of Fruits and Vegetables in India, 1988. Sec. B-2, pp.8-16.

Uota, M. (1957) Evaluation of polythene film liners for packaging Emperor grapes for storage. Proc. Amer. Soc. Horti. Sci., 70: 197.

Current Plant Science and Biotechnology in Agriculture

1. H.J. Evans, P.J. Bottomley and W.E. Newton (eds.): *Nitrogen Fixation Research Progress.* Proceedings of the 6th International Symposium on Nitrogen Fixation (Corvallis, Oregon, 1985). 1985 ISBN 90-247-3255-7
2. R.H. Zimmerman, R.J. Griesbach, F.A. Hammerschlag and R.H. Lawson (eds.): *Tissue Culture as a Plant Production System for Horticultural Crops.* Proceedings of a Conference (Beltsville, Maryland, 1985). 1986 ISBN 90-247-3378-2
3. D.P.S. Verma and N. Brisson (eds.): *Molecular Genetics of Plant-microbe Interactions.* Proceedings of the 3rd International Symposium on this subject (Montréal, Québec, 1986). 1987 ISBN 90-247-3426-6
4. E.L. Civerolo, A. Collmer, R.E. Davis and A.G. Gillaspie (eds.): *Plant Pathogenic Bacteria.* Proceedings of the 6th International Conference on this subject (College Park, Maryland, 1985). 1987 ISBN 90-247-3476-2
5. R.J. Summerfield (ed.): *World Crops: Cool Season Food Legumes.* A Global Perspective of the Problems and Prospects for Crop Improvement in Pea, Lentil, Faba Bean and Chickpea. Proceedings of the International Food Legume Research Conference (Spokane, Washington, 1986). 1988 ISBN 90-247-3641-2
6. P. Gepts (ed.): *Genetic Resources of* Phaseolus *Beans.* Their Maintenance, Domestication, Evolution, and Utilization. 1988 ISBN 90-247-3685-4
7. K.J. Puite, J.J.M. Dons, H.J. Huizing, A.J. Kool, M. Koorneef and F.A. Krens (eds.): *Progress in Plant Protoplast Research.* Proceedings of the 7th International Protoplast Symposium (Wageningen, The Netherlands, 1987). 1988 ISBN 90-247-3688-9
8. R.S. Sangwan and B.S. Sangwan-Norreel (eds.): *The Impact of Biotechnology in Agriculture.* Proceedings of the International Conference 'The Meeting Point between Fundamental and Applied *in vitro* Culture Research' (Amiens, France, 1989). 1990. ISBN 0-7923-0741-0
9. H.J.J. Nijkamp, L.H.W. van der Plas and J. van Aartrijk (eds.): *Progress in Plant Cellular and Molecular Biology.* Proceedings of the 8th International Congress on Plant Tissue and Cell Culture (Amsterdam, The Netherlands, 1990). 1990 ISBN 0-7923-0873-5
10. H. Hennecke and D.P.S. Verma (eds.): *Advances in Molecular Genetics of Plant–Microbe Interactions.* Volume 1. 1991 ISBN 0-7923-1082-9
11. J. Harding, F. Singh and J.N.M. Mol (eds.): *Genetics and Breeding of Ornamental Species.* 1991 ISBN 0-7923-1094-2
12. J. Prakash and R.L.M. Pierik (eds.): *Horticulture – New Technologies and Applications.* Proceedings of the International Seminar on New Frontiers in Horticulture (Bangalore, India, 1990). 1991 ISBN 0-7923-1279-1

KLUWER ACADEMIC PUBLISHERS – DORDRECHT / BOSTON / LONDON

CPSIA information can be obtained
at www.ICGtesting.com
Printed in the USA
LVHW051745170520
655861LV00004B/233